Evolutionary Ecology of Parasites

Evolutionary Ecology of Parasites

second edition

Robert Poulin

PRINCETON UNIVERSITY PRESS • PRINCETON AND OXFORD

Published by Princeton University Press, 41 William Street, Princeton, New Jersey 08540

In the United Kingdom: Princeton University Press, 3 Market Place, Woodstock, Oxfordshire OX20 1SY

Library of Congress Cataloging-in-Publication Data

Poulin, Robert, 1963–
   Evolutionary ecology of parasites / Robert Poulin. — 2nd ed.
     p. cm.
   Includes bibliographical references (p.   ) and index.
   ISBN-13: 978-0-691-12084-3 (alk. paper)
   ISBN-10: 0-691-12084-6 (alk. paper)
   ISBN-13: 978-0-691-12085-0 (pbk. : alk. paper)
   ISBN-10: 0-691-12085-4 (pbk. : alk. paper)
    1. Parasites.   2. Parasites—Ecology.   3. Parasites—Evolution.   I. Title.

QL757.P68 2007
591.7'857—dc22   2006050364

British Library Cataloging-in-Publication Data is available

Contents

Preface

A reviewer of the first edition of this book suggested that it was perhaps a bit premature, coming at a time when most questions about parasite evolutionary ecology were still being asked. How times have changed! The recent popularity of parasites has seen them joining the mainstream of ecology and evolutionary biology. After years on a dusty shelf under the label "gross and disgusting," parasites are now the subject of renewed interest. This has happened in part because they have turned out to be excellent models for a range of fundamental studies, and in part because there is a pressing need to find out more about their ecology and evolution at a time when emerging diseases threaten wildlife and human health. This new edition covers recent developments in the evolutionary and ecological study of parasites. It is not a basic parasitology text providing a phylum-by-phylum description of parasite life cycles; instead, it is a synthesis of the most exciting aspects of their evolutionary ecology. Fuller than its predecessor, it is still a concise book, aimed at advanced students and practicing researchers in parasitology.

A couple of warnings are necessary at this stage. First, I have tried to present the big picture instead of getting lost in details. My aim is to address big questions, even if only tentative answers are currently possible, rather than providing exact answers to specific problems. The book deals with the ultimate or evolutionary level of biology, rather than the proximate or mechanistic level. Thus, most questions relate to "why" parasites do certain things rather than "how" they do these things. As a rule, I look for general patterns and ignore exceptions, unless the latter can shed further light on the problem at hand. Because there is now much empirical evidence in support of the various patterns, I had to be selective in my choice of examples, focusing on those that serve to illustrate the conceptual framework instead of presenting exhaustive lists of case studies.

Second, this book is unashamedly biased toward parasites themselves rather than the host-parasite relationship as a whole. The influence of host biology on parasite evolution and ecology is a central theme of the book; in contrast, the impact of parasites on host biology is not discussed, unless it allows insight into parasite biology. With the growing recognition that parasites are omnipresent agents of natural selection as well as causes of morbidity in wildlife populations, they are increasingly seen as evolutionary and ecological forces, rather than as organisms in their own right. I aim to redress the balance.

The book is intended as a reference for practicing parasitologists, although hopefully it will also be useful to other evolutionary ecologists. Its goal is also to set the research agenda for the next several years, by pointing out promising directions for future investigations. I have assumed a basic knowledge and understanding of ecology, evolution, and parasite biology. The book could thus only be used as a textbook for an advanced undergraduate course or, more likely, for a graduate course in parasitology. My ambition is to stimulate both young and established researchers to use parasites as model organisms in evolutionary ecology studies.

Many people have contributed in one way or another to make this book possible; without them there would be no book. Over the years, my various students and collaborators have sharpened my thoughts about parasite evolution and ecology, forcing me to reconsider some of my craziest ideas or to revisit conclusions that I took for granted. Through our collaborative work or merely in more casual discussions, they have unwittingly helped to shape my views about parasite biology. In particular I would like to acknowledge, and thank, Claude Combes, Tom Cribb, Brian Fredensborg, Dave Latham, François Lefebvre, Janice Moore, Serge Morand, David Mouillot, Kim Mouritsen, Klaus Rohde, Frédéric Thomas, Dan Tompkins, Tellervo Valtonen, and Bill Vickery. I am especially grateful to Dale Clayton and Janice Moore, who read the entire manuscript and gave me constructive suggestions for its improvement. Of course, none of the above people are to blame for the remaining errors in the book: any unorthodox interpretation in the following pages is strictly my own. Sam Elworthy, of Princeton University Press, was enthusiastic about the project from the very start, and provided excellent publishing support. The Royal Society of New Zealand, via a James Cook Research Fellowship, freed some of my time to get the ball rolling in the first place. My colleagues and friends in the Department of Zoology, University of Otago, create the kind of stimulating and supportive working environment that was needed to complete such a project. Finally, I thank Diane, Eric, and Alex, who provided love, laughter, distraction, encouragement, support, and regular reality checks.

1 | Introduction

Ecology is the scientific study of interactions between organisms of the same or different species, and between organisms and their nonliving environment. One of the main goals of ecologists is to explain the abundance and distribution of organisms over space and time. The scope of ecology includes all sorts of interactions, from the most intimate, permanent associations to the briefest of encounters. Although parasitism qualifies as the sort of interaction of interest to ecologists, it has somehow become the focus for another branch of science, parasitology, which uses a multidisciplinary approach to investigate host-parasite interactions. Because of the intimate and intricate nature of the association between host and parasite, a broadly trained parasitologist using techniques ranging from molecular biology all the way to ecology seemed the most appropriate investigator. The problem with this takeover of parasitism by parasitologists, however, is that parasites have been ignored by ecologists for a long time. Sections on parasitism have only recently begun to appear in ecology textbooks (for instance, Begon et al. 2005 and earlier editions), and these bear mainly on the ecological impacts of parasites on free-living organisms. Studies on the population or community ecology of parasites themselves are practically absent from the ecology literature; they are almost exclusively restricted to parasitology journals, and often devoid of any references to important ecological studies on free-living animals.

Similarly, parasite evolution has until recently been studied by parasitologists rather than by evolutionary biologists. As with ecological studies, evolutionary investigations of host-parasite interactions undertaken by nonparasitologists are relatively recent (Poulin 1995a), but rapidly growing in number. While the quality of studies on the ecology or evolution of parasitism performed by parasitologists is not in doubt, it is unfortunate that for many years there have been few exchanges of ideas between parasitology and either ecology or evolutionary biology. Researchers in these disciplines attend different meetings and read different journals. Ecologists and parasitologists have even developed their own jargon; although they use similar terms, they assign different meanings to the same words, which hinders effective communication between the two disciplines (Bush et al.

1997). More importantly, the separation between parasitology, ecology, and evolutionary biology has lead to philosophical differences between these branches of biology, some of which are important. For instance, parasitologists have long known that parasites can affect host population dynamics, but ecologists took some time to realize this. Similarly, ecological parasitology developed as a discipline, in the mid- to late twentieth century, with strong doses of natural history and field biology; this period is beautifully captured, with a hint of nostalgia, by Esch's (2004) historical essays. During that time, however, most theoretical advances in ecology and evolutionary biology were ignored by parasitologists; those that were eventually adopted by parasitologists were only applied to parasites years after their introduction in other fields. For example, until recently and in part because of the influence of medical science on parasitology, many parasitologists accepted that evolution led to a decrease in parasite virulence, whereas modern evolutionary theory would have predicted a greater range of outcomes (Anderson and May 1982, 1991; Ewald 1994). These disagreements could have been avoided had there been a better integration of ecology, evolution, and knowledge of parasite biology by students of parasitism.

My purpose in this book is generally the same as in the first edition: to present an evolutionary ecologist's view of the biology of parasites. I want to discuss various aspects of the biology of parasites using an approach compatible with current theory in evolutionary biology and ecology. Many studies of parasite ecology or evolution published in the past fifty years were thorough descriptive investigations but failed to test the general hypotheses put forward by ecologists and evolutionary biologists; here I will try to rectify this by emphasizing the link with theory. In this book I approach parasites as an evolutionary ecologist would approach any other group of organisms, while recognizing the special attributes of parasites. The book focuses on parasites themselves rather than on the interaction with hosts. Instead, hosts are seen as a key part of the parasite's environment and as a major source of selective pressures. The influence of parasites on host biology has been dealt with extensively in recent reviews; it will only be covered here if it relates to the ecology or evolution of parasites.

My other objective is to suggest a research agenda for the next several years. Where relevant, I point out the gaps in our knowledge, and try to suggest ways of filling these gaps. With this book, I want to capture the present state of parasite ecology and evolution, and propose directions for its future growth.

1.1 The Evolutionary Ecology Approach

Organisms interact with one another and with the nonliving environment on an ecological time scale, measured in days, months, or years. These interactions, however, are the product of natural selection acting over evolutionary time, over thousands and millions of years, to produce organisms well suited to their environment. Evolutionary ecology is the study of the selective pressures imposed by the environment, and the evolutionary responses to these pressures. Natural selection has shaped not only the traits of individual organisms, but also the properties of populations and species assemblages. The subject matter of evolutionary ecology, therefore, includes topics such as the trade-off between the size and number of offspring produced by individual animals, the proportion of males and females in animal populations, and the composition of animal communities. All these phenomena can be studied on a human time scale but to understand the differences observed among organisms, one must consider the forces and constraints that have acted during their evolutionary history.

The study of evolutionary responses is not always as straightforward as that of phenomena occurring on shorter time scales. A major goal of science is to demonstrate causality; it can be inferred that an event causes a response if the response always follows the event in an experimental situation. For example, exsheathment of many nematode larvae and hatching of many cestode eggs always follow their exposure to the conditions encountered in their host's gut in in vitro experiments, therefore it can safely be inferred that these conditions cause exsheathment or hatching, at least in a proximate sense. In evolutionary ecology, the manipulation of variables in controlled experiments is often impossible. Instead, we must rely heavily on comparisons between species that have been exposed to different selective pressures. If species under a given selective regime have consistently evolved a certain combination of traits, these "natural" experiments can be used to draw conclusions about the effects of certain factors over evolutionary time. Obviously, similarity between species can be the result of inheritance of traits from a common ancestor as much as the product of independent lines of convergent evolution. A careful distinction must be made between phylogenetic influences and the true action of natural selection (Brooks and McLennan 1991; Harvey and Pagel 1991). In the absence of other evidence, comparisons across taxa can help to identify true adaptations, defined here as genetically determined traits that have spread or are spreading through a population because they confer greater fitness on their bearers.

Although only applied recently to parasitological problems, the comparative approach can shed much light on parasite evolution. This approach can do more than identify relationships between species traits. It can also be used to test evolutionary hypotheses even though it does not follow the classical experimental approach consisting of the manipulation of independent variables in controlled conditions (see Brandon 1994). Different parasite lineages leading to extant species can be viewed as different evolutionary experiments, in which the ancestor represents the initial experimental conditions and the current phenotype represents the experimental outcome. Comparing lineages evolving under different selective pressures (e.g., in different types of hosts) is like comparing the responses of subjects exposed to different experimental conditions, or their responses to the manipulation of selected variables. In this context, controlling for phylogenetic influences corresponds to avoiding pseudoreplication.

Proper comparative analyses are powerful tools for hypothesis testing in evolutionary ecology (Harvey and Pagel 1991). They are not, however, a panacea for the study of adaptation. Used in isolation from other kinds of evidence, comparative studies provide limited insights into evolutionary mechanisms and the causal links between biological traits (Doughty 1996). On the other hand, the comparative approach is the most useful to identify general patterns that can guide further research. Although the results of controlled experiments or field observations are used as tests of theory wherever possible throughout the book, much of the evidence presented in this book relies on the explanation of variability among species using a comparative approach.

At the same time as the comparative approach became a major tool for evolutionary biologists, a parallel development provided ecologists with a new way of tackling the complexity of natural systems. Macroecology has emerged as a research program aimed at trying to infer the laws of nature from the statistical patterns among its constituent parts (Brown 1995; Gaston and Blackburn 2000). Macroecology consists of the empirical detection of patterns, the formulation of mechanistic hypotheses to account for these patterns, and the empirical testing of the hypotheses. As a whole, macroecology has been remarkably successful at finding general patterns and likely explanations for these patterns (Brown 1995, 1999; Lawton 1999; Gaston and Blackburn 2000; Blackburn and Gaston 2001). Much of the information on parasite ecology presented in this book borrows strongly from the macroecological approach.

The comparative approach in both evolutionary biology and macroecology focuses on large-scale, general phenomena rather than on detail. As a rule, they provide broad, though sometimes tentative, answers to important questions instead of definitive answers to very specific questions. An effort is made throughout this book to combine evidence from comparative or macroecological studies with that from experimental studies, to provide different perspectives on the same problems.

This book is not an elementary treatise of evolutionary ecology. The reader who wants a more general overview of the theory and mathematical models at the core of modern evolutionary ecology can read any of several recent texts on the subject (e.g., Cockburn 1991; Bulmer 1994; Pianka 1994; Fox et al. 2001). This book applies many ideas from evolutionary ecology specifically to parasites, and aims to foster the use of evolutionary thinking in the study of parasite ecology. Some of the questions that will be addressed include: Why do some parasites have more complex life cycles than others? Why are some parasites more host-specific than others? Why are some parasites much more fecund than others? Why are some parasites much more virulent than others? Why are some parasites more highly aggregated among their hosts than others? Why is there greater gene flow among populations in some parasite species than in others? Why are some parasite communities richer than others? These questions have been addressed before by parasitologists, but usually not in an evolutionary context or not with appropriate comparative methods.

1.2 Scope and Overview

Because of its vague definition, the term *parasite* has been applied to a wide range of plant and animal taxa. One can even argue that parasites *sensu lato* greatly outnumber free-living organisms (Windsor 1998). The most widely accepted definition of a parasite is that it is an organism living in or on another organism, the host—feeding on it, showing some degree of structural adaptation to it, and causing it some harm (when the harm incurred by the host invariably leads to its death, the parasite is often referred to as a parasitoid). Interpretations of this definition vary among authors (see Zelmer 1998). Price (1980) included phytophagous insects as parasites, but excluded blood-sucking flies. Barnard (1990) included behavioral parasites, such as many birds that are not physiologically dependent on their host but exploit it in other ways, for example by stealing food from the host. Combes (1995, 2001) even included strands of DNA among parasitic entities. To avoid confusion, it is therefore necessary to specify the taxonomic scope of this book, which will focus exclusively on protozoan and metazoan parasites of animals. These include several diverse taxa of parasites (table 1.1) that have had several independent evolutionary origins. Because helminths and arthropods have been the subject of the majority of relevant studies, they will provide most examples. The general biology and life cycles of these parasites are described in detail in any basic parasitology text (e.g., Noble et al. 1989; Schmidt and Roberts 1989; Cox 1993; Roberts and Janovy 1996; Kearn 1998; Bush et al. 2001), and it will be assumed that the reader is at least superficially familiar with them. Some of the patterns discussed here and some of the conclusions they suggest may also apply to other groups of parasites, but these are beyond the scope of this book.

The evolutionary ecology of parasites can be studied at several hierarchical levels. The smallest unit of study in ecology is the individual organism, but ecologists also deal with populations of individuals of the same species, and with communities made up of several populations of different species. This book first examines how ecological traits of individual parasites have evolved, and then considers population and community characteristics. Chapter 2 begins with a discussion of how organisms that made a transition to parasitism from a free-living ancestral lifestyle have undergone changes in their biology; it also considers how historical events and selective pressures have shaped complex life cycles, and how these life cycles in turn have influenced the ecology of the parasites adopting them. Chapter 3 explores the reasons why some parasites have evolved the ability to exploit a wide range of hosts whereas others are restricted to a single host species. For organisms often thought to be small, degenerate egg-production machines, parasites also show a tremendous range of life-history traits. Chapter 4 discusses how much of this variation is explained by selective pressures from the host or the physical environment, and how much is due to phylogenetic constraints. The final characteristic of individual parasites that will be considered is their ability to harm or manipulate the host. Far from evolving to become benign commensals, parasites can be selected to become highly virulent exploiters of host resources, or they can evolve the ability to control the physiology and behavior of their host for their own benefit. Chapter 5 explores the conditions under which host exploitation strategies can evolve toward these extremes.

One of the most easily described properties of animal populations is their distribution in space. Parasite populations are typically aggregated among their host individuals, but the degree of aggregation varies greatly over time and among populations and species of parasites. The opening chapter on parasite population ecology will examine the causes of aggregation, and some of its potential evolutionary consequences (chapter 6).

Table 1.1 Diversity of some of the major taxa of metazoan parasites of animals. Estimates of species numbers are meant to be realistic minimum numbers. (Modified from Poulin and Morand 2004)

Parasite taxon	No. of parasitic species	Definitive host[a]	Life cycle[b]	Habitat[c]
Phylum Mesozoa	>80	Endo, I	S	M
Phylum Myxozoa	>1350	Endo, V	C	M, FW
Phylum Cnidaria	1(?)	Endo, V	C(?)	FW
Phylum Platyhelminthes				
Class Trematoda	>15000	Endo, V	C	M, FW, T
Class Monogenea	>20000	Ecto, V	S	M, FW
Class Cestoidea	>5000	Endo, V	C	M, FW, T
Phylum Nemertinea	>10	Ecto, I	S	M
Phylum Acanthocephala	>1200	Endo, V	C	M, FW, T
Phylum Nematomorpha	>350	Endo, I	S & C	FW, T
Phylum Nematoda	>10500	Endo, I & V	S & C	M, FW, T
Phylum Mollusca				
Class Bivalvia	>600	Ecto, V	S	FW
Class Gastropoda	>5000	Ecto, V	S	M
Phylum Annelida				
Class Hirudinea	>400	Ecto, V	S	M, FW
Class Polychaeta	>20	Endo, I	S	M
Phylum Pentastomida	>100	Endo, V	C	FW, T
Phylum Arthropoda				
Subphylum Chelicerata				
Class Arachnida				
Subclass Ixodida	>800	Ecto, V	S & C	T
Subclass Acari	>30000	Ecto, I & V	S	M, FW, T
Subphylum Crustacea				
Class Branchiura	>150	Ecto, V	S	M, FW
Class Copepoda	>4000	Ecto, I & V	S	M, FW
Class Cirripedia				
Subclass Ascothoracida	>100	Endo, I	S	M
Subclass Rhizocephala	>260	Ecto, I	S	M
Class Malacostraca				
Order Isopoda	>600	Ecto, I & V	S	M
Order Amphipoda	>250	Ecto, I & V	S	M
Subphylum Uniramia				
Class Insecta				
Order Diptera	>2300	Ecto, V	S	T
Order Phthiraptera	>4000	Ecto, V	S	T
Order Siphonaptera	>2500	Ecto, V	S	T
Order Strepsiptera	>600	Endo, I	S	T

[a] Parasites are classified as either endoparasitic (Endo) or ectoparasitic (Ecto) on either invertebrate (I) or vertebrate (V) definitive hosts.

[b] Life cycles are simple (S) if a single host is required and complex (C) if two or more hosts are required for the completion of the cycle.

[c] The habitat of parasites and their hosts can be marine (M), freshwater (FW) or terrestrial (T).

Parasite individuals in a population are not distributed evenly in space, and their numbers also fluctuate in time. The basic models of parasite population dynamics are reviewed in chapter 7, along with a discussion of how evolution may have shaped various population processes. This chapter also includes a synthesis of recent developments in parasite population genetics, from which much has been learned about the spatial dispersion of parasites at various scales, within or among populations.

In nature, any parasite population is likely to coexist with populations of other parasite species. The transition to parasite community ecology will be made by examining how populations of different parasite species interact and how parasites have responded to interspecific competition (chapter 8). Parasites of different species occurring in the same host individual form a community, which itself is only a small subset of the larger community comprising all parasite species found in the host population. In turn, this is a subset of the fauna of parasite species known from the combined populations of that host species. At all these levels of organization, assemblages of parasite species may be structured or they may be random, that is, they may be predictable sets of species drawn from the pool of available species, or an assemblage formed by chance events. This and other issues are addressed at all levels of parasite community organization in chapters 9 and 10.

This book is about the biology of individual parasites such as their transmission pattern and life cycle, the biology of parasite populations including their distribution among hosts, and the biology of parasite communities with emphasis on structure and richness. All these themes are linked to one another and set within an evolutionary or phylogenetic framework. The evolutionary ecology of parasites is still a young discipline and many questions remain unanswered. Throughout the text, areas in which further research is required are highlighted. Hopefully, these suggestions will lead to more investigations and a narrowing of the gap between parasite evolutionary ecology and the evolutionary ecology of free-living organisms. In chapter 11, general guidelines are offered for future studies, and the global importance of evolutionary studies of parasites is discussed.

2 | Origins of Parasitism and Complex Life Cycles

Parasites have obviously evolved from free-living ancestors—there first had to be animals around for parasites to exploit. Parasitism has originated independently in several animal taxa, and has sometimes arisen more than once in a given taxon. The origins of parasites typically go back several million years, as indicated by the classical fossil evidence (Conway Morris 1981) supplemented lately by the discovery of well-preserved specimens fossilized in amber (e.g., Poinar 1999; Klompen and Grimaldi 2001). Since these early days parasites have diversified greatly (Poulin and Morand 2000a, 2004), and modern species now display a wide variety of life cycles and adaptations to their parasitic existence. This chapter presents scenarios for the early origins and evolution of parasites and the subsequent increases in the complexity of their life cycles. Other aspects of the life history of parasites, such as the evolution of body size and reproductive output, will be covered in chapter 4.

2.1 Transitions to Parasitism

Two organisms of widely different sizes may come into contact for some time without the small one exploiting or harming the large one, and without the large one eating the small one. This is probably common in nature. If these encounters are frequent in a given pair of species, an opportunity exists for a more permanent association to develop. However, opportunity is only one of the requirements for the establishment of an intimate interspecific association. The parasite-to-be must possess some pre-adaptations for survival, feeding, and reproduction on the host, and its reproductive success as a parasite must be greater than its success as a free-living animal. Without pre-adaptations the parasite cannot begin to exploit the host, and without fitness benefits, host exploitation will not be favored by natural selection.

The need for pre-adaptation of parasite precursors is an old idea (Rothschild and Clay 1952). Specialization for a parasitic existence may be gradual, but the initial stages of host exploitation must be possible at the onset. Selection would not gradually shape feeding and attachment organs from one generation to the next if the parasite could not derive benefits from the host, that is, fitness gains must precede specialization. Thus the animals that can become parasites are the ones capable of remaining in or on the host for certain periods of time, during which they can feed on the host and achieve greater fitness than their conspecifics that do not exploit hosts. The mouthparts and other structures necessary to feed on dead animals, or to burrow into plant stems, or to suck the sap of plants, are all examples of pre-adaptation to parasitism on animals. Some gastropod molluscs (Bouchet and Perrine 1996) and ostracod crustaceans (Stepien and Brusca 1985), not usually known to be parasitic, have been documented feeding at night on fish resting near the seafloor. Crabs are usually free living but larval and adult crabs can enter the gill cavity of fish caught in traps on the seafloor, and feed on their gill filaments (Williams and Bunkley-Williams 1994). Such opportunistic foragers using a feeding apparatus designed for a different purpose could eventually evolve into specialized parasites forming more permanent and intimate associations with their fish hosts. Cymothoid isopods, for instance, are obligate ectoparasites of fish evolved from ancestors that were only facultative fish parasites (Brusca 1981). The closest relatives of the Cymothoidae, for example, Aegidae and Corallanidae, are still free-living predators of invertebrates capable of an occasional blood meal on a fish (Bunkley-Williams and Williams 1998).

Many small invertebrates attach to the external surfaces of larger animals for a limited time to disperse into new areas. This phenomenon, called phoresy, is seen as a common step toward a more permanent, parasitic association. For instance, both mites and nematodes include phoretic and parasitic species. In these taxa, parasitic species may have evolved from phoretic ancestors (Anderson 1984, 2000; Athias-Binche and Morand 1993; Walter and Proctor 1999). In particular, the dauer larvae of nematodes possess the necessary characteristics to use other animals as transport hosts, being resistant to a range of conditions including accidental passage through an animal's gut. Phoresis may be more beneficial in terrestrial than in aquatic environments; water currents facilitate dispersal in the latter. Not surprisingly, despite the rich fauna of free-living nematodes in aquatic habitats, nematodes parasitic of animals most probably all originated in terrestrial habitats from phoretic ancestors, and the relatively few nematodes parasitic in aquatic hosts have terrestrial ancestors (Anderson 1984, 1996). Whatever its exact nature, some form of pre-adaptation must have been present in the precursors of parasites to allow the association to get started; a strict physiological dependence of the parasite on the host and efficient methods of host-to-host transmission would have evolved subsequently.

As an alternative to phoresy and dispersal, small animals may also minimize the variance in environmental factors that they would normally experience by associating with larger animals. Hairston and Bohonak (1998) point out that there are no copepod taxa that exhibit both diapause and parasitism, even though both traits have evolved repeatedly and independently in all major copepod orders. Diapause is a form of arrested development in copepods thought to be an adaptation to cope with environmental variation. Hairston and Bohonak (1998) suggest that parasitism is an alternative strategy serving to reduce temporal variation in food availability and other environmental conditions.

Whether or not this is true, the ability to attach to and feed on hosts must have existed *before* any transition to parasitism in copepod lineages that subsequently specialized as parasites for the association to have been viable in the very beginning.

The opportunity to study a transition to parasitism "in progress" would be invaluable, and there may be some happening right now. For instance, pearlfishes (family Carapidae) are an unusual group of fish that live symbiotically inside sea cucumbers or starfish. Recent studies of stable isotope ratios suggest that some pearlfish species do not feed on their host's tissues at all, whereas others do (Parmentier and Das 2004). They form a continuum ranging from commensalism to parasitism. This range of strategies suggests that the association may begin with the fish first using the host strictly as a refuge in which to undergo metamorphosis and then subsequently adapting to feed on its tissues. More information is needed about this association to draw firm conclusions, but this example shows that parasitism can evolve in a stepwise manner.

The preceding discussion has started with the assumption that the parasite precursor had to be smaller than its host-to-be, and that it was not part of the latter's diet. There are of course exceptions to this general scenario. For instance, in some groups of organisms, parasites and their hosts are closely related—the parasites have evolved from their hosts, sometimes on several independent occasions within one lineage (Ronquist 1994; Goff et al. 1997; Coats 1999). Clearly, at the onset, the parasites were of roughly equal body size to their host, and only later evolved toward smaller sizes. It is also possible for a prey to become a parasite of its predator. For example, some free-living ciliates respond to the appearance of their predators, larval treehole mosquitos, by dividing and transforming into parasitic cells that attach to and penetrate the cuticle of the mosquitos, ultimately killing them (Washburn et al. 1988). This facultative parasitism is the ultimate anti-predator strategy, and it may explain the evolution of obligate parasitism in systems where predators are not just present seasonally, but continuously. There are so many different kinds of parasites that there must be many different ways in which parasitism can evolve.

How often has parasitism arisen as a mode of existence in the history of life on Earth? We cannot answer this question yet. We would need a fully resolved animal phylogeny just to start guessing; without one, we might easily underestimate how frequent the transitions have been. Indeed, some parasite taxa long thought to be monophyletic may in fact be polyphyletic, that is, parasitism may have evolved more than once in the group (e.g., in lice: see Johnson et al. 2004; Murrell and Barker 2005). Even with a complete phylogeny of extant groups, we could never get an accurate answer because we do not know how many parasitic lineages have gone extinct without leaving a trace. Some fossil evidence indicates that host-parasite associations that existed millions of years ago have no present-day counterparts (e.g., Poinar et al. 1997), suggesting that at least certain parasite lineages no longer exist.

Having pointed out the caveats, we can still look at the available evidence. It is clear that transitions from a free-living existence to parasitism have been very common (De Meeüs and Renaud 2002). Morphological convergence between related but distinct parasite lineages can mask their independent origins, and molecular data can shed light on the evolutionary history of parasitism. For example, although the parasitoid habit per se arose only a single time within the order Hymenoptera (i.e., wasps), endoparasitism, or the ability to lay eggs inside the host, has evolved independently more than once (Whitfield 1998); however, convergence among the morphological adaptations of endoparasitoids can obscure their independent origins (Quicke and Belshaw 1999).

The same is true for classical parasites, the ones covered in this book, and recent phylogenetic studies have often revealed multiple origins of parasitism within given groups. Larval parasitism in vertebrate flesh arose independently several times among blowflies (Stevens 2003). Among nematodes, parasitism of animals has evolved on many independent occasions (Anderson 1984; Clarck 1994), at least four times according to a molecular phylogeny of the phylum (Blaxter et al. 1998). Parasitism evolved only once in the ancestor of all living acanthocephalans (Herlyn et al. 2003). Digeneans (or trematodes), monogeneans, and cestodes are members of a monophyletic group presumably descended from a single evolutionary transition to parasitism by a common ancestor (Littlewood et al. 1999a, b); however, associations with hosts have had multiple independent origins in turbellarians, another group of parasitic flatworms (Rohde 1994a). In crustaceans, several taxa include both parasitic and free-living species, and parasitism has arisen on several different occasions in copepods (Poulin 1995b), isopods (Poulin 1995c; but see Dreyer and Wagele 2001), and amphipods (Poulin and Hamilton 1995). Considering all taxa in which there are parasitic lineages (table 1.1), it is probably safe to say that living metazoan parasites of animals are the product of at least sixty separate evolutionary transitions from a free-living existence to one of obligate parasitism. One of these transitions, in the phylum Cnidaria, is responsible for a single known parasite species, *Polypodium hydriforme*, a jellyfish relative that parasitises the eggs of sturgeons (Raikova 1994). Most passages to parasitism, however, have been followed by moderate to extensive diversification, giving rise to the vast diversity of parasite species in existence today.

2.2 Specialization of Parasites

Discussions of parasite evolution often revolve around the idea that parasites are evolutionarily retrogressive or degenerate. This perception comes from comparisons made between parasites and their hosts; obviously, worms are morphologically less complex than their vertebrate hosts, whatever definition of complexity we use. This superficial and inappropriate comparison leads to the conclusion that parasites lost sense organs and other structures during their evolution until they became the simple life forms we observe today. Such a process has been called sacculinization (after a parasite, of course, the crustacean *Sacculina*!) and is still associated with parasites in the contemporary literature.

Because hosts do not provide a suitable benchmark for comparisons, a more informative contrast would pit parasites against their closest free-living relatives. In some cases, the loss of complexity is evident. For instance, the microscopic endoparasites that comprise the phylum Myxozoa show unparalleled levels of morphological simplification when compared with any of the taxa now thought to be their closest free-living relatives (Okamura and Canning 2003; Canning and Okamura 2004). In other cases, no reduction in morphological complexity is apparent. Parasitic nematodes do not look much different from their free-living, soil-dwelling counterparts. Parasitic polychaetes provide another striking example. Polychaetes from the family Oenonidae (formerly Arabellidae) are endoparasites of other, free-living polychaetes (e.g., Pettibone 1957; Uebelacker 1978; Hernández-Alcántara and Solis-Weiss 1998); none of their morphological features suggest they are parasitic, and most specialists can tell that they are parasites only when finding them inside another polychaete. The loss of morphological complexity and structural specialization often associated with the parasitic mode of life are therefore not universal features of parasitic taxa.

The apparent loss of morphological complexity, however, can be deceptive, because it may or may not be mirrored by changes in genomic complexity. The measurement of complexity using genome size is not straightforward, because only a fraction of an organism's DNA actually codes for something, and larger genomes may not necessarily mean more coding base pairs (Cavalier-Smith 1985). We know practically nothing about how much genetic information is needed to encode a particular morphological structure. With these caveats in mind, it is still interesting to look at the results of the few studies on genome size in parasites. Bacteria that are obligate intracellular parasites typically have smaller genomes than free-living bacteria, a phenomenon that is apparently the outcome of genome degradation following a transition to a parasitic mode of life (Frank et al. 2002). In contrast, studies using DNA reassociation techniques have indicated that the genome of parasitic helminths is often larger and more complex that that of their free-living relatives, not what we would expect if parasites evolve to become simpler. For example, the parasitic nematodes *Ascaris* and *Trichinella* have much larger genomes than those of free-living nematodes, and the cestode *Hymenolepis* has a genome twice as complex as that of the free-living flatworm *Dugesia* (Searcy and MacInnis 1970). Ironically, the genome of the cestode *Bothriocephalus gregarius* is twice the size of that of its vertebrate host (Verneau et al. 1991). Thus, the morphological simplicity of parasites does not always reflect their genomic complexity. Perhaps the complexity gets partitioned among the different stages in the life cycle, so that a complex genome codes for an ontogenetic succession of relatively simple forms, with various genes turned on or off during development. Viewing the whole life cycle, and not just the adult form, as the unit of selection gives a more accurate perspective on parasite complexity.

As Combes (1995, 2001) points out, some functions such as digestion and locomotion are often left entirely to the host; in these situations the loss of structures or organs without active roles only makes sense. The economy of energy and resources thus achieved may have allowed other structures to develop in response to selection. With the aid of modern electron-microscopic techniques, a wide array of new sensory receptors has now been discovered in parasitic worms (Rohde 1989, 1994a). Some sense organs such as eyes may have been lost by parasitic lineages only to be replaced by more appropriate structures. Brooks and McLennan (1993a) examined the rates of character loss and character innovation in parasitic flatworms, based on the presence or absence of ancestral and derived characters in many taxa. They found that the majority of evolutionary changes in morphological characters were innovations rather than losses. This does not support the view that parasites are structurally simplified and degenerate. This approach

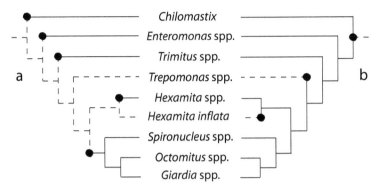

Figure 2.1 Two scenarios for the evolution of parasitism in the protist taxon Diplomonadida. Free-living lineages are indicated by broken lines, parasitic lineages by solid lines; black circles illustrate the proposed transitions between lifestyles. The phylogeny is based on the analysis of twenty-three ultrastructural characters; *Chilomastix* is the outgroup. In scenario (a), the ancestor was free living, and parasitism arose on five separate occasions. In scenario (b), parasitism was the ancestral state in the Diplomonadida, and there were two reversals to a free-living lifestyle. (Modified from Siddall et al. 1993)

would be more convincing if comparisons were made between the rates of character changes in parasitic flatworms and their free-living sister group. Nevertheless, all this evidence suggests that parasites are no more simple than free-living animals, only specialized for a different kind of existence.

The high degree of morphological specialization displayed by parasites also suggests that a transition to parasitism is irreversible. The evolution of parasitism could provide support for Dollo's Law, which states that progress and specialization are unidirectional. Early parasites can evolve a dependence upon a host organism, but once they are dependent on the host there is no going back. In other words, early specialization for a parasitic life commits a lineage forever. This may indeed be the case for obligate intracellular parasites, which have lost certain functional categories of genes, making them irreversibly specialized for parasitism (Moran and Wernegreen 2000). The Diplomonadida, a group of flagellated protists containing many species of obligate gut parasites as well as free-living species, suggests a different scenario (Siddall et al. 1993). When the mode of life of the different species is mapped onto their phylogeny, the most parsimonious evolutionary explanation for the observed pattern requires two independent reversals to a free-living lifestyle from parasitic ancestors (fig. 2.1). Are such reversals also possible in metazoan parasites? The nematodes would be likely candidates, because they include taxa alternating between free-living and parasitic cycles, and show little morphological specialization for parasitism, but a fully resolved phylogeny (Blaxter et al. 1998) suggests that reversals have not occurred in this group. However they have in other metazoan groups where parasitic species have retained some morphological resemblance with their free-living counterparts. A return to a free-living lifestyle has been documented for a mite species belonging to a family otherwise consisting strictly of parasitic species (Radovsky et al. 1997). In fact, loss of the parasitic lifestyle may have occurred on numerous independent occasions in the phylogenetic history of water mites parasitic on aquatic insects (Smith 1998). Recent phylogenetic reconstructions of the evolutionary history of leeches support blood feeding on vertebrate hosts as an ancestral characteristic that has then been lost repeatedly, as some leech lineages switched to a predatory life style (Apakupakul et al. 1999; Light and Siddall 1999; Borda and Siddall 2004). Similarly, nonparasitic lampreys are widely believed to have derived from parasitic ancestors not too different from extant parasitic lampreys (Salewski 2003). Thus reversals to a free-living existence are possible, at least in parasite taxa where morphological specialization is not too extreme.

2.3 Complex Life Cycles: Historical Contingency or Adaptation?

Organisms that have just completed a transition from a free-living lifestyle to a parasitic existence exploit a single host species. They begin by fine-tuning their exploitation of that host and their methods of transmission to other hosts of that species. During the course of evolution, however, these simple, direct life cycles can become much more complex. This does not happen because of some intrinsic tendency to evolve toward complexity; parasite evolution is neither more nor less progressive than that of other organisms. Rather, because of historical accidents and/or because of selective pressures to ensure high transmission efficiency, other host species may be added to the initial cycle and sometimes later dropped. The end result of this process is the panoply of life cycles observed among living parasites, from the most basic and simple ones to fantastically complicated cycles involving many host species, habitats, and parasite life stages.

Life-cycle changes must have an adaptive component and cannot be purely accidental. They are the product of selection acting on new variants resulting from random mutations. Changes in the complexity of the life cycle, however, can be either necessary responses to changes in external environmental conditions, or the outcome of the differential success of alternative transmission strategies under stable external conditions. The driving force behind the evolution of life cycles is thus either external events (contingency) or just plain old natural selection among heterogeneous individuals (adaptation). Of course, these two possibilities are not mutually exclusive.

2.3.1 Increases in Life-Cycle Complexity

The simplest explanation for the evolution of complex life cycles from simple ones is that new hosts were added to the cycle following historical events that affected the transmission of the parasite or the survival of its host. Selection favored those parasites capable of adjusting to the new conditions, and the lineages that survived were the ones able to make the best of a difficult new situation. Consider the case of an ancestral parasite with a one-host life cycle in which the host is an invertebrate. Long after the association between this parasite and its invertebrate host became established, vertebrates appeared on the scene and began feeding on the host. As vertebrates became more abundant, the probability of any invertebrate parasites ending up in a vertebrate gut increased. Natural selection would have strongly favored any parasite with the ability to survive in the gastrointestinal tract of vertebrates, and those able to do so would become the ancestors of parasites with a two-host cycle. As landing in a vertebrate's gut became more and more likely, the parasites made adjustments to their developmental schedule so that the passage to adulthood became linked with the ingestion by a vertebrate host. Whatever the exact sequence of events, many features of parasite life cycles can be explained by such adjustments to historical contingencies. Other scenarios trace a slightly different picture, in which the transmission patterns may have become established because they initially benefited the host (e.g., Smith Trail 1980). Such explanations are less parsimonious than the ones involving only historical accidents, but they also assume that the evolution of parasite life cycles was entirely at the mercy of external factors.

Viewing such complicated patterns of transmission as mere adjustments to chance events may be simplistic, however. Fitness differences between parasites with or without the ability to survive in a new potential host, and with or without the ability to use alternative transmission routes, have surely driven the evolution of the life cycles observed today. Bonner (1993) has rightly pointed out that an organism is not just the adult form but the whole life cycle from the fertilized egg to late in the adult life. We may think of a fluke as a flatworm living inside a vertebrate, but natural selection has acted on all stages of its development to shape the life cycle. In fact, it may be more adequate to view the life cycle itself as the unit of selection rather than the individual parasite. Or, in gene-centered language, the units of selection are the genes coding for the life cycle and not simply for the adult parasite. In this light, we may ask whether life cycles are adaptations rather than mere accidents.

In theory, there are two ways in which a simple, one-host life cycle can become more complex with the addition of a second host (fig. 2.2). Both can be favored by natural selection under a range of realistic conditions that have been formalized mathematically by Parker et al. (2003a). The first way is by upward incorporation, by adding a new host after the original host, with the latter subsequently becoming an intermediate host. If the original host is regularly eaten by a predator of larger size, mutants able to survive and reproduce in the predator will be favored if the cost of the ability to survive in this new host is small. Because parasite fecundity generally increases with size (see chapter 4), survivorship and further growth in the predator host generate selection for delayed maturity and enhanced reproduction in this new host (fig. 2.3). The relative increase in lifetime reproductive success obtained by delaying maturity is likely to be substantially greater than the one illustrated in figure 2.3, thus offsetting the fact that it is dependent upon successful transmission. An additional benefit of adding a predator of the original host as a new definitive host is that it allows the parasites dispersed across several individual hosts to be concentrated in the definitive hosts, a process that facilitates mating and cross-fertilization (Brown et al. 2001a).

The second way in which a simple one-host life cycle can become more complex is by downward incorporation, when a regular prey of the original host happens to commonly ingest the parasite's eggs (fig. 2.2). When the ability to survive in the prey host is not too

a. Upward incorporation

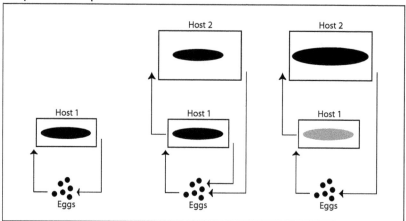

b. Downward incorporation

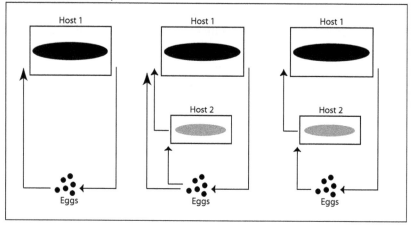

Figure 2.2 Transition from a one- to a two-host life cycle (a) by upward incorporation of a definitive host, and (b) by downward incorporation of an intermediate host. The different steps involved are ordered from left to right. In upward incorporation, the single host in the initial life cycle is ingested increasingly frequently by potential host 2, resulting in a flexible two-host cycle with parasites reproducing in both hosts. Eventually, reproduction in host 1 is suppressed and the latter becomes an intermediate host. In downward incorporation, propagules leaving the single host in the initial cycle regularly enter potential host 2, which is a frequent prey of host 1. Infection of host 1 can then be either direct or indirect, until the eventual loss of the direct transmission route. Adult parasites in definitive hosts are shown in black, immature parasites in intermediate hosts are shown in gray. (Modified from Parker et al. 2003a)

costly, this new host may become an intermediate host if it enhances transmission to the original host (which is now a definitive host). In an established two-host cycle, a further intermediate host can be added if it sufficiently increases transmission rates between the two hosts, resulting in three-host cycles (Parker et al. 2003a; see also Choisy et al. 2003). The end results of upward and downward incorporation of new hosts are identical, and determining which route was taken by any given taxon requires detailed knowledge of its phylogenetic history.

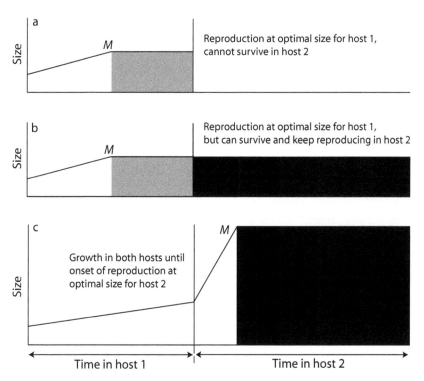

Figure 2.3 Parasite growth and reproduction in upward incorporation of a new definitive host. Here it is assumed that parasite growth stops at maturity, *M*, when the parasite begins to reproduce. (a) In the initial one-host cycle, the parasite matures at a size optimal for the space and other resources afforded by host 1, and its lifetime reproductive success is proportional to its size-dependent reproductive rate times its life span (the shaded area). (b) In the transitional cycle, the parasite can survive in host 2 and carries on reproducing in this host, achieving a greater reproductive success (the sum of the shaded and black areas). (c) The two-host cycle is achieved when reproduction is suppressed in host 1 to allow the parasite to achieve a much greater size, and therefore a higher reproductive output, in host 2; its expected lifetime reproductive success is also higher (compare the black area to those of earlier steps in the evolution of the cycle). (Modified from Parker et al. 2003a)

In acanthocephalans, phylogenetic evidence suggests that the life cycle became more complex via upward incorporation of a new host. The ancestor of living acanthocephalans was originally an endoparasite of a marine arthropod, and only subsequently added a vertebrate predator as a second host, to form the basic two-host life cycle inherited by extant acanthocephalans (Near et al. 1998; Garcia-Varela et al. 2000; Herlyn et al. 2003).

In contrast, nematodes provide examples of increases in life-cycle complexity by downward incorporation of a new host. Complex life cycles have evolved independently in several large and distantly related groups of parasitic nematodes (Anderson 1984; Adamson 1986; Clarck 1994; Durette-Desset et al. 1994; Blaxter et al. 1998). Although other scenarios are not impossible, it seems likely that nematodes parasitic in vertebrates began with one-host cycles involving only the vertebrate host. Intermediate hosts, either invertebrates or vertebrates, were added later. In some cases the term *paratenic host* may be more appropriate; paratenic hosts serve to transfer infective stages from one host to another but parasites do not develop in paratenic hosts. Whether the dependence of the parasite on a particular host is strictly ecological and not physiological as well is not important in the present discussion. Some of the advantages of using intermediate or paratenic hosts include protection from the external environment and the channeling of the parasite toward the final host if the intermediate host is a prey of the definitive host (Anderson 1984, 2000). The same advantages apply if the intermediate host is itself a parasite of the definitive host rather than one of its prey: several nematodes parasitic in fish apparently use the ectoparasitic crustacean *Argulus* as intermediate hosts, from which they derive nutrients for their continued development and transport to a suitable definitive host (Moravec 1994). In light of these benefits, the common addition of intermediate hosts to the life cycle of nematodes suggests that it was adaptive rather than accidental.

Downward extension of the life cycle also explains the evolution of complex life cycles in parasitic flatworms. The most parsimonious reconstruction of the evolution of parasitism in the phylum Platyhelminthes is that the lineage that gave rise to digeneans and cestodes first became parasitic on vertebrates (Littlewood et al. 1999a). Subsequently, cestodes added an arthropod intermediate host, well before diversifying into all extant species (Mariaux 1998; Olson et al. 2001), whereas digeneans added a molluscan intermediate host (Cribb et al. 2003; see also Brooks and McLennan 1993a). The digenean life cycle further increased in complexity when an additional intermediate host was inserted between the mollusc and the vertebrate hosts. This happened several independent times, in separate branches of the digenean phylogenetic tree (Cribb et al. 2003), resulting in the vast panoply of life cycles and host associations observed among living digeneans (Shoop 1988; Gibson and Bray 1994; Rohde 1994a; Cribb et al. 2003; Galaktionov and Dobrovolskij 2003). The inclusion of a second intermediate host allowed the prolongation of the infective life of the digenean cercarial stage, which encysts as a metacercaria in the second intermediate host, and thus increased its probability of ingestion by the definitive host. In addition, the use of a second intermediate host by digeneans can allow the mixture of different clones produced asexually in the first intermediate host, thus reducing the risk of mating between genetically identical individuals after transmission to the definitive host (Rauch et al. 2005).

The life cycles of trophically transmitted parasites have become embedded in food webs, with parasites passed upward from prey to predators via predation. Although textbooks often present parasite life cycles in an ecological vacuum, there are many predators other than suitable definitive hosts that may ingest a prey and the parasites it contains. This is not necessarily the end of the road for the parasites, however. For instance, in intertidal systems, some predatory gastropods can feed on live or recently dead molluscs and crustaceans, and at the same time ingest the larval stages of digeneans, nematodes, or acanthocephalans in these prey. If the definitive hosts of these parasites can also feed on the gastropods, the parasites might still reach their final destination, albeit after a detour (e.g., McCarthy et al. 1999; McFarland et al. 2003; Latham et al. 2003). Similarly, the two-host life cycle of nematomorphs, involving an aquatic and a terrestrial insect, can possibly involve spurious paratenic hosts, such as snails and crustaceans, and still be completed (Hanelt and Janovy 2003, 2004). In the complex network of trophic relationships within a natural food web, many secondary, even fortuitous, transmission routes can all lead to a definitive host. Life cycles can thus be even more complex than they appear.

The complex life cycles of helminth parasites generally comprise at least one instance of trophic transmission, and the scenarios depicted in figure 2.2 can account for their evolution. However, the complex life cycles of some ectoparasites will require different explanations. Many ticks require two or three individual hosts, usually of different species, to complete their life cycle (Oliver 1989). Similarly, ectoparasitic copepods of the family Pennellidae require two hosts. After attaching to the first fish host, the young copepods develop into adults. Depending on the genus, they either mate and then detach from the host, or detach first and then mate while free in the water column. After mating, males die whereas females spend some time in the plankton, and then attach to a new fish host, on which they usually undergo profound morphological changes and massive increases in size prior to egg laying. In some genera, such as *Lernaeenicus*, the two hosts are of the same fish species, whereas in other genera, such as *Lernaeocera*, the two fish hosts are of different species (see Anstensrud and Schram 1988; Whitfield et al. 1988). In ticks, the successive hosts are of increasing sizes, but not in pennellid copepods. There is little doubt that the female copepod could achieve the same resource uptake and the same growth on the initial host, and there are currently no good explanations for the evolution of complex life cycles in these parasites.

Mathematical models tend to emphasize transmission or reproductive benefits as the main drivers behind the evolution of complex life cycles (Dobson and Merenlender 1991; Brown et al. 2001a; Choisy et al. 2003; Parker et al. 2003). Several other studies have used different adaptive arguments to explain the evolution of complex life cycles (Combes 1995, 2001; Ewald 1995; Morand 1996a). Clearly, if fitness benefits can be gained by adding steps to the life cycle, natural selection should favor the parasite individuals able to follow the most rewarding cycle. Ewald (1995) proposed that benefits associated with specialization on different hosts for different resources could have driven the increase in complexity of parasite life cycles. If the fitness of the parasite is greater when one host is used as a food base and another host as an agent of dispersion in time and space, than when a single host is used for both purposes, then a two-host cycle will be favored. Ewald (1995) supports his idea with data on the virulence of several helminth parasites in their intermediate and definitive hosts that show that parasites have more severe effects on the intermediate (food base) host than on the definitive (transport) host. The arguments in support of this specialization are not entirely convincing; for instance, few parasites convert tissues from their intermediate hosts into parasite tissues, so that the role of intermediate hosts as food bases is not clear. In any event, data on parasite transmission success and how it varies as a function of the complexity of the life cycle are what we need to evaluate life cycles as adaptations.

The only way to assess whether more complex life cycles actually lead to improved fitness would be to compare the fitness of pairs of related species that differ only in the complexity of their life cycle. An example is provided by two sympatric species of the marine cestode genus *Bothriocephalus*. The two species differ only with respect to their life cycle (Robert et al. 1988). One species, *B. barbatus*, has a two-host life cycle; larval stages

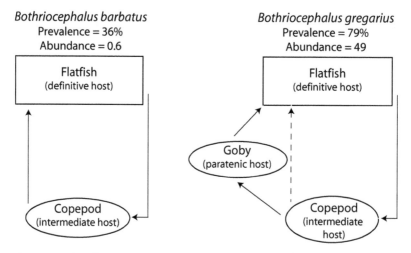

Figure 2.4 Life cycles of the cestodes *Bothriocephalus barbatus* and *B. gregarius*. Values for prevalence and abundance of infection in the definitive host are from populations in the Mediterranean Sea. (Modified from Robert et al. 1988; Morand et al. 1995)

live in a planktonic copepod, which is ingested by the final host, a flatfish. The second species, *B. gregarius*, can be transmitted in the same way, but usually goes through a paratenic host (a gobiid fish), which feeds on copepods and is itself eaten by the flatfish definitive host. In the Mediterranean Sea where the two cestodes occur, *B. gregarius* is more prevalent and much more abundant in the definitive host than its relative with the simpler life cycle (fig. 2.4). This may suggest that the addition of the paratenic host resulted in greater transmission success. Prevalence and abundance, however, are population measures and cannot be used to assess the success of individuals, on which selection acts. A mathematical model devised for the life cycles of the two species of *Bothriocephalus* suggests that individuals with the more complex life cycles do in fact achieve greater transmission success (Morand et al. 1995). This is due to a larger proportion of infective stages reaching the definitive host when a paratenic host is used. In general, the conditions for the addition of hosts to a life cycle that are obtained from mathematical models indicate that the additional mortality incurred because of the new transmission step must be compensated by increases in net transmission efficiency or reductions in the mortality of free-living stages (Dobson and Merenlender 1991; Choisy et al. 2003; Parker et al. 2003).

Many more contrasts between related taxa that differ only in the complexity of their life cycles would be necessary before we know whether complexity is the outcome of selection or not. General trends associated with evolutionary additions of hosts to life cycles may be difficult to extract from these comparisons, however, because only a few of them are possible. Despite the many thousand living species of helminths displaying a multi-host life cycle, the radiations from which these parasites are issued follow just a handful of transitions from simple to complex life cycles. Acanthocephalans, for instance, are a monophyletic group; all living species and their typical two-host cycles are most likely descended from a single common ancestor in which the cycle evolved (Herlyn et al. 2003).

Complex cycles have had a few independent origins among nematodes and platy-helminths parasitic in animals (Anderson 1984; Adamson 1986; Brooks and McLennan 1993a; Blaxter et al. 1998; Littlewood et al. 1999a). Similarly, the life cycles of parasitic protozoans have evolved toward greater complexity on separate occasions, even within particular groups (Barta 1989; Smith et al. 2000; Slapeta et al. 2001; Su et al. 2003). Still, the relative rarity of these changes in the complexity of life cycles within any given phylum may prevent any robust comparative analysis. Comparisons between individuals of the same species that differ in life cycle would be even better, though they would be even more difficult to carry out since not many species have facultative steps in their life cycle.

2.3.2 Abbreviation of Complex Life Cycles

If the addition of hosts to simple life cycles can be advantageous in some cases, surely the loss of hosts from complex cycles can also be advantageous in other situations. For example, blood flukes of the families Sanguinicolidae, Spirorchidae, and Schistosomatidae display a two-host cycle most likely derived from an ancestral three-host cycle, typical of the majority of digenean groups (Shoop 1988; Brooks and McLennan 1993a; Cribb et al. 2003). Is the abbreviated two-host life cycle an adaptation? Cercariae of blood flukes penetrate the definitive host directly after leaving their mollusc intermediate host; the "current" definitive host probably used to be the second intermediate host in the ancestral life cycle, and became the host in which the parasite matures after the original definitive host was lost (Shoop 1988; Cribb et al. 2003). Why it was lost we do not know; however, we can speculate that transmission rates to the former definitive host must have dropped to the point where it paid off to abandon it. If this return to a simpler life cycle is adaptive we might expect it to have occurred in other digenean lineages.

A shortening of the life cycle has indeed taken place in other digenean groups, on numerous independent occasions (Poulin and Cribb 2002). Truncations of the life cycle of some sort have been reported in over thirty digenean families, and because they are often observed in only one or a few species per genus or family, they appear to be the product of at least twenty independent evolutionary events. For instance, some species of the genus *Alloglossidium* have a normal three-host digenean life cycle, but other species in the same genus have a two-host cycle (Carney and Brooks 1991; Smythe and Font 2001). A phylogenetic analysis of the genus suggests that the three-host cycle was the ancestral condition and that species with a two-host cycle are all derived from a common ancestor, the loss of the vertebrate host having happened only once in the lineage (fig. 2.5).

There are many ways in which digeneans can achieve simpler life cycles (fig. 2.6). The first is to adopt progenetic development: following encystment as a metacercaria in the second definitive host, the worm develops precociously into an egg-producing adult. What is normally the last transmission event in the cycle—ingestion of the second intermediate host by the definitive host—is bypassed. Progenesis is the most common type of life cycle abbreviation in digeneans (Poulin and Cribb 2002; Lefebvre and Poulin 2005). The second type of life-cycle truncation consists of also using the molluscan first

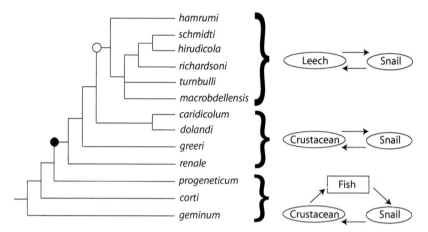

Figure 2.5 Phylogeny of species in the digenean genus *Alloglossidium* showing the evolution-ary changes in life-cycle patterns. The three-host cycle is the ancestral condition and is still present in three species. The fish host was lost only once in the evolution of the lineage (black circle), and there was one host switch in which the crustacean host was replaced by a leech (open circle). Worms using only two hosts reach adulthood in either the crustacean or the leech. (Modified from Smythe and Font 2001)

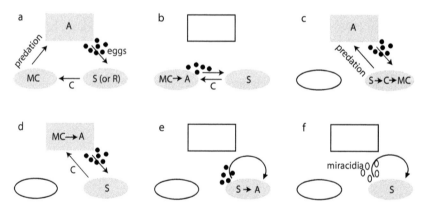

Figure 2.6 Typical three-host digenean life cycle and the ways in which it can be abbrevi-ated. (a) In the three-host cycle, adults, A, living inside definitive hosts release eggs into the environment. Molluscan first intermediate hosts are infected by eating the eggs or by the miracidia that hatch from the eggs. In the mollusc, following asexual multiplication of lar-val stages in sporocysts, S, or sometimes rediae, R, the cercariae, C, leave the mollusc to en-cyst as metacercariae, MC, in the second intermediate host. The cycle is completed when this second intermediate host is ingested by the definitive host. The five different types of life cycle truncation are: (b) progenesis in the second intermediate host; (c) first intermedi-ate host used as second intermediate host; (d) second intermediate host also used as defini-tive host; (e) sexually reproducing adult develops in first intermediate host; and (f) sporo-cysts in the first intermediate host produce miracidia directly. (Modified from Poulin and Cribb 2002)

intermediate host as the second intermediate host; this way, cercariae do not need to emerge from the mollusc, they simply turn into metacercariae on site. The third form of life-cycle truncation involves using the second intermediate host as a definitive host as well. For example, cercariae of the macroderoidid *Haplometra cylindracea* encyst as metacercariae in the mouth of frogs; after a few days, the cysts burst and the worms migrate to the same frogs' lungs where they mature into adults (Grabda-Kazubska 1976).

The fourth and fifth types of life-cycle truncation are the most extreme, consisting of one-host life cycles derived from a three-host cycle. In the fourth type, sexually mature adult worms develop directly inside the mollusc first intermediate host, before producing eggs that hatch into miracidia (e.g., Barger and Esch 2000). Finally, the fifth type of life-cycle truncation is observed in species of the genus *Mesostephanus* (family Cyathocotylidae). As in other digeneans, miracidia of these species penetrate the snail intermediate host in which they multiply asexually. Instead of producing cercariae, the next stage in the life cycle responsible for infecting the second intermediate host, they can directly produce and release new miracidia, which can presumably reinfect other snails (Mohandas 1975; Barker and Cribb 1993). The parasite skips two hosts and one round of sexual reproduction each time it does this.

Abbreviations of complex life cycle are not restricted to digeneans. Some species in the cestode genus *Hymenolepis* can truncate their cycle from two hosts to a single one, at least when immunodeficient definitive hosts are available (Andreassen et al. 2004). Progenesis and the ability to bypass the definitive host have also been reported in other cestodes (Poddubnaya et al. 2003) and in some nematodes (e.g., Jackson et al. 1997). In addition, the nematode *Camallanus cotti*, which normally has a two-host life cycle, is capable of abandoning the need for a copepod intermediate host when its fish definitive host is reared in aquaria from which copepod hosts are absent (Levsen and Jakobsen 2002). The fact that the life cycles of many helminth groups have repeatedly and independently evolved toward greater simplicity well after complex life cycles became established is a strong indication that parasites have retained the genetic variation necessary to adjust to changing ecological conditions.

What factors promote the truncation of parasite life cycles? A few hypotheses have been proposed to explain this phenomenon (Poulin and Cribb 2002). Their common theme is that shorter life cycles are favored by selection whenever parasites experience, on a permanent or regular basis, low or highly variable probabilities of transmission at one or more stages in their life cycle. In other words, truncation is advantageous when a host species normally required for the completion of the life cycle is temporarily absent or going extinct locally. Abbreviation of the life cycle is facultative in most parasite taxa where it has been reported, and its occurrence often varies seasonally following the availability of certain host species (Poulin and Cribb 2002). In the opecoelid digenean *Coitocaecum parvum*, metacercariae inside the crustacean second intermediate host are more likely to develop precociously into adults when their crustacean host is reared in the absence of fish odors than when it is reared in fish-conditioned water (Poulin 2003a). This parasite can switch to a two-host life cycle when it obtains cues from its crustacean host that there are no fish definitive hosts around (see also Thomas et al. 2002a). In this and many other parasite species with facultative truncation of the life cycle, the truncation serves as an insurance against host bottlenecks. There are some costs associated with shorter life cycles; for instance, parasites maturing in what would normally be an intermediate host may experience lower fecundity, and, because they often cannot mate with conspecifics, they may have offspring with reduced genetic heterogeneity (Poulin and Cribb 2002; Lefebvre and Poulin 2005). Still, in many situations, the ability to complete a complex life cycle by truncation confers greater fitness to its bearer.

On longer evolutionary time scales, if a change in the complexity of the life cycle is adaptive, it may lead to higher rates of speciation because it could allow the colonization of new niches and lower rates of extinction. Brooks and McLennan (1993b) compared the species richness of the major groups of parasitic platyhelminths. They concluded that of the three groups (Digenea, Monogenea and Eucestoda) displaying an independently derived high species richness, an adaptive radiation has only occurred in the Monogenea. They attributed this to the appearance of a suite of key innovations involving the loss of one host from the life cycle and the reversal to a direct one-host cycle. This would support the claim that changes to the complexity of the life cycle are adaptive and driven by selection. The conclusions of Brooks and McLennan (1993b) are only valid, however, if they used the correct phylogeny to derive contrasts in richness among platyhelminth taxa, and if their estimates of species richness are accurate. Rohde (1996) has questioned the analysis of Brooks and McLennan (1993b), and convincingly shown, among other things, that the results are entirely dependent on a phylogenetic hypothesis that may be flawed. As for estimates of species diversity, there are indications that we only have identified a fraction of existing species (Poulin 1996a; Rohde 1996; Poulin and Morand 2004), so that any attempt to compare the richness of lineages with different life cycles may be premature.

In conclusion, the adaptiveness of parasite life cycles will prove difficult to demonstrate. The possibility that they are adaptive is enticing, and at least forces us to consider alternatives to classical ideas founded solely on the constraints of history. A final example illustrates how challenging old ideas brings a fresh new perspective on parasite evolution. The larvae of many nematodes parasitic in vertebrates undergo extensive migrations through the tissues of their host. Some migrating species infect the host by penetrating the skin, and have no choice but to migrate through the host to reach the host's gut, where they spend their adult life. Other migrating species, however, are acquired by ingestion; they start their migration from the host's gut only to end in the same place. The prevailing explanation for migration in these latter species has long been that it is an evolutionary legacy, inherited from a skin-penetrating ancestor (e.g., Adamson 1986). Comparisons between migrating nematode taxa and their nonmigrating sister groups, however, show that migrating parasites are on the whole significantly larger, and presumably more fecund, than their nonmigrating relatives (Read and Skorping 1995). Migrating worms achieve faster growth rates, possibly because of some nutritional advantage over worms developing entirely in the gut. Migrating worms may also benefit by evading the immune attacks they would face if they spent their entire life associated with the mucosal surfaces of the host's gut (Mulcahy et al. 2005). In the strongylid nematodes at least, skin penetration and tissue migration appears to be the ancestral condition (Sukhdeo

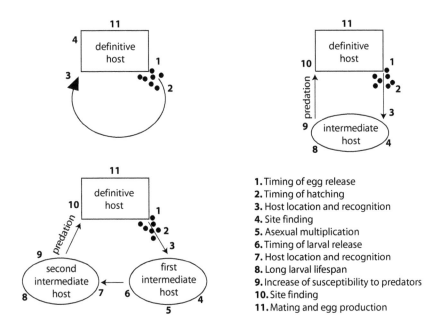

1. Timing of egg release
2. Timing of hatching
3. Host location and recognition
4. Site finding
5. Asexual multiplication
6. Timing of larval release
7. Host location and recognition
8. Long larval lifespan
9. Increase of susceptibility to predators
10. Site finding
11. Mating and egg production

Figure 2.7 Generalized one-host (e.g., monogenean), two-host (e.g., acanthocephalan or cestode), and three-host (e.g., digenean) life cycles. Adaptations favored by the cycle itself are indicated at each step in the transmission from one host to the other.

et al. 1997); the loss of skin penetration and the retention of migration as the nematodes adopted an oral infection route suggests that migration is a legacy, but it could have been maintained and favored by selection if it benefited the worms. Thus, nematode migrations appear to be more than just vestigial behavior patterns, and perhaps a similar adaptive explanation applies to parasite life cycles in general.

2.4   Evolutionary Consequences of Complex Life Cycles

Whether or not complex life cycles are the products of selection, they have in turn imposed selective pressures on parasites and lead to drastic adjustments in the parasites' biology. In the course of a complex life cycle, parasites will inhabit taxonomically unrelated hosts, visit different physical habitats, and need means of getting from one host or habitat to the next. They will assume widely different shapes, each fitted to a given part of the cycle. An adult digenean, a miracidium, and a cercaria are very distinct in morphology, but are all vehicles of the same genetic information. At each stage of the cycle, selection has favored morphologically different carriers of the parasite's genes in response to pressures from the cycle itself.

The addition of a step in transmission may or may not lead to fitness benefits, but it entails a new set of challenges that must be met by the parasite. Each stage is accompanied by an array of adaptations aimed at facilitating the completion of that stage (fig. 2.7). Some of these same adaptations are also expected in parasites with simple one-host cycles and these parasites will also be used as examples here. From the moment eggs are released by the adult parasite in the definitive host, a cascade of events must take place in the proper sequence for the successful return of the offspring to the right definitive host. The constraints and pressures on egg production by adult parasites will be dealt with in detail in chapter 4. This section focuses on the hurdles and challenges facing parasites from the moment eggs are released by the adult, and on how these have been overcome. Adaptations intrinsic to parasitism itself, such as immune evasion or the extraction of nutrients from hosts, are not considered in detail here; the focus is on the traits evolved in response to the increased complexity of life cycles. The following discussion bears on metazoan parasites that release eggs from their definitive hosts; other life cycles, such as those of vector-borne protozoans, are also associated with a panoply of adaptations. Combes (1995, 2001) provides a more comprehensive review of these and other adaptations to complex life cycles.

2.4.1 Transmission and Infection

At each stage in the cycle, selection favors those parasites better at infecting the next host in the cycle, the target host. Because target hosts, whatever they are, are patchily distributed in space and time, increasing the likelihood of finding them can begin with a precisely timed release of eggs or infective stages. The cestode *Triaenophorus crassus*, for example, stores up to a few million eggs in its body during its yearlong life in the pike, *Esox lucius*, the cestode's definitive host. Pike only enter shallow water once a year, to spawn. The cestode's first intermediate host is a planktonic copepod inhabiting shallow water. Egg release by the cestode coincides precisely with the pike's spawning activity: after a year during which eggs are accumulated, they are released in a brief instant the moment the parasite finds itself in the microhabitat of its next target host (Shostak and Dick 1989). An even more striking example of well-timed egg release comes from some polystomatid monogeneans, in which the temporal window of opportunity for transmission is even narrower. Polystomatids, like other monogeneans, are transmitted by a swimming ciliated larva; however, unlike other monogeneans, many polystomatids exploit vertebrates that do not spend their entire life in water. One species, *Pseudodiplorchis americanus*, lives in the urinary bladder of spadefoot toads in Arizona (Tinsley 1990, 1999). The toads live in deserts and hibernate in burrows for nine to ten months per year. They become active following the annual summer rains, and spawn in temporary pools. Spawning takes place over one to three nights, and individual toads may spend, in total, less than twenty-four hours per year in water. The parasite has synchronized its egg release with these brief visits to water, abruptly expelling a whole year's production of offspring into the host's urine when the host enters water.

The preceding examples are convincing, but they do not tell the whole story. Variability in the timing of egg release among related species exploiting the same host species suggests that other factors are also important. For instance, two of the three most abundant species of trichostrongyle nematodes parasitic in reindeer in northern Norway show peaks of egg release in summer, a pattern common among other trichostrongyles in temperate regions. In contrast, the third species has its highest egg output in the winter, when cold temperatures limit egg and larval development (Irvine et al. 2000). What promoted these differences remains to be determined, but clearly factors other than the probability of immediate transmission must influence the evolution of patterns of egg release in parasites.

The timing of egg release with the period of greatest availability of the target host is only the first step toward ensuring that a target host will be reached. In many parasites eggs do not hatch until they have been ingested by the target host. Since the time between release from the definitive host and ingestion by the target host may be highly variable, selection has often favored eggs that can remain viable for very long periods of time. Nematode eggs, for instance, possess a thick, resistant shell that protects the larvae against environmental hazards such as desiccation, and the larvae themselves can tolerate sub-zero temperatures; these adaptations allow them to survive for up to several years in some cases (Wharton 1986, 1999; Perry 1999).

In other parasites, eggs hatch to release infective larvae that must locate and infect the target host. In some parasites, such as the fish ectoparasite *Argulus coregoni*, egg hatching within a single clutch of eggs ranged over several months, with peaks corresponding to temperature fluctuations (Hakalahti et al. 2004). This fits well with a bet-hedging strategy in which the parasite copes with the temporal variability in host availability by spreading egg hatching over months. In other types of ectoparasites, however, because parasite larvae typically do not feed and are therefore short-lived, hatching of the eggs has evolved to be fully synchronized with the presence of target hosts if these are transients and if their availability is variable. Again, monogeneans provide good examples. Hatching of monogenean eggs is typically finely tuned either to environmental cycles or to stimuli of direct host origins (Kearn 1986; Tinsley 1990). In species exploiting hosts with a clear-cut circadian behavior pattern leading to day-night differences in their susceptibility to infection, egg hatching often shows circadian rhythms timed with photoperiod (fig. 2.8). In other species exploiting hosts with unpredictable availability, eggs remain viable for long periods and hatching is only triggered by specific cues. These include chemical substances specific to the host mucus, as well as less specific stimuli such as passing shadows or physical disturbances in the surrounding water (Kearn 1986). Such precise hatching synchrony illustrates how selection has finely tuned the match between the parasite's biology and that of the host, as a response from pressures generated by the life cycle itself.

The release of eggs by the adult parasite in the definitive host may not be the only place in the cycle where infective stages are released to find a target host. In digeneans, the life cycle also includes the release of cercariae from the first (mollusc) intermediate host and the subsequent infection of a second intermediate host (fig. 2.7). Here the life cycle also favours the parasite genotypes that allow a perfect match between the timing of cercarial release and the susceptibility of the target host. Cercariae are typically short-lived, having a life span of a few to several hours (Combes et al. 1994). Because the availability of the next host in the life cycle is unpredictable, cercarial production declines with the age of infection in some digenean species, a strategy that keeps the snail host alive for longer and extends the period of cercarial production (Karvonen et al. 2004b). On shorter time scales, for instance, over twenty-four-hour periods, cercarial release often peaks

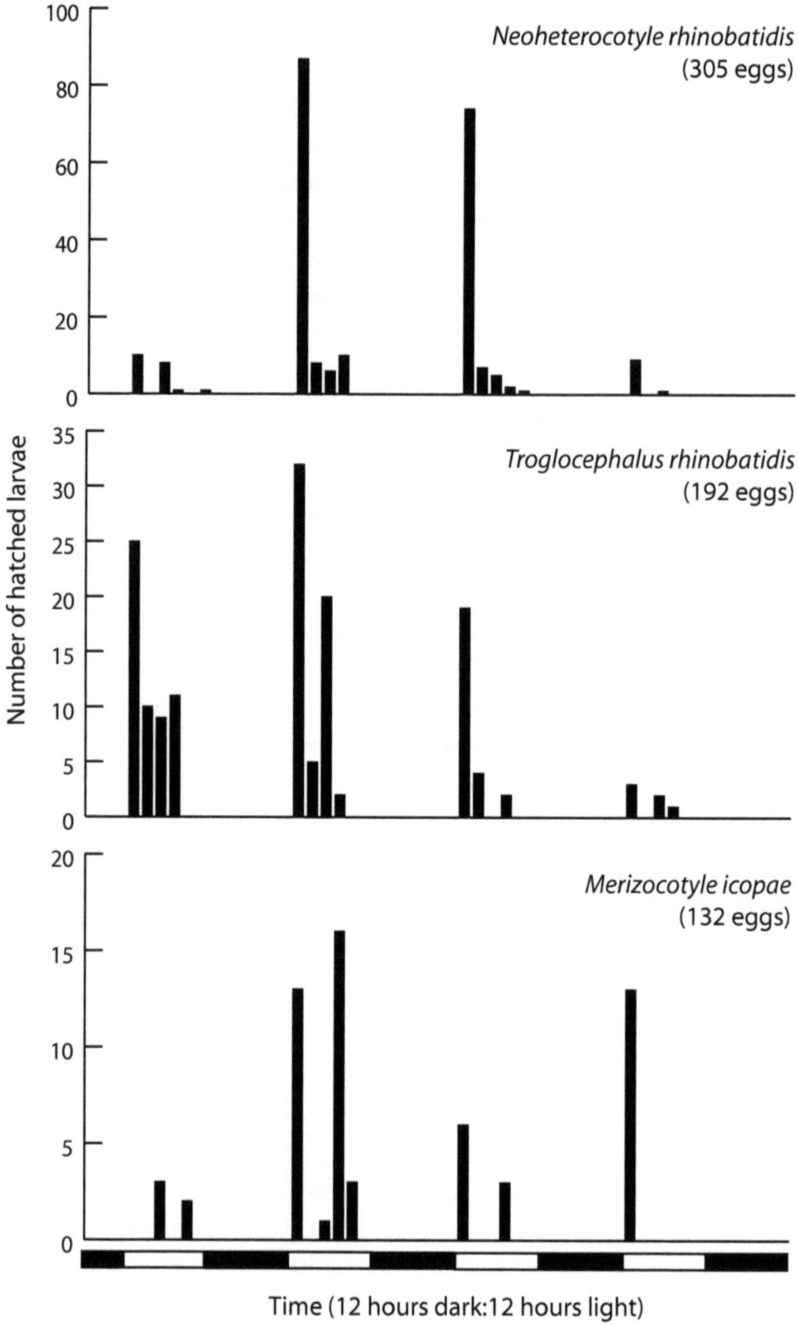

Figure 2.8 Temporal patterns of egg hatching in three species of the monogenean family Monocotylidae, all parasitic on the shovelnose ray *Rhinobatos typus*. All three species show a clear-cut circadian rhythm, with hatching only occurring during daylight hours. (Modified from Chisholm and Whittington 2000)

when hosts are most likely to be available. The main factor synchronizing cercarial shedding from mollusc hosts is the ambient photoperiod, although other factors can be important. For instance, in intertidal habitats, tides are also key external agents synchronizing the release of cercariae, sometimes indirectly by allowing increased light levels and higher temperatures in shallow tidal pools where many snails aggregate (Mouritsen 2002a; Fingerut et al. 2003a). Patterns of emergence vary greatly across species, and often appear to correspond to temporal patterns of activity or susceptibility of the target host (e.g., Théron 1984; Combes et al. 1994; McCarthy et al. 2002). Among schistosomes, cercarial emergence coincides with the periods during which target hosts are most likely to visit water (fig. 2.9). Other environmental influences such as predator avoidance may sometimes play a role in the evolution of the diurnal pattern of cercarial emergence (Shostak and Esch 1990). But in many cases, the precise fit between the activity of the target host and the timing of cercarial emergence is strong evidence that emergence patterns are adaptations for transmission. In addition, experimental crosses between populations of the same schistosome species that differ in the timing of cercarial emergence indicate that cercarial release is under genetic control (Combes et al. 1994).

Maximizing temporal overlap with the target host is one of two types of adaptation required by the infective stages of a parasite in order to complete their phase of the life cycle. The other challenge is to find the host in space (fig. 2.7). The free-living stages of most metazoan parasites are short-lived and highly susceptible to environmental conditions (Pietrock and Marcogliese 2003); without finely tuned host location mechanisms, their infection success would be nil. During the infective stages, most parasites do not feed, and thus have finite energy reserves. Typically, whatever the type of parasites, the proportion of infective larvae surviving over time, following hatching or release from the previous host, decreases following a sigmoidal function (fig. 2.10). Although mortality may be negligible initially, it increases rapidly afterward. Environmental temperature plays a key role here. Because the metabolic rate of the larvae is proportional to temperature, they utilise their finite energy stores faster at higher temperatures, and incur a shorter lifespan as a consequence (fig. 2.10).

In most cases there appears to be a trade-off between energy depletion and host encounter rates (Fenton and Rands 2004). The optimal infection strategy ranges from a passive ambushing strategy, in which energy is conserved while the parasite waits for a host to come by, to a more active cruising strategy in which the parasite spends its limited energy at a higher rate by searching for hosts (Fenton and Rands 2004). Which strategy is favored by natural selection depends mainly on the local probability of encountering a host. Whether ambushing or cruising, parasite infective stages rely heavily on sensory input to detect hosts. The infective stages of parasites with direct life cycles, such as monogeneans, copepods, or rhizocephalans, respond to a variety of stimuli that tend to bring them within proximity of their hosts (e.g., Rothsey and Rohde 2002; Pasternak et al. 2004). In digeneans, there are usually two free-living infective stages, miracidia and cercariae, and both often display a range of responses to simple cues that facilitate host location (see Saladin 1979; Sukhdeo and Mettrick 1987; Combes et al. 1994). These infective stages can either respond to cues from the physical environment that bring them within

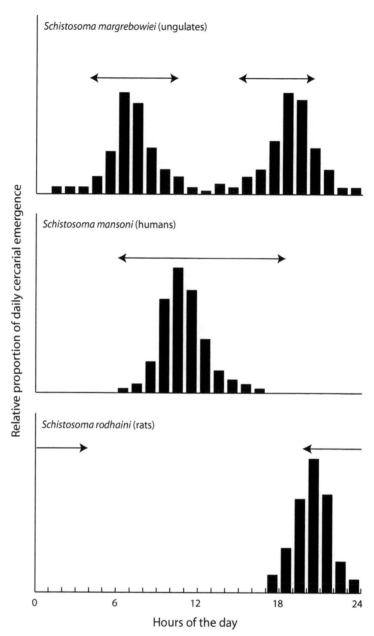

Figure 2.9 Temporal patterns of cercarial emergence in three species of schisto-somes with different definitive hosts. The columns represent the relative number of cercariae released by snail intermediate hosts for each hour of the day. The definitive hosts are indicated in parentheses, and their approximate periods of visits to water are indicated by arrows. (Modified from Combes et al. 1994)

the area inhabited by the target host, or on a smaller scale they can respond to specific cues produced by the target host (Combes 1991a; Combes et al. 1994).

Miracidia searching for snail hosts are quite specific in their choice of host species. They can identify specific macromolecules released by their hosts and respond by altering their swimming behavior: they increase the frequency of turning, which maintains

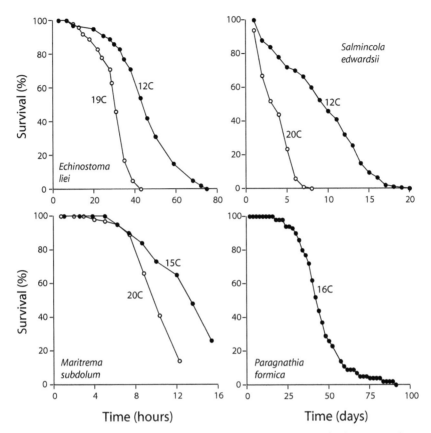

Figure 2.10 Survival of infective larval stages as a function of time, in the digeneans *Echinostoma liei* and *Maritrema subdolum*, the ectoparasitic copepod *Salmincola edwardsii*, and the ectoparasitic isopod *Paragnathia formica*. Digenean cercariae released from snail first intermediate hosts, and crustacean larvae hatched from eggs, are nonfeeding stages that must find and attach to or penetrate a host in order to pursue their life cycle. Their life span decreases with increasing temperature, as shown for the digeneans and the copepod. (Data from Evans 1985; Mouritsen 2002a; Conley and Curtis 1993; Tinsley and Reilly 2002)

them in the small area where host chemicals have been detected (Haas et al. 1995). Cercarial host location seems to operate mostly on a larger scale. Cercariae of different species, for instance, show different responses to light or gravity and consequently migrate up or down the water column in an attempt to match the spatial distribution of their target host (fig. 2.11). The swimming abilities of cercariae cannot be underestimated. For instance, even in turbulent coastal waters, active swimming by cercariae of the echinostomatid *Himasthla rhigedana*, guided by appropriate responses to environmental cues, can overpower water flow and eddies under all but storm conditions and bring the cercariae within the vicinity of their next host (Fingerut et al. 2003b). On a finer spatial scale, the cercariae of many, but not all, digenean species are able to respond to host chemicals (Combes et al. 1994; Haas 1994; Haas et al. 1995). Interestingly, the host finding of miracidia and cercariae of the same species appear to operate independently: in *Echinostoma caproni*, both miracidia and cercariae search for the same snail host, since this digenean uses the same snail as first and second intermediate host, and yet

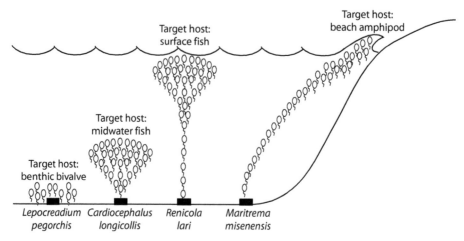

Figure 2.11 Spatial dispersion patterns of cercariae of four species of digeneans, each with a different target second intermediate host, after their emergence from their benthic molluscan host in a marine littoral habitat. (Modified from Combes et al. 1994)

each infective stage relies on different chemical signals to locate the host (Haberl et al. 2000). In any event, cercarial host finding is often precise and sophisticated. For example, certain chemical cues stimulate the cercariae of *Diplostomum spathaceum* to remain in contact with a surface following contact, but a different set of chemical signals is required to induce the cercariae to penetrate host skin (Haas et al. 2002). The unique profile of cues necessary to complete the infection sequence limits the number of mistakes made by the searching cercariae. Mortality of cercariae in this species is age dependent, suggesting that cercariae begin their short life (<36 hours) with roughly equal amounts of resources and die as these are exhausted (Karvonen et al. 2003); efficient host recognition mechanisms are essential as second chances are likely to be rare. Cercariae of many digenean species often have a broader range of acceptable host species than miracidia, but the lack of specific host-recognition mechanisms in some species results in mistakes and the death of the cercariae. The dermatitis caused by avian schistosomes accidentally penetrating human skin is an example of such mistakes. That the ability to recognize suitable hosts has not evolved in cercariae of all species is therefore surprising.

It must be pointed out that not all digeneans rely on their cercariae finding a suitable host: many rely on the host finding the cercariae. For instance, cercariae of some species form dense clusters instead of dispersing individually following their emergence from snail hosts. These aggregates are supposedly large enough to attract suitable target hosts, such as fish, to feed on them. Whether or not transmission does indeed occur by predation of cercarial clusters in these species, this strategy must be efficient because it has apparently evolved independently in several digenean lineages (see Hendrickson and Kingston 1974).

At each transmission stage in the life cycle, the probability of finding a suitable host is very small indeed. The addition of new steps in the cycle has apparently stacked the odds against the completion of the cycle by any individual larva. Whatever the reasons for the increase in life-cycle complexity, the pressure has been on parasites to come up with ways to compensate for the heavy losses incurred during successive transmission events. In a three-host cycle (see fig. 2.7), with two stages involving the release of propagules and the infection of hosts by mobile larvae, larval mortality can make the completion of the cycle unlikely. If the first intermediate host is used only as a food source for the transformation of the first mobile larva into a single new larva for the next stage, and if the probability of success in both transmission events is .001, then out of a million eggs produced by the adult on average only a single one would reach the second intermediate host. And the cycle would not yet be completed, as the parasite must reach the definitive host, survive, grow, and mature successfully inside the host. Although these numbers are arbitrary, they illustrate the sort of odds faced by parasites with complex life cycles.

There are at least three obvious solutions to this problem, and parasites typically resort to more than one of them. First, the adult can be selected to produce more eggs; this is seen as a typical adaptation of parasitic helminths and will be discussed in chapter 4. Second, selection could opt to increase the success rate and favor the production of larvae increasingly better equipped for survival and host location. The costs of producing such larvae may impose limits on how many could be produced; again, chapter 4 will return to potential trade-offs in life history strategies like this one. Third, selection could favor the production of a fresh new wave of infective stages by the original larvae that were successful in the first infection event. This strategy is a characteristic of digeneans: each miracidium successful at infecting a mollusc intermediate host will multiply asexually to produce large numbers of cercariae ready to embark on the next leg of the cycle. For example, in schistosomes, thousands and even tens or hundreds of thousands of cercariae may be produced from one miracidium (Loker 1983). Similar numbers are observed in other digenean groups (e.g., Shostak and Esch 1990; Combes 1995, 2001). Digeneans produce such staggering numbers of cercariae by achieving high daily output rates in combination with a long tenure in the snail intermediate host. For instance, in the long-lived estuarine snail *Ilyanassa obsoleta*, infections by different species of digeneans persist for several years, possibly even decades (Curtis 2003).

Since all digeneans share this trait, it is most likely the outcome of a single evolutionary event, that is, it appeared once in the ancestor of all digeneans. The independent appearance of a trait in several lineages under similar selective pressures would provide strong evidence of adaptive convergence. Digeneans are not the only taxa capable of asexual multiplication in intermediate hosts: the trait has evolved independently in several cestode lineages (Moore 1981; Moore and Brooks 1987). It is therefore likely that this amplification of numbers is a trait strongly favored by selection but restricted to lineages in which it can evolve because of sufficient genetic variation. The nature of the intermediate host used can also constrain the evolution of asexual amplification; for instance, many arthropods used as intermediate hosts by helminths may not provide sufficient resources for asexual multiplication, or may not be able to survive it.

The use of mobile infective stages is not the only way in which parasites are transmitted from one host to another. Predation of an intermediate host by the definitive host (fig. 2.7) is common in the life cycles of parasitic protozoans, digeneans, cestodes, nematodes, and acanthocephalans. Many of these parasites exploit existing food chains and ride

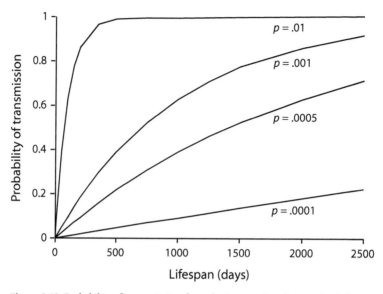

Figure 2.12 Probability of transmission from the intermediate host to the definitive host as a function of the parasite's life span inside the intermediate host. Given p, the probability of transmission by predation on any given day, the function is obtained from the standard formula for the probability of at least one successful outcome from a series of trials, where the overall probability of transmission is equal to $1 - (1 - p)^t$, in which t is life span in days. The function is illustrated for four values of p.

along predator-prey interactions, a fact that leads many to believe that life cycles are merely the accidental by-products of the evolution of predation. In any event, in cycles depending on such rare events as the capture of infected prey by suitable predators, selection could favor at least three different strategies. First, it could favor the truncation of the life cycle, such that the need for predation is abolished; this has been discussed in section 2.3.2. Second, selection could favor parasites capable of increasing the frequency of rare predation events (Poulin 1994a). Many parasites induce modifications to the appearance or behavior of their intermediate hosts. These modifications may often be adaptive manipulations by the parasite of the intermediate host's susceptibility to predation by the definitive host (see chapter 5 for further discussion). Third, parasites could accept the rarity of predation and wait patiently for it to happen by increasing their life span in the intermediate host. Consider a parasite for which the daily probability of transmission from the intermediate to the definitive host, corresponding to the risk of predation by the latter on the former, is .001; this is a realistic number indicating that on an average day, there is one chance in a thousand that the intermediate host in which a given parasite is located will be captured and eaten by a suitable definitive host. The overall probability of successful transmission will increase asymptotically with an increase of the parasite's life span inside the intermediate host (fig. 2.12). If it can survive for a year inside the intermediate host, the parasite will have a probability of transmission of about 0.3, if it survives for three years its probability of transmission will reach 0.66, and so on. An extended life span in the intermediate host is therefore one way of beating the odds, assuming that the intermediate host itself lives longer than the parasite. For example, plerocercoids of the cestode *Triaenophorus crassus* can survive for several years in their fish intermediate host waiting to be ingested by a predatory pike, the definitive host (Rosen and Dick 1984). Other cestodes not known to alter the behavior of their vertebrate intermediate host can survive in them for years (Moore 1981). In these parasites the life cycle is interrupted by long periods without growth or development spent waiting for an improbable event. Truncation, long life in the intermediate host, and manipulation of the intermediate host are possibly alternative strategies, either incompatible or too costly to all be selected in the same parasite (Poulin 1994a; see chapter 5). This again illustrates the sort of dilemma imposed on parasites by complex life cycles.

Reaching the definitive host does not signal the end of the cycle; the parasite still has two tasks to complete. First, it must find its site of infection, the site where its adult phenotype can survive and reproduce. This site is often very specific, and may have very precise boundaries even within a large organ like the vertebrate intestine. Fitness of the parasites is dependent on how close they get to the optimal attachment site. For example, the precise location of worms in the host's intestine determines the likelihood of mating and subsequent fecundity in the acanthocephalan *Moniliformis moniliformis* (Lawlor et al. 1990) and female fecundity in the nematode *Trichinella spiralis* (Sukhdeo 1990a). During the evolution of the life cycle, the route of entry into the definitive host may have been changed entirely. For instance, parasites that used to penetrate the host through the skin may have evolved to use the oral route of infection. Changes like this one during the evolution of the life cycle would have put great pressures on parasites to adjust their way of getting to their final site of infection. Sukhdeo (1990b) suggested that the host internal environment is much more predictable than the environment at large, because all conspecific hosts are almost identical in construction and function, the same organ secreting the same chemicals in all hosts, and so forth. In these predictable conditions, any behavior increasing the chance of arriving at the correct site of infection would be favored and would quickly spread to fixation through the parasite population (Sukhdeo 1990b, 1997; Sukhdeo and Sukhdeo 1994, 2002). The rapid evolution of fixed, simple behavior patterns could make adjustments in site finding relatively easy following alterations to the life cycle affecting the route of entry into the definitive host. The second and final task necessary to complete the cycle is finding a mate and the act of mating itself.

2.4.2 Sexual Reproduction

Negative frequency-dependent selection is probably the most important force maintaining sexual reproduction in organisms in general: because some components of the environment keep changing, gene combinations that are adaptive in one generation may be much less favorable in the next generation (Maynard Smith 1978). In the case of parasites, host populations can develop protective immunity against specific parasite strains, and genetic resistance against host immune responses rapidly become obsolete. For long-lived parasites such as many helminths, sexual reproduction remains the best way to generate genetic diversity among offspring and stay ahead in the coevolutionary race with the host (Galvani et al. 2003). Thus, even in parasite taxa with asexual reproduction at one stage in the life cycle, sexual reproduction also features, usually in the definitive host.

Because reaching the definitive host after going through the whole cycle is a challenging task, the co-occurrence and physical encounter in the same host at the same time of a male and a female parasite of the same species is unlikely in many situations. One remarkable adaptation provides a solution to the low frequencies of encounters between sexes: the fusion of two individuals. In several families of parasitic copepods, such as the Chondracanthidae, the small male attaches permanently to the female's genital region upon first encountering a female (Raibaut and Trilles 1993). A single encounter is enough to guarantee lifetime mating success for both members of the couple. Similarly, in the monogenean family Diplozoidae and in some members of the digenean family Didymozoidae, two pre-adult worms become physically fused for life in the days following their first meeting (Kearn 1998). In the latter example, the worms are hermaphrodites, and even without fusion of individuals, hermaphrodism itself offers a solution to the low frequency of encounters between potential mates. Hermaphrodism is an ancestral trait of platyhelminths (turbellarians, monogeneans, digeneans, and cestodes); these parasites do not have to worry about finding a mate of the opposite sex, any conspecific will do. Self-fertilization is also common, especially in tapeworms (e.g., Haag et al. 1999), and eliminates the need to find a partner altogether. Things are not that simple, however, and hermaphroditic parasites face a few dilemmas. For instance, when co-occurring with conspecifics in the same definitive host, how should a hermaphrodite partition its reproductive effort between selfing and outcrossing? When outcrossing, what features should it look for when choosing a mating partner?

Answers to these and related questions are available for the cestode *Schistocephalus solidus*, which has been extensively studied in this context. This simultaneous hermaphrodite normally reproduces in the gut of birds, but it can be reared and studied in vitro. Selfing, or self-fertilization, is normally associated with inbreeding depression, or the accumulation of deleterious mutations and a reduction in the genetic variability of offspring (Charlesworth and Charlesworth 1987; Jarne and Charlesworth 1993). In *S. solidus*, worms kept alone and reproducing by selfing produced eggs at a higher rate than worms placed in pairs (Wedekind et al. 1998). Over the worms' lifespan, there was no difference in total egg output because paired worms compensated for their lower rate of egg production by producing eggs for longer than selfing worms (Schärer and Wedekind 1999). The lower rate of egg production could still represent a time cost of outcrossing, resulting from pairing and gamete transfer. However, selfed eggs and the embryos they contained were smaller than those of paired worms, and they also hatched at lower rates (fig. 2.13). In competitive situations, offspring of paired worms had a higher infection success and faster growth in copepods, the cestode's first intermediate host, than selfed offspring (Christen et al. 2002; see also Wedekind and Rüetschi 2000). Thus, the lower genetic quality of selfed offspring, which is an expected consequence of inbreeding depression, probably nullifies any advantage selfing worms may incur from their higher rates of egg production. Still, even when placed in a group of conspecifics, *S. solidus*

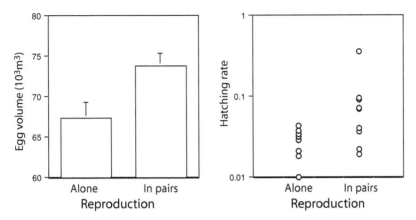

Figure 2.13 Mean (+ standard error) egg volume and hatching rates of several egg clutches of the cestode *Schistocephalus solidus* reproducing either alone (by selfing) or in pairs (by outcrossing). (Modified from Wedekind et al. 1998)

worms always self-fertilize at least some of their eggs; outcrossing rates increase with group size and decreasing size asymmetries among pairing worms (Schärer and Wedekind 2001; Lüscher and Milinski 2003). In *S. solidus*, larger worms allocate more to the female function and produce more eggs; when given a choice of partners, worms clearly prefer to mate with other individuals that are larger than they are (Lüscher and Wedekind 2002). Outcrossing is apparently the best option, but selfing remains profitable when a worm finds itself alone in a definitive host, or in the presence of smaller conspecifics with few eggs to offer.

The number of conspecific individuals occurring with a hermaphroditic parasite plays an important role in other species as well. In the simultaneous hermaphroditic digenean *Echinostoma caproni*, allocation to the male function is relatively low when the worms occur alone in a definitive host, that is, single worms have small testes (Trouvé et al. 1999). As the number of worms in a host increases, so does local mate competition, and in response the worms allocate relatively more to the male function, developing larger testes. This result suggests that the worms are capable of assessing the number of conspecifics occurring with them, and thus the opportunities for outcrossing and the intensity of competition for access to mates. In the parasitic cymothoid isopod *Ichthyoxenus fushanensis*, the situation is more extreme. The isopods only occur as single males or as male-female pairs in their fish hosts. Fecundity is proportional to size in females, whereas small males can fertilize even the largest females. Selection has favored protandrous hermaphrodism in this and other isopod parasites (Tsai et al. 1999). All individuals begin their life as males seeking fish hosts. If a juvenile male infects a fish already harboring a single male, the previously established male becomes a female and pairs with the newly arrived small male. If the juvenile male infects a previously unparasitized fish, it settles and grows while waiting for a partner. The male-first hermaphroditic system ensures that the female in the pair will always achieve the largest size, and therefore the highest fecundity possible (Tsai et al. 1999).

In the digenean family Schistosomatidae, individuals in any given species are dioecious, that is, they have separate sexes. In schistosomes, dioecy is a derived trait that evolved from a hermaphroditic condition (Basch 1990; Platt and Brooks 1997). Adult males and females have different body sizes, and they also appear to have different transmission strategies and survivorship at each step in their life cycle (Boissier et al. 1999; Moné and Boissier 2004). Why did selection favor the adoption of dioecy and sexual dimorphism in schistosomes? It may have something to do with a division of labor between the sexes leading to a higher fitness for both participants than simple hermaphrodism. Schistosomes live in the venous system of vertebrates, and precision egg-laying is necessary to ensure the passage of eggs to the external environment. There is evidence suggesting that female schistosomes have evolved slender bodies, narrower than those of their closest relatives in the families Spirorchidae and Sanguinicolidae, to fit into the smallest blood vessels where egg release must take place (Platt and Brooks 1997). In addition to fertilizing the females, males provide muscle: they transport females to the mesenteric venules and help females feed on red blood cells. Note that some of the advantages of simultaneous hermaphrodism are retained, because the typical mating pattern of schistosomes is one of lifelong monogamy, with females and males remaining physically attached for life. This is not a universal feature of schistosomes, though. In otherwise monogamous species, changes of partners can be frequent (Pica-Mattoccia et al. 2000). Also, although the possession of few testes and a groove (the gynecophoric canal) in which a female can be fitted for long-term pairing are ancestral characteristics of male schistosomes, reductions of the groove and increases in the number of testes have occurred repeatedly in the evolutionary history of the family (Morand and Müller-Graf 2000). Thus some schistosome lineages have adopted a male strategy consisting of investing in the production of sperm and fertilizing a high number of females instead of monogamy.

It has also been shown that parthenogenesis has evolved, possibly more than once, in schistosomes (Jourdane et al. 1995). When a female schistosome is paired with a male from a distantly related species, offspring are produced through parthenogenesis; in at least one species, *Schistosoma douthitti*, parthenogenesis occurs regularly even when conspecific males are available, possibly an advantage when population density is low. It is intriguing that schistosomes, after switching to dioecy, show a tendency to revert to a mode of reproduction not involving pairing. Although this latest ability appears advantageous, we can only speculate that the life-cycle conditions under which dioecy was favored were different from the ones under which parthenogenesis has later appeared.

In most other major groups of metazoan parasites (nematodes, acanthocephalans, arthropods, etc.), separate sexes are the norm, and two individuals must meet for egg fertilization. The few exceptions are interesting. Hermaphrodism has apparently evolved independently in the rhabditid nematode genera *Steinernema* and *Heterorhabditis* (Griffin et al. 2001). This convergence may be related to their similar life cycles, involving penetration of insect hosts by an infective stage and the subsequent use of symbiotic bacteria to kill the host and make its tissues available for the developing worms produced by the original worm (Poinar 1993). Infective nematodes in these genera are characteristically alone on invasion of a new host, and thus the ability to self-fertilize should be highly advantageous. In other nematodes, some ancestrally dioecious lineages have evolved toward parthenogenesis and eliminated the need for eggs to be fertilized. For example, in the nematode *Strongyloides ratti*, adult worms in the small intestine of the rat host are always parthenogenetic females. Eggs produced by these females and passing to the external environment develop either directly as infective larvae, or as free-living dioecious adults that will reproduce sexually by mating in the external environment. The offspring of these sexual adults develop into infective larvae that will turn into parthenogenetic females once inside a host. The proportion of larvae that develop into sexual adults is under both genetic and environmental control (Viney 1996; Harvey et al. 2000), with simple developmental switches activated by external conditions determining the fate of the larvae. In particular, the immune status of the host appears important: larvae issued from hosts that have acquired immune protection against *S. ratti* are more likely to develop into sexual adults than larvae issued from naive hosts (Gemmill et al. 1997). Even host immunity against other nematode species is sufficient to promote the development of sexual adult worms (West et al. 2001a). Other forms of stress experienced by worms inside the host, such as sublethal doses of anthelmintics, can have opposite effects (Crook and Viney 2005). Different relative frequencies of both developmental strategies will be favored depending on the prevailing environmental conditions, both inside and outside the host (Fenton et al. 2004). Thus in this species facultative sex has been retained following the adoption of parthenogenesis as the main mode of reproduction, probably to both increase the number of infective larvae per cycle and facilitate rapid adaptation during periods when the host environment is changing.

In protozoan parasites also, sexual reproduction competes with other forms of replication. The phylum Apicomplexa includes several well-known pathogens, such as the malaria parasites *Plasmodium* spp., which must balance asexual multiplication with sexual reproduction to maximize their transmission to new hosts. Inside the host, asexual reproduction within host cells results in the production of large numbers of merozoites; some of these merozoites develop into sexual stages known as male and female gametocytes, while the rest can embark on a new cycle of asexual reproduction. In the genus *Plasmodium* and in other apicomplexans, gametocytes are not only responsible for sexual reproduction, but also for the invasion of the mosquito vector while the latter feeds on host blood. Several factors, such as the physiological and immunological status of the host, can influence the parasite's commitment to the production of gametocytes, and therefore its commitment to sexual reproduction (Taylor and Read 1997; Dyer and Day 2000). Interestingly, probing by mosquito vectors can accelerate the switch to gametocyte production in *P. chabaudi* inside the rodent host (Billingsley et al. 2005; but see Shutler et al. 2005). The parasite apparently displays considerable plasticity in its reproductive strategy in response to immediate transmission opportunities. Further asexual multiplication extends the period of transmissibility by ensuring a continuing source of gametocytes, but it does so at the expense of current transmission success by limiting the current density of circulating gametocytes. Natural selection must maintain a balance between asexual and sexual reproduction, a balance no doubt influenced by the identity of the host and vector and the characteristics of the life cycle.

Everything in the biology of parasites, from larval development to adult reproduction, has been shaped by the nature of the life cycle. The complexity of the transmission process from a definitive host and back to another definitive host has forced parasites to adopt many guises, each a different manifestation of the same genotype. The present section was only a brief survey of the many influences of the life cycle, but is intended to demonstrate the key role of the cycle in parasite evolution.

2.5 Conclusion

From simple beginnings, parasite lineages have evolved complex adaptations to their way of life. The ephemeral nature of the host as a habitat has forced parasites to find ways of constantly colonizing new hosts. Individuals that are better at doing so because they go through additional steps in their life cycle are favored by selection, and thus evolve complex life cycles. In turn, these products of selection act as strong selective pressures leading to adaptive adjustments in all aspects of the parasite's biology.

The variety of parasite life cycles observed today is evidence that parasite evolution does not follow a single road. There are however striking similarities between the life cycles of very distantly related groups in which parasitism had independent origins. In endoparasitic platyhelminths, nematodes, and acanthocephalans, complex life cycles involving transmission by predation from an invertebrate intermediate host to a vertebrate definitive host are extremely common. Adaptations to such a life cycle also appear to have evolved in parallel in these three groups, providing good examples of convergent adaptations. Therefore ancient events in the phylogeny of parasites can commit a lineage to a particular evolutionary path, and make difficult the distinction between constraint and adaptation. Contrasts between the transmission rates of sister taxa differing only in one aspect of the life cycle would be one way to achieve this distinction. Another would be to exploit model systems that are amenable to experimental investigations and in which some facets of the life cycle are variable or easily manipulated. Complex life cycles remain one of the most baffling features of parasites, and there is still much to be learned about their evolution.

3 | Host Specificity

Over evolutionary time, parasites have added hosts to their life cycles by adding steps to the cycle and thus increasing its complexity. Hosts can also be added to the life cycle in parallel rather than in series (Combes 1991b, 1995, 2001). The spectrum of potential hosts that can be used at any step in the cycle can be broadened without an increase in the number of steps. Instead, selection simply adds alternative pathways through the cycle. A complex life cycle allows parasites to specialize on two or more hosts by partitioning specialization to different times during the cycle (Thompson 1994); however, at each step in the temporal sequence of the cycle, the specialization can be relaxed to include more than one host species. The ability to exploit several host species at a given stage of the life cycle may be the adaptive outcome of natural selection, and studying host specificity is equivalent to studying resource specialization in free-living organisms. This chapter will extend the previous chapter's discussion of the evolution of complexity in parasite life cycles by discussing the evolution of host specificity at the different stages in the life cycle and the wide differences in specificity displayed by extant parasites.

3.1 Measuring Host Specificity

The term *host specificity* has been used in different contexts and has many interpretations. According to its most widely accepted definition, host specificity is the extent to which a parasite taxon is restricted in the number of host species used at a given stage in the life cycle. Highly host-specific parasites are restricted to one host species, and specificity declines as the number of suitable host species increases. Before moving on to more sophisticated definitions of host specificity, some issues are worth discussing.

When specificity is measured as the number of host species used, it can be estimated by summing up the number of known host species from published records of parasite occurrence. There is an obvious danger in using these estimates, however.

Consider two parasite species, the first, which has been the subject of a single survey in which it was described from one host species at one location, and the second, which has been regularly reported from several populations of the same host species over a large geographical area. Both these parasite species have therefore a single known host species. But can we be sure that the first species is not in fact exploiting a broad range of host species in which it is yet to be found? High host specificity can be an artefact of inadequate sampling. Among species of parasites of freshwater fish, sampling effort explains much of the variability in host specificity (Poulin 1992a, 1997a): the number of known host species is strongly, positively correlated with the number of times a parasite species has been recorded in the literature (fig. 3.1). The same is true among tick species, and the distinction between highly specific and less specific ticks may really be a distinction between rarely and frequently collected species (Klompen et al. 1996). If the number of host species used is characteristically underestimated in poorly studied parasite species, then it is not a very adequate measure of host specificity. Corrections for sampling effort, however, can make this measure somewhat more reliable (Poulin 1992a).

Another problem of using lists of published records is that this method does not provide an accurate measure of the specificity of parasites in one population, only that of their species as a whole. If a species of parasite is known to exploit seven host species, there is no reason to believe that individuals in a given population of that species are capable of infecting all seven hosts. Members of the population may only be adapted to locally available host species, and thus be more host specific than their species as a whole.

Incorrect species identification can also influence estimates of host specificity. On the one hand, a species of parasite known to exploit n host species in a given area can in fact prove to be a complex of n species of superficially identical, highly host-specific parasites. With the recent application of molecular techniques to parasite systematics, several groups of cryptic species have been recognized where it was once thought there was a single species exploiting several host species (see examples in Combes 1995; Thompson and Lymbery 1996; Chilton et al. 1999; Hung et al. 1999; Jousson et al. 2000; Blouin 2002; Leignel et al. 2002). On the other hand, what appears to be n related species of parasites exploiting n different host species can prove to be a single parasite species with low host specificity in which the phenotype is influenced by the identity of the host species, with a resulting confusion in taxonomy. This has been compounded by the trend among systematists in the past to propose a new species each time a parasite genus is found for the first time in a new host species. There are probably many instances in which "different" parasite species are in fact one and the same (e.g., Dallas et al. 2001; Bell et al. 2002), and these synonymies can also affect estimates of host specificity. What follows ignores the above problems by assuming they can be accounted for.

The number of host species used by a parasite has been referred to as the host range, in recognition of the fact that it is only a crude measure of host specificity (Lymbery 1989). It assumes that all host species used by a parasite are equal, whereas in fact they generally differ on two fundamental levels, and the mere number of host species used fails to capture these differences. First, from an ecological perspective, some host species

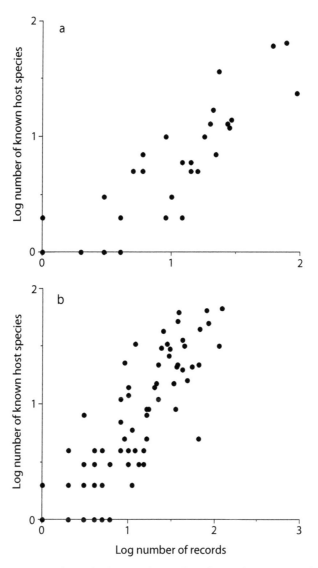

Figure 3.1 Relationship between the number of known host species and the number of published records mentioning a parasite genus among metazoan genera of (a) ectoparasites and (b) endoparasites of Canadian freshwater fish. More hosts tend to be known for intensely studied parasite species. (Data from Poulin 1997a)

are used more intensely than others. The prevalence, intensity, or abundance of infection by a particular parasite usually varies widely among its host species, even within the same locality. A true measure of specificity should take into account how heavily and how frequently the various host species are infected by a given parasite. There is no doubt that information on whether a parasite species utilizes its various host species equally or whether it concentrates on only one, would be valuable for any attempt to examine patterns of host use. Rohde (1980) developed an index of specificity, S, based on the number of parasite individuals found in each host species. The index can only be used for a large sample including many species of hosts collected at the same time and location, and is computed as follows:

$$S = \frac{\sum (x_i/n_i h_i)}{\sum (x_i/n_i)}$$

where x_i is the number of parasite individuals on the ith host species, n_i is the number of host individuals examined of the ith host species, x_i/n_i is the abundance of parasites on host species i, and h_i is the rank of host species i (the host species with greatest abundance of parasites has rank 1). Rohde (1980, 1993, 1994b) showed that the index could be modified to use prevalence of infection instead of abundance as a measure of host utilization by parasites. Prevalence and abundance are generally positively related, and using one or the other to compute the index probably produces similar results. However, as abundance is a better indicator of the success of individual parasites at infecting hosts of different species, it is the most appropriate measure to use in the computation of Rohde's index.

The value of the index tends toward one when the parasites concentrate predominantly in one host species, and is expected to tend toward zero when parasites are equally distributed among host species. However, the value of the index when host species are used evenly approaches zero only when the number of host species included in the computations is much greater than twenty. The majority of parasite species utilize fewer than ten host species and for such species a high value of S does not necessarily imply that parasites concentrate on a single host species. Rohde and Rohde (2005) have proposed a modified version of the index S in which the raw value of the index is corrected for the minimum possible index value, making it reliable for comparisons of parasite species using different numbers of host species.

Most parasites show biases in the way they utilize suitable host species, sometimes occurring on different host species at abundances that vary by orders of magnitude (fig. 3.2). An ecological index of host use is thus necessary to measure host specificity. Rankings of host species according to parasite abundance could also serve to assess which hosts are preferred, that is, used more extensively, among the spectrum of suitable species (fig. 3.2). Robust estimates of abundance require the examination of many hosts, and rankings of host species are only meaningful if all species have been sampled more or less equally. Even when host species are well sampled, however, ranking them with respect to parasite abundance to determine which species are more important to the parasite population may not be informative because different host species are not equally available to parasites. The relative abundance of hosts of different species may determine what proportions of the parasite population they harbor. For instance, if parasites are twice as abundant on host species A than on host species B, but host species B is twice as common in the environment as host species A, then both host species harbor the same proportion of the overall parasite population. Ideal indices of host specificity should take into account both usage and availability of hosts (Lymbery 1989), and would

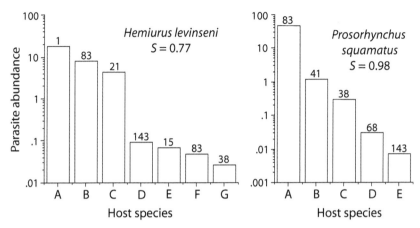

Figure 3.2 Parasite abundance (total number of parasites found divided by the number of hosts examined) in different host species for two species of digeneans parasitic in marine fish. Note that abundance is on a log scale. The numbers of fish examined in each host species are indicated above the bars, and so is S, Rohde's index of host specificity. (Data from Rohde 1994b)

therefore be difficult to use in natural situations. A final issue concerning ecological differences among host species and how they affect the quantification of host specificity is that not all host species are equally suitable for parasite development and reproduction (Kennedy 1975). For instance, in the acanthocephalan *Acanthocephalus tumescens*, which uses six fish species in an Argentinian lake, the definitive host species in which parasite abundance is highest is not the one in which parasite egg production is highest (Rauque et al. 2003). As a rule, data on egg production in different host species are not readily available, and an index of host specificity that takes into account prevalence or abundance is still preferable to the mere use of the number of host species exploited.

Differences in parasite infection levels among host species are only one way in which the different host species used by a parasite cannot be considered to be equal. The second becomes apparent from a phylogenetic perspective: some of the host species used by a parasite are likely to be closely related phylogenetically, whereas others are only distantly related. Consider two parasite species, A and B, each exploiting four host species. If all four host species of parasite species A are congeners, whereas those of parasite species B all belong to different families, we have a situation in which one parasite (species A) is much more specific than the other because its hosts represent only a single host genus. Parasites with low host specificity are those capable of broad taxonomic "jumps" during their evolutionary history, regularly switching from one host species to a distantly related one. A useful measure of host specificity needs to take host relationships into account (see Caira et al. 2003; Poulin and Mouillot 2003a).

The specificity index, S_{TD}, proposed by Poulin and Mouillot (2003a), measures the average taxonomic distinctness of all host species used by a parasite species. When these host species are placed within a taxonomic hierarchy, based on the Linnean classification into phyla, classes, orders, families, genera, and species, the average taxonomic distinctness is simply the mean number of steps up the hierarchy that must be taken to reach a

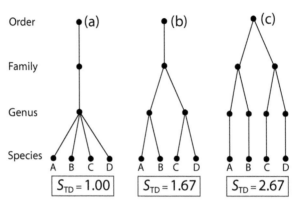

Figure 3.3 Hypothetical taxonomic trees for the hosts of three species of parasites. Each parasite species is found on four host species, A to D. In (a), all four host species are congeners, and thus they average a single step toward a common node, i.e., $S_{TD} = 1$. In (b), the four hosts belong to two different genera but to the same family, with some species pairs, such as A-B, requiring one step to reach a common node, and others, such as A-C, requiring two steps. Example (c) shows even greater taxonomic distinctness, with the four hosts belonging to different families within the same order. Note that the maximum possible value of S_{TD} in these examples is 3, because there are three taxonomic levels above species. (Modified from Poulin and Mouillot 2003a)

taxon common to two host species, computed across all possible pairs of host species (fig. 3.3). Thus, if two host species are congeners, one step (species-to-genus) is necessary to reach a common node in the taxonomic tree; if the two species belong to different genera but the same family, two steps will be necessary (species-to-genus, and genus-to-family); and so on, with these numbers of steps averaged across all host species pairs. Equal step lengths of one unit are postulated between each level in the taxonomic hierarchy, but actual phylogenetic distances could also be used if these were known. The value of the index S_{TD} is actually inversely proportional to specificity: a high index value means that on average the hosts of a parasite species are not closely related. Formally, the index is computed as follows (see Poulin and Mouillot 2003a):

$$S_{TD} = 2 \frac{\sum \sum_{i<j} \omega_{ij}}{s(s-1)}$$

where s is the number of host species used by a parasite, the double summation is over the set $\{i = 1, \ldots s; j = 1, \ldots s,$ such that $i < j\}$, and ω_{ij} is the taxonomic distinctness between host species i and j, or the number of taxonomic steps required to reach a node common to both (see fig. 3.3). Using the standard five taxonomic levels above species—genus, family, order, class, and phylum—the maximum value that the index S_{TD} can take (when all host species belong to different classes) is 5, and its lowest value (when all hosts are congeners) is 1. Given that few (if any) parasite species, at a given stage in their life cycle, infect hosts belonging to different phyla, this range is sufficient for all practical purposes. The index S_{TD} measures the average taxonomic distinctness between host species, but it is also possible to compute its variance (Poulin and Mouillot 2003a) to obtain separate information of how much taxonomic heterogeneity there is among a group of host species. One weakness of S_{TD} is that it cannot be applied to parasite species infecting only one host species: there is no pair of host species in these cases from which an average taxonomic distance can be computed. Such highly host-specific parasites can either be excluded from any comparative analysis in which S_{TD} is used, or they can be assigned a default S_{TD} value of 1, since by definition "all" host species for such specialized parasites belong to the same genus. Overall, the index S_{TD} provides a measure of host specificity from an evolutionary perspective, one that is independent of the number of published records about given parasites (see Poulin and Mouillot 2003a) and easy to compute.

Ideally, though, we need an index that combines both the ecological and evolutionary facets of host specificity. Poulin and Mouillot (2005) have recently proposed such an index. Their index, $S_{TD}*$, measures the average taxonomic distinctness of all host species used by a parasite species, weighted by the prevalence of the parasite in these different hosts. The taxonomic side of the index is similar to the earlier index S_{TD} of Poulin and Mouillot (2003a). Taxonomic distinctness is computed as before for all possible pairs of host species, and is then weighted by the product of the parasite's prevalence in each host species in a pair. The weighting factor has a maximum value of 1 when prevalence is 100 percent in both host species in a pair, and converges toward 0 when the prevalence in both host species is very low. Thus, more weight is given to the taxonomic distance between two host species if the parasite achieves high prevalence in these hosts than if the parasite occurs infrequently in these hosts. This way, the average weighted taxonomic distinctness reflects more strongly the taxonomic distances among the main hosts of the parasite. The index $S_{TD}*$ is the ratio of the sum of the weighted taxonomic distinctness values to the sum of the weighting factors. More formally, the index is computed as follows:

$$S_{TD}* = \frac{\sum\sum_{i<j} \omega_{ij}(p_i p_j)}{\sum\sum_{i<j} (p_i p_j)}$$

where the double summation is over the set $\{i = 1, \dots s; j = 1, \dots s$, such that $i < j$, and s is the number of host species used by the parasite$\}$, ω_{ij} is the taxonomic distinctness between host species i and j, or the number of taxonomic steps required to reach a node common to both, and p_i and p_j are the prevalences of the parasite in host species i and j, respectively. Fallon et al. (2005) have also independently derived a very similar index.

Figure 3.4 shows two hypothetical sets of host species, for two different parasite species. In each case the taxonomic tree for the host species is the same, and the prevalence values are the same; however, the distribution of prevalence values among host species differs between the two examples. The value of $S_{TD}*$ increases as the taxonomic distinctness between the high-prevalence hosts increases (fig. 3.4). As for S_{TD}, the value of the index $S_{TD}*$ is therefore inversely proportional to specificity: high value means that

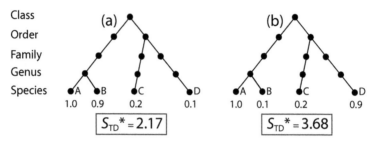

Figure 3.4 Taxonomic structure of the sets of hosts for two hypothetical parasites, with prevalence in each host indicated below. There are four host species, A to D, in each example. The taxonomic tree of host species is the same in both cases, as are the prevalence values; however, the distribution of prevalence values among host species differs between the two examples. The index S_{TD}^* achieves a higher value in (b) than (a) because of the greater taxonomic distance among host species with high prevalence. (Modified from Poulin and Mouillot 2005)

on average the host species most frequently used by a parasite are not closely related. The range of possible values for S_{TD}^* are also the same as for S_{TD}, and parasite species exploiting a single host species again need to be assigned a default value or excluded, because S_{TD}^* cannot be computed for a single host (see Poulin and Mouillot 2005).

The preceding discussion makes it clear that measuring host specificity is not as straightforward as it may look at first glance. Detailed analyses of patterns of host species use are complicated by the lack of information about the relative abundance of different host species, their unequal use by (and suitability for) any given parasite, and uneven study effort across parasite species. These problems should be kept in mind when reading the later section on observed patterns of specificity. Although the indices presented above have been designed primarily for comparative purposes, the comparative studies summarized later in this chapter rely almost exclusively on the number of host species exploited as a measure of host specificity; they must therefore be interpreted accordingly.

3.2 Host-Parasite Coevolution and Host Specificity

Host specificity has the main hallmark of a true species trait: it varies significantly less among populations of the same parasite species than among different parasite species, independently of how it is measured (see Krasnov et al. 2004a). The specificity of a parasite for its host species can be seen as the outcome of both historical events and current ecological conditions. This section will examine how this fundamental trait has evolved, first by looking at historical patterns of host-parasite associations, and then at smaller-scale phenomena determining whether new host species can be colonized.

3.2.1 Macroevolutionary Patterns

To understand why some parasites are very host specific and others are not, we must first understand how parasites and hosts have coevolved since the origin of their association.

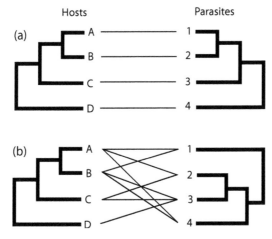

Figure 3.5 Phylogenetic trees of hypothetical host and parasite species. Thin lines indicate host-parasite associations. (a) Host and parasite trees are perfectly congruent, and parasites are strictly host specific. (b) Host and parasite trees are incongruent, and the host specificity of parasites varies among species. (Modified from Poulin and Morand 2004)

Recent years have seen much progress in efforts to reconstruct the phylogeny of parasites based on morphological or molecular information. The many available parasite phylogenies have made possible comparisons between the phylogeny of a group of parasites and that of their hosts. Differences and similarities between host and parasite phylogenies can shed light on the history of their association. Recent developments in the field of host-parasite coevolution as it pertains to the evolution of host specificity are reviewed here.

Consider a speciation event taking place in an ancestral host population harboring one species of parasite. If the barrier to gene flow that isolates the two allopatric subpopulations of hosts also prevents gene flow between the two newly created subpopulations of parasites, then the parasite will cospeciate with its host. If this process of "association by descent" is repeated several times in the daughter species, it will result in n species of hosts and n species of parasites. Each parasite species will be strictly host specific and the phylogeny of the parasite species will be a mirror image of that of their host species (fig. 3.5a). The perfect congruence of host and parasite phylogenies is a pattern known to parasitologists as Fahrenholz's Rule, and it can serve as a null model of host-parasite coevolution against which other evolutionary scenarios can be tested. Typically, host and parasite phylogenies will not be entirely congruent and parasites will not show strict host specificity (fig. 3.5b). The occurrence of one parasite species on more than one host species can indicate the maintenance of a single parasite species following host speciation, resulting from the continuation of gene flow between the parasite populations on different hosts (e.g., Brant and Orti 2003). It can also result from host switching, or the colonization of new host species. The opportunities for host switching, and the frequency of such events in the phylogeny of parasites will determine the number of host species they can exploit.

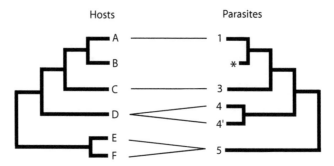

Hosts   Parasites

Figure 3.6 Illustration of some possible evolutionary events that can lead to false incongruence between host and parasite phylogenies: parasite extinction (*), intrahost speciation, such as the duplication of parasite lineage 4 without a corresponding speciation event in host lineage D, and failure of the parasite to speciate in response to host speciation (parasite 5 in hosts E and F). (Modified from Poulin and Morand 2004)

To reconstruct the history of cospeciation with the host and host switching among hosts in a parasite lineage, we must begin by comparing host and parasite phylogenetic trees and assessing their congruence. This poses some problems. False congruence between trees can arise from a sequential colonization of hosts by parasites that coincidentally mirrors the pattern of host radiation (Brooks and McLennan 1991). False incongruence between host and parasite trees is also possible, and can result from duplications and losses of parasites independent of host phylogeny (Page 1993, 2003; Paterson and Gray 1997). The absence of parasite species from host species in which they should occur if both lineages had cospeciated can be the result of parasite extinction following a cospeciation event (fig. 3.6). Parasites may also have missed the boat, may simply have been absent from the founder population of hosts during a host speciation event. The aggregated distribution of parasites among hosts (chapter 6) should result in many small founder host populations being parasite free (Paterson and Gray 1997). Missing the boat appears to be a relatively common event, at least for lice parasitic on birds (Paterson et al. 1999). The duplication of a parasite lineage without a corresponding speciation in the host lineage will also create minor discrepancies between the branching patterns of host and parasite phylogenies, as will the failure of parasites to speciate following host speciation (Johnson et al. 2003) (fig. 3.6). Sampling error can also lead to false incongruence between the trees. If the abundance of parasites is very low, they may escape detection (Paterson and Gray 1997). In addition, the presence of members of a parasite species on an unusual host species may be a case of "straggling," that is, the accidental occurrence of a parasite on a host, rather than either a successful host switching event or one in progress (Rózsa 1993; Whiteman et al. 2004).

With these caveats in mind, one can still derive meaningful information from a comparison of host and parasite phylogenies. There are at present two main analytical methods for measuring the congruence between host and parasite phylogenies. One of them, Brooks's parsimony analysis or BPA, treats parasites and their phylogeny as host characters and then compares the most parsimonious host cladogram derived from the parasite data with the original host phylogeny (Brooks 1988; Brooks and McLennan 1991; Wojcicki and Brooks 2004). Incongruence is taken as evidence for one or more host switching events, which can lead to an inflated estimate of the number of host switches (Paterson and Gray 1997). The other method, reconciliation analysis in its various guises, contrasts independently derived host and parasite phylogenies, and determines whether their congruence is greater than expected by chance (Page 1990, 1993; Charleston 1998). Reconciliation analysis assumes that host switching has not occurred and explains incongruence with parasite extinctions or duplications such that all incongruence is false and cospeciation is the rule. Despite the different assumptions behind the two approaches and their inherent biases, both can lead to similar general conclusions when applied to the same data set (Hoberg et al. 1997; Paterson and Gray 1997). Recent modifications to reconciliation analysis allow for host switches as a process to explain incongruence (Page 1994, 2003), and the differential weighing of different evolutionary events (Charleston 1998, 2003; Paterson and Banks 2001). This apparent convergence of the assumptions of reconciliation analysis and BPA suggests that a unified method for the study of host-parasite coevolution should be conceivable. However, the ongoing and bitter debate between the proponents of the two methods suggests otherwise (Dowling 2002; Dowling et al. 2003; Siddall and Perkins 2003; Brooks et al. 2004; Siddall 2004, 2005). No distinction will be made between the two methods in the discussion that follows, although they can lead to different results under certain circumstances.

What are the general patterns of coevolutionary history observed to date? There does not seem to be a common scenario, even among related host-parasite assemblages. Lice and their hosts, for instance, illustrate how patterns of coevolution are variable. Because lice are transmitted by contact between hosts, one might expect a tendency toward cospeciation with few opportunities for host switching. A large-scale examination of patterns of diversification in lice and their mammalian hosts, across all taxa, suggests that cospeciation has not been important (Taylor and Purvis 2003), but studies on a finer scale are necessary to determine the extent of cospeciation. In the classical example, the evolutionary history of several species of two related genera of chewing lice and their hosts, members of the rodent family Geomyidae, was shown to be one of rather strict cospeciation with host switching playing a very minor role (Hafner and Nadler 1988, 1990; Hafner and Page 1995). Not only is there a clear congruence between the branching patterns of host and parasite phylogenies, but the timing of speciation events in both host and parasite lineages coincides remarkably well based on evidence from rates of molecular change. Not surprisingly, these lice species display strict host specificity. In contrast, host-switching events appear to have been very common and cospeciation almost nonexistent in lice parasitic on several species of one genus of rock wallabies in Australia (Barker 1991). To some extent the discrepancies between these studies may be due to differences of opinion among researchers, with the same evidence being interpreted differently by different people (Barker 1994, 1996; Page et al. 1996), but biological differences between the host-lice systems in these studies no doubt account for most of their contrasting results.

In recent years, studies on avian lice have shed new light on patterns of host-parasite coevolution. In the first rigorous studies, cospeciation appeared as the main explanatory process in the evolution of lice on seabirds (Paterson et al. 1993; Paterson and Gray 1997). Host-switching events are more important in other bird-lice associations (e.g., Clayton et al. 1996; Page et al. 1998; Johnson et al. 2002a), though cospeciation remains the dominant process. Ecological factors seem to play a central role in bird-lice coevolution by determining the likelihood of host switching. Transfer experiments have shown that host-specific lice can survive and reproduce on other bird species; what prevents them from exploiting these birds in nature is a lack of opportunities because of ecological barriers preventing their dispersal (Clayton et al. 2004). The ability of lice to survive on other bird hosts is beautifully illustrated by the lice fauna of the brown-headed cowbird *Molothrus ater*, a brood-parasitic bird that lays its eggs in the nests of other bird species. Lice are normally only transmitted via bodily contacts between hosts, something that usually only happens among birds of the same species. The cowbird nestlings have a unique opportunity to encounter lice species normally found only on their host parents' species. As a consequence, cowbird fledglings harbor about as many lice species as all their potential host species combined (Hahn et al. 2000). If none of these numerous lice species are in fact specific to cowbirds (see Clayton and Johnson 2001), then the rich lice fauna on cowbirds would suggest that strict host specificity and tight cospeciation are only maintained because of a lack of opportunities for host switching. Some taxa of avian lice can disperse and contact new host species in other ways. Predation can provide opportunities for dispersal and host switches that vary even among lice species using the same bird host (Whiteman et al. 2004). In addition, several species of lice are known to use hippoboscid flies for transport (i.e., phoresis) and dispersal to new hosts. This has been shown to lead to frequent host switching to locally available bird species, even if they are unrelated to the original host, resulting in very little congruence between host and parasite phylogenies (Johnson et al. 2002a; Clayton and Johnson 2003; Weckstein 2004). We thus observe the full spectrum of possibilities in lice parasitic on vertebrates, from tight cospeciation and congruent host and parasite phylogenies, to frequent host switching and incongruence. All seems to depend on opportunities for host switches (Clayton et al. 2004).

The patterns uncovered in other kinds of host-parasite associations also cover the full gamut of possibilities. In principle, transmission mode should dictate the likelihood of host switching, with contact- or vertically (parent-to-offspring) transmitted parasites having fewer opportunities than trophically transmitted parasites. In reality, these expectations are not always met. At one extreme, a high degree of cospeciation between hosts and parasites has been reported from studies on ectoparasites with direct life cycles, such as feather mites and their bird hosts (Dabert et al. 2001) and copepods and their fish hosts (Paterson and Poulin 1999). Cospeciation has also been important, but to a lesser extent, in some protozoan parasites, such as microsporidians with their insect hosts (Baker et al. 1998). In apicomplexans such as *Plasmodium*, *Haemoproteus*, or *Hepatozoon*, one analysis suggests that cospeciation with the vertebrate host has been important (Ricklefs and Fallon 2002), whereas another suggests that if cospeciation has played a role, it occurred between the parasites and their invertebrate vectors, not their vertebrate hosts (Jakes et al. 2003). In contrast, host switching appears to have been a more common event, and to have played a greater role, in the evolution of many helminth parasites. Rampant host switching is a better hypothesis to explain the coevolutionary history of many nematodes with their hosts (e.g., Brant and Gardner 2000; Beveridge and Chilton 2001; Perlman et al. 2003). In monogeneans, which are notorious for their strict host specificity, host switching can nevertheless be frequent and cospeciation less so (e.g., Desdevises et al. 2002a; Huyse and Volckaert 2005), or intrahost parasite speciation can produce clusters of congeneric and closely related species all highly specific to the same host species, as in the genus *Dactylogyrus* (Simkova et al. 2004). The same is true for onchobothriid cestodes: they are remarkably specific for their elasmobranch hosts, and yet show little evidence of having cospeciated with them (Caira and Jensen 2001). Frequent host switching also appears to have been the main theme in the evolutionary history of other cestodes (e.g., Skerikova et al. 2001). Indeed, the cestode genus *Taenia* twice independently colonized early humans in Africa, switching from carnivore hosts to primates as our ancestors shifted their dietary habits (Hoberg et al. 2001). Finally, the same coevolutionary pattern, dominated by several host switches, emerges from studies of digeneans, whether their phylogeny is contrasted with that of their definitive hosts (Snyder and Loker 2000; Radtke et al. 2002) or that of their snail intermediate hosts (Morgan et al. 2002; Donald et al. 2004). The latter finding is puzzling in light of the high host specificity of digeneans for their snail host, which one would normally expect to be associated with tight host-parasite cospeciation.

Another way to examine the evolutionary origins of existing host-parasite associations is to focus on the community of parasite species of given host taxa, rather than focusing on the phylogenetic relationships of a parasite taxon. For instance, an investigation of the digenean faunas of groupers indicates that these marine fish have acquired their parasites mainly via localized episodes of host switching: their digenean faunas are typically nothing but a selection of the locally available pool of digenean species (Cribb et al. 2002). Similarly, the yellow perch, *Perca flavescens*, has a rich fauna of helminth parasites, yet it appears that most if not all of them were acquired by host switching from unrelated but sympatric North American fish species (Carney and Dick 2000a). The relationships among the parasites of *P. flavescens* and those of its European relatives are not consistent with any scenario involving cospeciation.

As new studies on host-parasite coevolution accumulate, there is still only one conclusion possible to date: all scenarios are possible, from strict cospeciation to rampant host switching. In addition, host specificity is not necessarily coupled with cospeciation, as was assumed traditionally. Indeed, several important parasites of humans appear to have expanded their host range by switching to humans, only to later abandon their initial vertebrate hosts to specialize on the newly colonized human hosts (e.g., Combes 1990; Waters et al. 1991). One fact should not be forgotten, however: despite most parasite species being quite host specific, many are today found in more than one species of host. If we assume that the same parasite species cannot have more than one origin (i.e., it cannot evolve independently more than once!), there are two ways in which low host specificity can evolve. First, it can be the result of host switching, where a parasite repeatedly colonizes host lineages related or unrelated to that of its original host. Second, it can result from multiple speciation events in the host phylogeny without any corresponding speciation in the associated parasite lineage. This second possibility requires that gene flow be interrupted between parts of the host population but not within the parasite population. Both processes are equally possible and both require that the parasite be able to disperse efficiently.

One final macroevolutionary issue needs to be addressed here. Has there been any directional trend in the evolution of host specificity? In other words, has specificity tended to increase or decrease over evolutionary time in various parasite lineages? On the one hand, extreme specialization is often thought to be evolutionarily irreversible. If any species that becomes strictly host specific never reverts to a generalist strategy, then host-specific species may accumulate over time. On the other hand, strict specificity increases the extinction probability of a parasite species; thus, generalist species should survive better and become overrepresented in any higher taxon. Recent studies on other animal groups have challenged the paradigm that specialization is both directional and irreversible. In his reevaluation of fifteen studies on evolutionary transitions between specialized and generalized host-plant use in phytophagous insects, Nosil (2002) found that generalist-to-specialist transitions were more frequent overall, but that in some groups of butterflies and bark beetles, the opposite was true. Specialization is thus not a dead end for phytophagous insects (Nosil and Mooers 2005). Furthermore, Stireman (2005) reported that transitions from specialist to generalist strategies have occurred more frequently than the reverse during the evolutionary history of tachinid flies, a group of endoparasitoids of insect hosts. The result is that generalist tachinid species tend to be the most derived, that is, they tend to occupy branch tips in the phylogeny of the group (Stireman 2005). The only test of directionality in the evolution of host specificity in parasites has revealed that among fleas parasitic on mammals, host specificity may have decreased over evolutionary time. In the phylogeny of flea species, ancestral or basal species tend to have stricter host specificity than derived or more recently evolved species (Poulin et al. 2006). These trends are weak, though, and insufficient to conclude that host specificity evolves preferentially in one direction and one direction only.

3.2.2 Microevolutionary Processes

The discussion above focused on the macroevolutionary history of host-parasite associations and host specificity. On a microevolutionary scale, many phenomena can facilitate host switching and subsequent decreases in host specificity, or, conversely, promote greater specialization on fewer host species. The physiological and ecological characteristics of hosts and parasites are obviously important and will be addressed in the next section. Here, the processes by which natural selection may favor changes in host specificity are discussed.

The fundamental issue concerns the direction of selection: should we expect natural selection to favor increases or decreases in host specificity? Of course there are no easy answers. Just as parasites may be selected to increase the range of hosts in which they can successfully develop, they may also sometimes face selection for greater specialization through a narrowing of their range of suitable hosts. The growth and fecundity of any given parasite vary among host species (Poulin 1996b). If selection can fine-tune the mechanisms of host infection to ensure that fewer host species are encountered, then one would predict that host species in which development is suboptimal will be excluded. This would result in a narrow host range comprising only host species on which parasite fitness is high. A related advantage of high host specificity is that it leads to the concentration of parasite individuals on fewer host individuals than if several host species were used. This should lead to greater genetic exchanges among individuals and thus higher genetic diversity of the parasites' offspring (Combes and Théron 2000). Although greater specialization on fewer host species can be advantageous, it also links the fate of parasites to that of their hosts and can make highly host-specific parasites more prone to local extinction.

In many cases, we might expect a trade-off between the ability to use many host species and the average fitness achieved in these hosts (Ward 1992). Close adaptation to one host species may only be achieved at the expense of adaptations to other host species. Given that different host species have different defense systems, investing in many counteradaptations should have a fitness cost for the parasite: a jack-of-all-trades may be a master of none. The performance of parasites in different host species can be estimated by their abundance in these hosts, which is a reflection of the parasite's ability to exploit host resources. Different parasite species may achieve greater overall fitness at different points along the continuum of strategies between the high-specificity-low-mean-abundance and low-specificity-high-mean-abundance extremes. This kind of trade-off is often used to explain the host specificity of phytophagous arthropods (Fry 1990).

Is there evidence for parasites that such a trade-off exists between host specificity and average fitness achieved on different host species? Among the metazoan parasite species of North American freshwater fish, a negative relationship was found between the number of host species used by a parasite and its average abundance on these hosts, suggesting a trade-off between how many hosts can be exploited and how well the parasite does on them (Poulin 1998a). However, this pattern was completely reversed in a similar study on fish parasites performed on a smaller geographical scale (Barger and Esch 2002). Among helminths parasitic in birds, there is no relationship between the number of host species used and the parasites' average abundance in their hosts, but a trade-off appears between host specificity and abundance once the index S_{TD} (see section 3.1) is used to measure specificity (Poulin 1999a; Poulin and Mouillot 2004).

In contrast, among flea species parasitic on small mammals, there are generally strong positive relationships between parasite abundance and either the number of host species used or the index S_{TD} (Krasnov et al. 2004b). In light of previous studies, this latest finding indicates that there is no general trade-off between how many host species a parasite can use and how well it does on them. In some groups of parasites, such as fleas, the opposite happens: whatever features of fleas make them successful on a host also allows them to colonize other host species. This is not because fleas are the ultimate generalist parasites: they incur fitness reductions on certain hosts compared to others (Krasnov et al. 2004c).

In particular, fleas and other parasites often achieve high abundance on host species closely related to their main host species, but almost never on phylogenetically distant host species (Krasnov et al. 2004d; Poulin 2005a). Therefore adding new hosts to its repertoire does not necessarily benefit a flea. Perhaps the population dynamics of different kinds of parasites can account for the contrasting results coming from studies of the potential trade-off between host specificity and average abundance on different host species. Mathematical models predict that for parasites with direct, one-host life cycle in which transmission is strongly dependent on host density, such as fleas, the more host species are exploited in a locality, the greater the probability that the parasite population will persist and spread (Dobson 2004). For indirectly transmitted parasites, however, the models predict the exact opposite (Dobson 2004). These different dynamical features of parasite populations with different modes of transmission may explain the complete absence of any trade-off in parasitic fleas.

The basis for the existence of a trade-off between how many host species can be used and how well a parasite does on them on average, is the potential fitness costs of maintaining adaptations to many host species. These costs may be extremely small, however. For instance, the generalist nematode *Howardula aoronymphium* parasitizes several species of *Drosophila* flies. Even after a twenty-five-generation selection experiment on one host species, *H. aoronymphium* did not lose any of its ability to exploit other host species, and it showed no trade-off in performance on its different host species (Jaenike and Dombeck 1998). This parasite appears to have little heritable genetic variation for adaptation to specific hosts, indicating it may possess a general-purpose genotype to go with its high likelihood of encountering many different host species.

If lower average abundance is not constraining parasites from expanding to new host species, then what is? On microevolutionary time scales, host specificity is mainly determined by opportunities for colonization and availability of suitable host species. Opportunities can arise in many ways. Hybridization between host species, for example, can create a genetic and ecological bridge between host species and allow the colonization of one host by parasites from the other (Floate and Whitham 1993). Le Brun et al. (1992) have documented such a situation between two related fish species. One of them, *Barbus meridionalis*, is host to the monogenean *Diplozoon gracile*, whereas the other, *B. barbus*, is never parasitized by *D. gracile* under natural conditions but is a perfectly suitable host in laboratory infections. The two congeneric fish species sometimes hybridize in nature, however, and the likelihood of *D. gracile* infections among hybrids is proportional to the percentage of *B. meridionalis* genes in the hybrid (fig. 3.7). The microhabitat preferences of hybrids are also related to which of the parent species they most resemble, and determine the encounter rates between parasites and fish. It is easy to see how this could lead to a gradual extension of the parasite's spatial range to cover the microhabitats of both parent host species. Monogeneans form generally one of the most highly host-specific groups of parasites (see section 3.4). The example above illustrates that host switches can easily occur even in the most specific of parasites.

Opportunities for host colonization are important, but only if there are suitable hosts to be colonized. Experimental infections are a powerful tool to determine whether certain

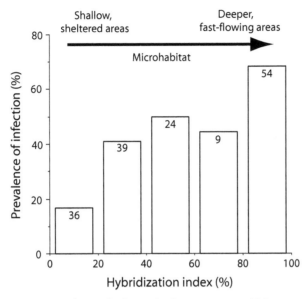

Figure 3.7 Prevalence of infection by the monogenean *Diplozoon gracilis* as a function of the degree of hybridization between the fish species *Barbus meridionalis* and *B. barbus*. A pure *B. barbus* fish has a score of 0 percent and a pure *B. meridionalis* specimen has a score of 100 percent; numbers of fish analyzed in each hybridization class are given in the bars. Only *B. meridionalis* is a natural host of the parasite; as hybrids get genetically closer to *B. meridionalis*, they are found in microhabitats preferred by *B. meridionalis* and are more likely to harbor parasites. (Data from Le Brun et al. 1992)

host-parasite combinations are compatible, and if not, they can also be used to pinpoint the reasons for the incompatibility. For example, in laboratory assays, several allantonematid nematode species, parasitic in *Drosophila* flies, have proven capable of successfully infecting many fly species in which they are never observed in nature, suggesting that they are fully compatible with these novel hosts but have not had opportunities to colonize them (Perlman and Jaenike 2003). Experiments on avian lice indicate that host defense systems may be a key component of host suitability. Species of the feather lice genus *Columbicola* exploit pigeons and doves, and are generally very host specific in nature. However, under experimental conditions, they can remain attached and feed successfully on other bird species taxonomically close to their original host (Clayton et al. 2003). What they cannot do, though, is escape from the main host defense against lice, preening, when the novel host differs in size from the original host. This is due to the fact that louse body size is adapted to its host's bill size and shape, and when placed on a novel host of different size, lice can easily be removed by the host (Clayton et al. 2003). In other lice taxa, the parasite can only attach firmly to feather barbules of a certain size: survival on a novel host depends greatly on whether or not its feathers have similar dimensions to those of the original host (Tompkins and Clayton 1999). Experimental studies in other taxa have also shed light on the maintenance of host specificity. For instance, the nematode *Strongyloides ratti*, a gastrointestinal parasite of rats, has only a limited attachment success and achieves reduced fecundity in mice, even in immunosuppressed mice (Gemmill et al. 2000). Experimental selection, achieved by serial passage in mice for eighteen generations, failed to improve the performance of *S. ratti* in this novel host, suggesting that factors stemming from the different physiologies of rats and mice are responsible for maintaining host specificity (Gemmill et al. 2000).

Strict host specificity is therefore not always easily overcome. Recent models offer reasons for this observation. In these models, adaptation to a particular host species occurs via the fixation of alleles whose beneficial effects are host specific; this is more rapid and more likely to occur in parasite populations restricted to that host species than in parasite populations spread among several host species (Kawecki 1997, 1998). These models predict that parasite species that begin as generalists gradually lose the ability to exploit seldom-encountered host species and eventually exclude them altogether from their range of suitable alternatives. The copepod *Lepeophtheirus europaensis*, parasitic on two species of flatfish with different habitats in the Mediterranean Sea, provides a possible example (De Meeüs et al. 1992). It infects flounder in brackish lagoons and brill in marine habitats. In what is believed to be a speciation event in progress, individual parasites taken from the two different host species show genetic differences as well as physiological adaptations to the salinity of their hosts' respective habitats. This is a case where an ancestral parasite species capable of exploiting two host species may give rise to two genetically isolated subspecies each specific to a single host species. In addition, *L. europaensis* populations infecting brill live in sympatry with *L. thompsoni*, a parasite of turbot. In the laboratory, the two congeneric species can meet and mate on turbot and produce viable hybrids, but in nature strict host specificity maintains the two species as distinct genetic entities (De Meeüs et al. 1990, 1995). The formation of genetically distinct host races has also been documented in other parasites. In the generalist tick *Ixodes uriae*, for instance, there is a clear genetic differentiation between individual parasites from different seabird host species, whether these host species are sympatric or geographically isolated (McCoy et al. 2001, 2003, 2005a). The same is true in some genera of avian lice, especially those with more limited dispersal abilities (Johnson et al. 2002b).

The selection of greater host specificity in parasites can even be seen on a finer scale, when comparing the specificity of different populations of the same parasite species exploiting different populations of the same host species. In some trematode species, for instance, there is evidence of adaptation to local host populations (fig. 3.8). Cercariae of *Diplostomum phoxini* are more successful at infecting local fish than fish of the same species but from allopatric populations (Ballabeni and Ward 1993). In cross-infection experiments, miracidia of *Microphallus* sp. are more successful at infecting snails from the local snail population than allopatric snails (Lively 1989), despite relatively high rates of gene flow among parasite populations (Dybdahl and Lively 1996). In this example, the trematode disproportionately infects host genotypes that are locally common, and thus local adaptation results from the parasite tracking the most frequent host genotypes (Lively and Dybdahl 2000). Exceptions to this pattern are known and expected if parasite genotypes track local host genotypes with a lag (Morand et al. 1996b). In a review of the

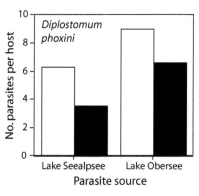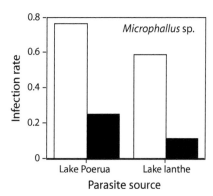

Figure 3.8 Examples of local adaptation by the digeneans *Diplostomum phoxini* in its fish second intermediate host in two Swiss lakes, and *Microphallus* sp. in its snail first intermediate host in two New Zealand lakes. The performance of parasites on sympatric or local hosts (open bars) and on allopatric hosts (black bars) was assessed experimentally under standardized conditions; performance was measured by the mean number of parasites establishing in hosts for *D. phoxini*, and by the proportion of hosts that became infected for *Microphallus* sp. In both examples, parasites achieve greater infection success on sympatric hosts than on allopatric hosts. (Data from Ballabeni and Ward 1993; Lively and Dybdahl 2000).

literature on local adaptation by parasites, Lajeunesse and Forbes (2002) found that local adaptation is more likely to be observed in parasite species that already show some host specificity, that is, parasites that only exploit few host species. This makes sense because generalist parasites exploiting many host species would have difficulty simultaneously tracking the changes in genotype frequencies in different local populations of their different host species. In general, though, after generations of isolation from other host genotypes, parasites may lose the ability to infect allopatric hosts in favor of a greater specialization for the local host genotypes. Alternatively, parasites can retain the ability to infect allopatric genotypes but achieve lower fitness when exploiting them. For example, a microsporidian protozoan parasite of the crustacean *Daphnia magna* achieves much higher spore production on local hosts than on hosts from distant populations (Ebert 1994).

Before moving on to the conditions that may predispose certain parasites for low or high host specificity, I want to extend the above ideas one step further. If parasites can become specialized for the local host population, why stop there? Why not specialize for the individual host itself? This is obviously not an option for the majority of parasites that do not spend more than one generation on the same host individual. For example, working with the nematode *Strongyloides ratti* and its rat host, Paterson (2005) tested all possible combinations of several host and parasite genotypes and found that no parasite genotype showed consistently greater performance in one particular host genotype. The host being only an ephemeral habitat, the parasites' offspring must disperse and locate new hosts; they must be adapted to infect hosts of different genotypes and not only the genotype of the host they just left. However, some rapidly dividing protozoan parasites exploiting long-lived hosts do spend several generations inside one individual host, after the host acquires an initial infection. Recent evidence suggests that over time, selection acts on the many dividing parasites inside the host and favors the parasite genotypes best suited for the particular conditions in that individual host (Seed and Sechelski 1996). A very similar phenomenon occurs in phytophagous arthropods, which can complete thousands of generations on a single host plant (Karban 1989; Mopper and Strauss 1998). Clearly parasites must retain the ability to colonize new host individuals and cannot become specialized exclusively for a certain host genotype. Rapidly dividing parasites, however, provide an example of a microevolutionary trend toward ultimate specificity.

The findings summarized above indicate that the evolution of host specificity may proceed in both directions, toward higher or lower specificity. Host specificity is just another continuous variable on which selection acts in no fixed direction; how it evolves depends on opportunities for host switching, availability of suitable hosts, and how host switching affects parasite fitness.

3.3 Determinants of Host Specificity

To illustrate the constraints and pressures acting on the evolution of host specificity, I will use Combes' (1991b) filter concept. Two filters determine how many animals can be used as hosts by a parasite, an encounter filter and a compatibility filter (fig. 3.9). The encounter filter excludes all animals that the parasite cannot meet for ethological or ecological reasons. The compatibility filter eliminates species in which the parasite cannot survive and develop, for morphological, physiological, or immunological reasons. Selection will act on these filters to increase or decrease their permeability and specificity; with respect to figure 3.9, selection acts to change the diameter of the encounter filter or the angle of the compatibility filter. For instance, in several species of allantonematid nematodes parasitic in *Drosophila* flies, many fly species are compatible hosts, as determined by laboratory infections, but not encountered in nature, and thus the encounter filter limits the realized host range of each nematode species (Perlman and Jaenike 2003). In contrast, the rhizocephalan parasite *Loxothylacus texanus* can encounter and even settle on many different crab species in the Gulf of Mexico, but it fails to survive and develop in crabs outside the genus *Callinectes*; in this case it is the compatibility filter that limits the host range (Boone et al. 2004). Other frameworks developed to illustrate the evolution of host specificity, including mathematical models (Garnick 1992a), lack the generality and simplicity of the filter concept. Here some of the factors that may influence the filters are discussed, leading to simple predictions that can be tested with comparative data.

Properties of the parasites can determine what range of potential hosts will be encountered (Price 1980). Selection has provided parasites with detailed genetic instructions to help them find a suitable host at each stage of the life cycle. Nevertheless, mistakes happen and are usually fatal. Several events can change the frequency of such mistakes through evolutionary time. For example, a new species with ecological attributes (diet, size, microhabitat preferences, etc.) identical to those of a parasite's only host, can invade the area occupied by the parasite and become more abundant than the host. Because suitable hosts would now become diluted in a sea of nonsuitable targets, mistakes could become more common. Selection would favor more sophisticated host-location

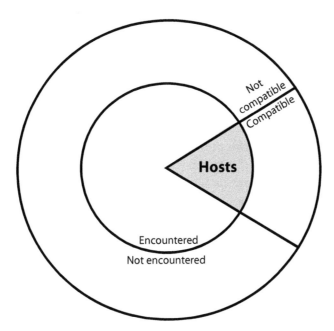

Figure 3.9 Schematic representation of the forces determining the range of host species used by a parasite. Of all potential host species (the outer circle), only a subset are actually encountered by the parasite. The encounter filter is represented by the smaller inner circle. Another filter, the compatibility filter represented by the triangular area, excludes all species in which the parasite cannot develop. Some compatible species are never encountered, and vice versa. Only species passing through both filters, i.e., those in the shaded area, can become hosts. (Modified from Combes 1991b)

mechanisms to minimize the possibility of mistakes. Alternatively, it could favor any mutant parasite capable of surviving in and exploiting the invading host species. The fitness of such parasites would not initially have to be equal to what could be achieved in the original host; it would just need to be greater than zero. Even if alternative targets are readily available, the likelihood of making mistakes during transmission and the opportunity for colonization of new hosts will depend on the mode of transmission. Therefore we might expect a link between transmission strategies and host specificity.

The infective larvae of ectoparasites such as monogeneans and copepods are equipped to locate and infect the right species of host. Similarly, the mobile infective stages of many endoparasites must locate, attach to, and penetrate the skin of their host, and possess several mechanisms for host location and recognition (see section 2.4). Although active choice of individual hosts may be possible in some species (McLachlan et al. 1999), in general, mechanisms of host location and recognition serve to discriminate against host species, and not host individuals. The mobility of many infective stages often suggests that they can potentially encounter a much greater number of host species than the ones that they actually exploit. Among related species, variation in the behavior (Snyder and Janovy 1996) or dispersal ability (Downes 1989) of mobile infective stages can explain interspecific differences in host specificity, but large-scale comparative studies are yet to confirm this tendency. Other parasites can only be transmitted via direct physical contact or during copulation between their host individual and another individual; since physical contact between members of different species is unusual, opportunities for transferring to another host species are limited. Parasites using any of the above transmission routes should be less subject to mistakes than infective stages entering the host by ingestion. This latter route of infection relies on hosts eating infective stages or infected intermediate hosts and does not prevent nonhost species from ingesting the parasites (Noble et al. 1989). All else being equal, we may predict that parasites entering the host through the mouth are less host specific than parasites using other routes of infection. In other words, the encounter filter (fig. 3.9) allows more host species to reach parasites using the oral route of infection. Complex life cycles, with their many steps and modes of transmission, may select for flexibility in the parasite's range of acceptable hosts. Noble et al. (1989) suggested that parasites with complex life cycles should be less host-specific than parasites with simple, direct cycles, and parasites with complex life cycles should be more specific in their choice of intermediate hosts than when settling on a definitive host.

Characteristics of the hosts will also influence opportunities for host switches and evolutionary changes in host specificity. Suitable but not yet colonized hosts are those that provide parasites with suitable living conditions and opportunities for the completion of the life cycle. They must therefore be similar to the hosts used by a parasite. The similarity can be the result of phylogenetic inheritance, if the suitable host is related to the parasite's current host, or convergence, if the suitable host has independently evolved to resemble the parasite's current host (Kennedy 1975). Obviously, in most cases similar hosts will be relatives, so we might predict that parasites of hosts belonging to species-rich taxa will be less host specific than parasites of hosts with no or few relatives, because of a difference in the availability of suitable hosts. Exploitation of a host from a speciose taxon means that the compatibility filter is wide open (fig. 3.9).

Two key host features that may influence the evolution of specificity are host body size or longevity, and host social behavior or spatial distribution. First, exploiting an ephemeral resource, such as a small and short-lived host, places limits on the reproductive output of parasites. Natural selection may not favor specialization on such hosts. In contrast, we might expect selection to favor the specialization of parasites that exploit large-bodied hosts, which are also generally long-lived and provide a stable, predictable resource base. Thus, we might predict a negative relationship between the average body size of a parasite's host species, and how many the parasite exploits. Second, opportunities to colonize new host species will depend heavily on the current host's tendency to interact with other species, and its spatial isolation from other potential host species. This applies mainly to contact-transmitted parasites with simple life cycles, of course. Such parasites exploiting spatially or socially isolated host species should be more host-specific than parasites of more uniformly distributed host species, if only because of a lack of opportunities for host switching.

Other host traits may also affect the evolution of host specificity. For instance, it is widely recognized that a high degree of host specificity can lead to a high risk of local extinction when host abundance drops below a critical level and the parasite is unable to use other available animals (Kennedy 1975; Bush and Kennedy 1994). Combes (1995, 2001) suggested that parasites exploiting hosts with unstable populations may be selected to maintain a wide host spectrum. The ability to exploit a wide range of hosts may come at a cost: a lower degree of specialization can mean lower fitness in general (see previous section). Because of this cost, perhaps a better long-term solution against the risk of extinction would be to maintain a wide-open encounter filter but a very exclusive compatibility filter (see fig 3.9 and Combes 1991b). Similarly, parasite populations on islands may be selected to exploit a wider range of host species because of the greater risk of extinction faced by host species in insular habitats (Freeland 1983).

Interesting as they may be, such considerations do not lead to any easily testable predictions. Other factors can mask the role of host or parasite features in the evolution of host specificity. For instance, changing environmental conditions can lead to a suspension of host specificity (e.g., Zander 1998). With this in mind, the next section will summarize how the available evidence supports the few predictions made above.

3.4 Observed Patterns of Host Specificity

Within any large parasite taxa, the majority of parasite species tend to be very host specific. Similarly, free-living animals in general tend to be limited in their distribution among potential habitats. For instance, the sizes of the geographical ranges of free-living species within any given taxon vary greatly, but their distribution is typically strongly right skewed. However, the observed distributions are usually lognormal, that is, when geographical range sizes are log transformed their distribution becomes normal (Gaston 1996). The predominance of small geographical ranges is an artefact of scale, and the skewness is not pronounced enough to survive the normalizing effect of a log transformation. Distributions of host specificities of parasite species within a taxon, on the other hand, are strongly right skewed before and even after log transformation (fig. 3.10). This indicates a very strong tendency for parasites to be highly host specific. Low host specificity is very much the exception. This, of course, applies to host specificity measured as the number of host species used by a parasite. Using the index S_{TD} instead, the right skew disappears, indicating that the taxonomic distances of the host species used by parasites are as likely to be high as they are to be low, within the range observed (fig. 3.11).

The key point is that there exists variability in host specificity between related parasite taxa. Can some of it be explained by differences in the route of transmission? Finding related parasite taxa that differ only in how they infect their host is not easy. Perhaps the best group for such a contrast would be digeneans, in which one lineage penetrates the vertebrate definitive host through the skin whereas other digeneans reach it via ingestion. Detailed information is not available to perform a robust comparison, but skin-penetrating schistosomes often have a wide repertoire of definitive hosts (Noble et al. 1989) whereas many surveys report that among other digeneans close to 50 percent of species occur in a single host species (Gregory et al. 1991; Poulin 1992a). Thus the

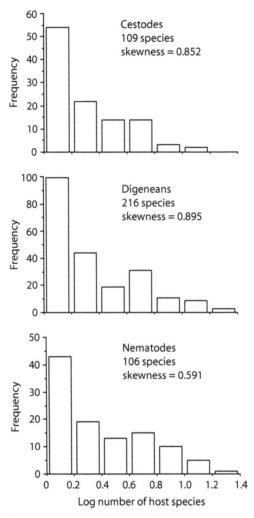

Figure 3.10 Frequency distribution of the number of known host species for cestodes, digeneans, and nematodes parasitic in birds. The index of skewness presented is the g_1 of Sokal and Rohlf (1995). Even though the distributions use log-transformed numbers of host species, all three are significantly right skewed and not lognormal. (Data from Gregory et al. 1991)

prediction that orally infecting parasites are less specific than those penetrating through the skin may not be supported across this parasite taxon.

Among helminth species parasitic in wild primates, however, host specificity and mode of transmission show a clear association (Pedersen et al. 2005). Parasites transmitted by close contact, sexual or other physical contact, are highly specific, with about two-thirds only known to infect a single host species and none capable of infecting host species belonging to different families. In contrast, parasites using intermediate hosts and transmitted via food are much less specific, with less than half restricted to a single host species, and more than a quarter exploiting hosts belonging to different mammalian

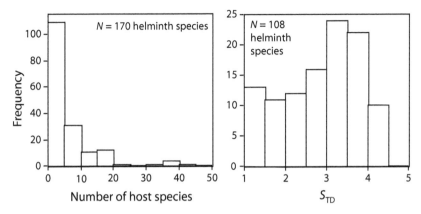

Figure 3.11 Frequency distribution of two measures of host specificity, i.e., the number of known host species and the index S_{TD}, among helminth parasites of Canadian freshwater fish. The data for both measures of specificity come from the same parasite species except that those for S_{TD} include only parasites with at least two known host species, as this index cannot be computed for single host species. (Data from Poulin and Mouillot 2003a).

orders. Vector-transmitted parasites show intermediate levels of host specificity (Pedersen et al. 2005). Although these comparisons do not account for phylogenetic influences on both host specificity and mode of transmission, they still suggest that how a parasite is transmitted may in part constrain its ability to exploit diverse host species.

Along those lines, comparisons between a large number of sexually transmitted parasites and their relatives that use a different route of infection has indicated that sexually transmitted parasites utilize a much narrower range of host species (Lockhart et al. 1996). Given that sexual contacts between hosts are almost exclusively intraspecific, this is not surprising. Yet, we need more comparative analyses of this nature, which attempt to control for phylogenetic influences to determine if the mode of transmission itself selects for a certain level of host specificity.

Monogeneans are widely regarded as highly host specific; typically, in any data set, much more than half of monogenean species are only known from a single host species (Kennedy 1975; Poulin 1992a; Rohde 1993). They are capable of evolutionary host switches (Guégan and Agnèse 1991) but generally are so specific that their mere presence is often enough to identify the host species (Lambert and El Gharbi 1995). Why are they much more host specific than their relatives, the digeneans and cestodes? Monogeneans attach to the skin of their aquatic vertebrate host rather than infecting it orally but also differ from digeneans and cestodes by having simple, one-host life cycles. Their route of infection, the simplicity of their life cycle, or some other characteristic such as the specialization of their attachment organs, may all have contributed to their specificity. For instance, at a proximate level, successful attachment of monogeneans on fish appears to depend on chemical affinities between the adhesive used by the parasite and the composition of fish mucus (Whittington et al. 2000). There is no way to tell at present which if any of these factors affected the evolution of host specificity in monogeneans.

There has only been one true test of the prediction that parasites with complex life cycles are less host specific than their relatives with direct life cycles. The average number of known vertebrate host species of 35 ascaridoid nematode species with simple life cycles was found not to differ from that of 334 ascaridoid species with complex life cycles (Morand 1996a). In this comparison, parasite species were treated as independent from one another regardless of their taxonomy, and thus phylogenetic influences were not removed. Still, this result provides no support for the prediction that host specificity is relaxed in parasites evolving complex life cycles.

Yet another prediction made in the previous section, that parasites with complex life cycles should display a higher specificity for their intermediate host than for their definitive host, remains mostly untested. The specificity of some helminths with complex life cycles for their intermediate host can prove lower than what is often believed when detailed experimental infections are performed (e.g., Dupont and Gabrion 1987). In digeneans, the specificity of miracidia for their snail hosts is notoriously rather strict. Experimental evidence indicates that it can be maintained by either a closed encounter filter or a narrow compatibility filter, or both. In some species, precise host-finding mechanisms allow miracidia to avoid all snail species except their only known host species; in other digeneans, miracidia readily attach to and penetrate novel snail species, but are subsequently killed by humoral immune factors in these snails (Sapp and Loker 2000). Various field surveys suggest that the host specificity of digeneans is much looser at later stages in their life cycle, but this has never been tested rigorously. Comparisons of host specificity between intermediate and definitive hosts, from a wide range of parasite species, are therefore needed before the prediction can be evaluated adequately.

The expansion of the number of host species used by the process of host switching should be facilitated when alternative species physiologically and ecologically similar to the host are available. Among parasites of Canadian freshwater fish, this certainly appears to be the case (fig. 3.12): parasites exploiting hosts belonging to species-rich taxa have colonized several of these other potential hosts closely related to the original host (Poulin 1992a). The same apparently occurred among digeneans exploiting coral-reef fish on the Great Barrier Reef (Barker et al. 1994). The average number of host species used by eighteen digenean species found among thirty-nine species of pomacentrid fish was almost five, with four parasite species each occurring in more than ten host species. In addition, nine of the eighteen digenean species were also found in hosts from other fish families. The large number of related fish species, living in sympatry at high densities, probably made host switching easy.

Other results also support the idea that parasites can expand their host range when potential host species related to their original hosts are locally present. Among fleas parasitic on small mammals, the taxonomic distances among the host species used by a parasite are generally no different from those of random subsets of species drawn from the pool of potential host species locally available. However, when they differ from random selections, they are consistently smaller, indicating that parasites preferentially use taxonomically related host species (Krasnov et al. 2004a). Under experimental conditions, allantonematid nematodes parasitic on *Drosophila* flies infect novel host species more readily if these are closely related to their original host; infection success approaches zero with

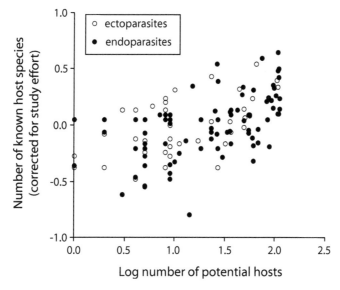

Figure 3.12 Relationship between the number of known host species and the number of potential hosts available among metazoan genera of ectoparasites (N = 41, r = 0.514, P < 0.001) and endoparasites (N = 90, r = 0.511, P < 0.001) of Canadian freshwater fish. The number of known hosts was corrected for study effort; the scores used are the residuals of the regression of log number of known hosts on log number of published records (see fig. 3.1). The number of potential hosts are obtained by summing up all fish species belonging to the family or families that include the parasite's known hosts. (Data from Poulin 1997a)

increasing genetic distance between the novel and original hosts (Perlman and Jaenike 2003). These several lines of evidence all suggest that host specificity should be lower for parasites exploiting species-rich host taxa.

Host switches between unrelated hosts can also take place, of course, but probably mainly under unusual circumstances. For instance, the colonizations of humans by parasites from unrelated hosts (e.g., schistosomes or the nematode *Onchocerca*; see discussion in Combes 1995, 2001) are probably the consequences of the rapid geographical expansion of humans, their modified behavior, and their impact on the environment. Similarly, animal species introduced to new geographical areas generally leave most of their parasites behind them because of founder effects and small initial population sizes (Torchin et al. 2003). However, following their establishment in the new area, they may acquire parasites from local host species even if they are not closely related to them. Salmonid fishes, for instance, have acquired parasite faunas in countries where they have been introduced that are taxonomically as diverse as those they possess in their areas of origin, but consisting exclusively of parasites specialized for other fish families (Poulin and Mouillot 2003b). Remarkably, the assembly of these diverse parasite faunas has occurred over just one hundred to one hundred fifty years, and not over evolutionary time. Thus, although host switching is probably easier among related host species, it is far from impossible among unrelated species.

The strength of host defenses is one general feature of locally available host species, in addition to their phylogenetic relatedness with a parasite's original host, that can facilitate their colonization by parasites. Møller et al. (2005) have found that avian flea species that parasitize a large number of bird species only exploit host species with weak immune responses, in contrast with host-specific fleas. They suggest that low host specificity in these fleas can be facilitated by a high availability of hosts with poor immune defenses, allowing easy colonization.

Two other host characteristics are expected to influence the evolution of specificity. Separate comparative analyses, based on different data, have both demonstrated that highly host-specific monogeneans tend to parasitize larger fish species, and that host specificity decreases as the average body size of the hosts decreases (Sasal et al. 1999; Desdevises et al. 2002b). Although similar evidence from other kinds of parasites would be welcome, these results support the hypothesis that specialization is favored on stable, predictable resources, such as large-bodied and long-lived host species. The other important host feature is their likelihood to interact with other potential host species, or show spatial overlap with other species. Among flea species parasitizing birds, those exploiting colonial birds have narrower host ranges, that is, are more host specific, than those infecting birds with a less pronounced degree of spatial clumping (Tripet et al. 2002). In this study, bird species classified as colonial nesters were those that form dense, monospecific breeding colonies, and therefore do not come into close proximity with other species. This finding supports the hypothesis that the specificity of directly transmitted parasites, such as fleas, is linked with the extent to which their hosts provide them with opportunities for host switching.

So what have we learned so far? Given sufficient genetic variation and potential fitness benefits, natural selection will favor host switching and adding new host species to a parasite's repertoire in situations where opportunities arise. Thus, in cases where a parasite's mode of transmission or its host's behavior exposes it to new host species, and if these other host species are related to the parasite's original host, host switching will typically result in parasites with lower host specificity. The examples above show this to be generally true for a wide range of parasite taxa.

There is still a great gap in our understanding of the evolution of host specificity. Few solid patterns have been reported, probably not because they do not exist but because no one has bothered looking for them. Some of the few known patterns are yet to receive a satisfactory explanation. For instance, there is a latitudinal gradient in the host specificity of digeneans parasitic in marine fish; digenean species from the tropics are restricted to fewer host species than their relatives in colder seas (Rohde 1993). Monogeneans, on the other hand, display consistently high host specificity from low to high latitudes. Why the two taxa follow different trends is not clear, and exemplifies how little we know about the evolution of host-parasite interactions and host specificity.

Certain patterns of host specificity are likely to defy explanation for some time. Species in the Myrmecolacidae, a family of the unusual endoparasitic insect order Strepsiptera, provide a superb example: the males parasitize ants, whereas the females parasitize grasshoppers and crickets (Kathirithamby 1989). This unusual type of host specificity would seem to require sex-specific mechanisms of host location and immune evasion. How could it have evolved?

3.5 Conclusion

The spectrum of host species used by a parasite at each stage in its life cycle is one of its most fundamental characteristics. This information is not available for most parasite species. We have it for several other species but are yet to understand it. One phenomenon renders impossible any quick and easy interpretation of large-scale patterns of host specificity: the host species used by a parasite in one locality are merely those suitable species that happen to be present locally. Although the number and taxonomic diversity of host species used by a parasite are constrained within limits set by its past history, they do vary on a geographic scale based on the presence and relative abundances of local host species (Krasnov et al. 2004a). This creates a situation in which the different populations of a given parasite species are each engaged in a slightly different coevolutionary association with local hosts (see Thompson 2005). This geographic perspective has been lacking from most previous investigations of host specificity, but we now need to consider it more seriously.

Three avenues are open for the investigation of the evolutionary forces shaping host specificity. First, experimental studies in which novel host-parasite combinations are created are excellent ways of determining what proximal mechanisms limit the host range of a given parasite species. Second, recent and future developments in cladistics and the reconstruction of parasite phylogenies should provide us with an increasing number of historical accounts of associations with hosts. Refinements in existing methods of reconciling host and parasite phylogenies may allow us to infer extinction or host-switching events and make predictions about the conditions favoring the colonization of new host lineages. The third approach should be to test these predictions with comparative analyses of the existing data in many different groups of parasites. As comparative biology gains in popularity among parasitologists, the next few years should see great strides taken toward elucidating the many aspects of host-parasite associations.

4 | Evolution of Parasite Life-History Strategies

Parasites must complete difficult journeys through several hosts as well as through the external environment. Evolution has sometimes favored longer and more complex journeys. Sometimes it has opened up alternative paths through the same cycle by relaxing host specificity. The length, complexity, and flexibility of the life cycle are the outcome of selection acting on the cycle as a unit of selection. But evolution has also shaped the organism itself, and this chapter now focuses on the parasite as an organism rather than as a life cycle.

Throughout the history of parasitology, several laws or rules have been proposed to describe the evolution of parasitic organisms, most of which have not survived the test of time and robust analysis (Brooks and McLennan 1993a; Poulin 1995d). Two of the most pervasive misconceptions about parasite evolution are that parasites evolve toward reduced body size and toward higher fecundity. Parasites are smaller than their hosts; that is a fact, but an irrelevant one when considering the evolution of parasite body size. The interesting comparison is not between the size of the parasite and that of its host, but between the size of the parasite and that of either its free-living ancestor or its closest free-living relative. Similarly, parasite fecundity is often pitted against that of free-living organisms. Many textbooks dazzle their readers with tables listing the egg output of some of the most fecund parasites (e.g., Esch and Fernández 1993). Again, the truly interesting questions relate to how these numbers compare to egg production by the parasites' free-living ancestors or closest free-living relatives, and also why some parasites exhibit less than prodigious fecundity. Taxa that contain both parasitic and free-living members and in which parasitism has appeared more than once, such as crustaceans and nematodes, are very promising for such investigations.

Life-history strategies are combinations of physiological and demographic traits such as body size, life span, age at maturity, fecundity, and offspring size that have been favored by selection because they result in higher fitness in particular environments than other possible combinations. The simultaneous maximization of several traits may be impossible because of phylogenetic, physiological, or physical constraints. Thus, the combinations of life-history traits that can evolve are limited, and are characterized by trade-offs between various pairs of traits (Stearns 1989). For instance, any investment into offspring size may come at the expense of fecundity. External pressures determine which way particular trade-offs will go, and the emerging set of life-history traits is expected to show some association with the organism's mode of life or habitat (Partridge and Harvey 1988; Southwood 1988). Other reviews give detailed explanations of life-history theory (Roff 1992; Stearns 1992) or how it applies to parasitic organisms in general (Poulin 1996b). This chapter will explore the evolution of life-history strategies in parasites, but with special emphasis on body size, age at maturity, and egg production.

4.1 Phenotypic Plasticity and Adaptation

In populations of parasites or free-living organisms, adjustments of life-history strategies to changing environmental conditions can take place through two distinct but mutually compatible processes (Stearns 1992). First, a given genotype may be capable of producing a variety of phenotypes under different conditions. In animals living in habitats that are highly variable spatially or over short time periods, a flexible developmental schedule would be highly advantageous. The set of possible phenotypes, referred to as a reaction norm, offered by a single genotype allows an organism the possibility to adjust to local and current conditions through small changes in development resulting in the life-history strategy best suited to those conditions. Phenotypic plasticity allows immediate responses to environmental changes and does not result in changes in genotypic frequencies in the population.

Second, if environmental changes persist in time, a true evolutionary response can occur, involving a shift in gene frequencies in the population. The genotypes producing sets of phenotypes better suited to the new conditions will be favored, and can spread through the population over time. In contrast to phenotypic plasticity, adaptive genetic responses can only take place over several generations and not within a single generation. The two processes, however, are compatible and can operate jointly. For instance, if new environmental conditions favor large body sizes, individual organisms in one generation can opt to grow to the upper limit set by their genotype, while selection acting over several generations can shift the mean or upper limit of sizes that individuals can reach toward greater values.

Parasites often display considerable phenotypic plasticity in life-history traits. For example, in *Triaenophorus crassus*, a cestode parasite of freshwater fish, mature and gravid individuals of the same age range in mass between 5.7 and 124 mg (Shostak and Dick 1987). The largest adult individual in the population is therefore over 20 times larger than the smallest. In the nematode *Raphidascaris acus*, also a parasite of freshwater fish, gravid female worms range in mass between 0.7 and 61.2 mg, the largest being almost ninety times the size of the smallest (Szalai and Dick 1989). The ranges of fecundity values among individuals of these two species are even greater. The relative scale of the phenotypic plasticity displayed by these worms greatly exceeds that shown by most free-living organisms. Such large differences in the adult size or fecundity of conspecific parasites from the same population may result from the distribution of parasites among their hosts (see chapter 6): the aggregation of many parasites in a few hosts means that many worms live in crowded conditions and may not be able to reach their maximum adult size. Parasites that do not share their host with conspecifics, on the other hand, are able to reach their maximum potential size. Thus growing conditions in the host are unpredictable and not all parasites in a population live in the same conditions. A flexible developmental program therefore affords individual parasites the opportunity to make the best of difficult situations: they get as close as they can to their maximum potential size but can still mature at small sizes if they have no other choice. Other conditions inside the host, such as immune responses, are genetically determined and vary from host to host; these can generate considerable phenotypic variation in body size or other traits among parasite individuals within a population (Stear et al. 1997; Davies and McKerrow 2003). In the above examples, the phenotypic plasticity in parasite life-history traits was observed within a single species of host. When parasites can use more than one host species at a given stage in their life cycle, the resulting variability can be even greater and more far reaching. For instance, redial development and cercarial production in the digenean *Fasciola hepatica* in its snail intermediate host are greatly influenced by the identity of the mammalian definitive host, among the few suitable species, from which the miracidium originated (Vignoles et al. 2004).

However, differences between individuals in a population are not entirely the product of phenotypic plasticity. Different individuals have different genotypes, and the source of differences in phenotypes can also have a genetic basis. True evolutionary change, involving changes in gene frequencies, can occur rapidly once a parasite population is subjected to selective pressures. For instance, life-history traits, such as patterns of egg production and parasite development rate within the host, differed significantly from those of the source population after only eleven generations in which selected lines of the nematode *Heligmosomoides polygyrus* were serially passaged through mice (Chehresa et al. 1997). Similarly, changes in life-history traits appear after as few as twelve generations under laboratory-rearing conditions in the entomopathogenic nematode *Steinernema glaseri* (Stuart and Gaugler 1996). The potential for rapid adaptation may be more limited in other parasite taxa. In the acanthocephalan *Moniliformis moniliformis*, for example, a laboratory population isolated for sixty generations did not show substantial life-history divergence from its wild population of origin (Reyda and Nickol 2001).

Many variables are associated with the expression of life-history traits in individual parasites (Poulin 1996b). Because adaptive responses are not easily distinguished from phenotypic plasticity at the level of the individual organism, I will mostly examine adaptive adjustments of life-history strategies using comparisons among species or higher taxa in the remainder of this chapter. At these levels, differences are surely the product of adaptive genetic changes. To compare species, however, one must use species-typical values for traits such as body size or fecundity. If these traits are as plastic within species as in the two above examples, how can we be sure that the values used are indeed representative of the whole species? The answer is that we cannot; one can only hope that averages provide good estimates of a species' strategy. Also, most studies to date have investigated average species values and only few have examined the variance in life-history traits. Variance of a trait can not only constrain evolutionary rates but it may also itself be subject to selection. Different parasite taxa may exhibit different levels of variability in life-history traits as a result of selection in hosts or habitats that differ in predictability, and further explorations of how this variability evolved would be particularly interesting.

4.2 Parasite Body Size

Many animal lineages undergo increases in average body size over evolutionary time, a phenomenon known as Cope's rule and supported by some fossil evidence (see Stanley 1973; Bonner 1988; Arnold et al. 1995; Jablonski 1996; Alroy 1998). The rule is not universal, however, and enough exceptions exist to suggest that body size may follow different evolutionary trajectories in different groups of organisms (Jablonski 1997; Knouft and Page 2003). In contrast, lineages making an evolutionary transition from a free-living existence to a parasitic mode of life are almost invariably assumed to evolve toward smaller body sizes (Price 1980; Hanken and Wake 1993). The physically restricted habitats of parasites are believed to have been the key factor promoting a reduction in their body size. The unstated and untested assumption behind this line of reasoning is that the free-living ancestors of modern parasites had larger body sizes than their parasitic descendants. Parasitism had multiple origins, and precursors of parasites came in all shapes and sizes. Evolution of parasite body size has no doubt proceeded differently in different lineages. This section begins with a comparison between the body size of parasites and their free-living relatives, and attempts to reconstruct the evolution of parasite body size to determine the direction of change following the adoption of parasitism as a lifestyle. The discussion then moves on to examine what other factors, either related to the host or the environment have played a role in the evolution of parasite body size. Finally, the forces creating a dimorphism in size between males and females in dioecious parasites will be explored.

4.2.1 Changes in Size as Adaptations to Parasitism

Within large taxa comprising both free-living and parasitic groups, quantitative comparisons between the sizes of these groups indicate clearly that parasites can be as large or larger than their free-living counterparts (figs. 4.1 and 4.2). There should therefore be no a priori reason to believe that parasites evolve to be smaller than free-living animals.

One way of contrasting the evolution of body size in parasites and their free-living relatives is to compare the shape of the frequency distributions of their body sizes. The shape of a body-size distribution can indicate which size classes have undergone extensive diversification and which remain species poor, and can thus reflect the relative selection pressures acting on body size within a large taxon. Almost universally, body-size distributions are right skewed, even on a log scale, such that the small-bodied classes,

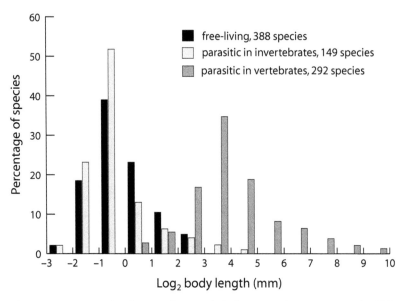

Figure 4.1 Frequency distribution of nematode body sizes with respect to mode of life and type of host. (Modified from Kirchner et al. 1980)

though not the very smallest, invariably include many more species than the large-bodied ones (Blackburn and Gaston 1994a). Several evolutionary mechanisms have been proposed to explain this ubiquitous pattern, most involving different rates of speciation and/or extinction among size classes (see Blackburn and Gaston 1994a; Johst and Brandl 1997; Kozlowski and Gawelczyk 2002). Many taxa of parasites, however, depart from the typical right-skew pattern (Poulin and Morand 1997). Log-normal distributions are common among ectoparasitic taxa such as monogeneans, copepods, and isopods parasitic on fish. In these parasites large size leads to higher fecundity but a greater likelihood of being dislodged from the external surfaces of the host, and selection may favor parasites of intermediate size. Ticks, however, show the usual right-skewed distribution of body sizes. Hosts of ticks, unlike fish, are capable of self-grooming, and this selective pressure may have pushed tick body sizes toward the smaller end of their spectrum (Poulin and Morand 1997). Internal parasites for which data were available, for instance, nematodes and digeneans, show right-skewed distributions of body sizes. The microhabitats available to these parasites may be physically constrained such that more niches are available for small-bodied endoparasites than for large ones. This would lead to higher speciation rates among small-bodied parasite taxa or to reduction in body size in the large-bodied taxa. Examinations of body-size distributions therefore suggest that evolution of body sizes proceed in different directions in different parasitic groups, and not always toward a reduction in size. In particular, the patterns of body-size distributions reported by Poulin and Morand (1997) appear to support a rough dichotomy between ectoparasites and endoparasites, similar to that proposed for ecto- and endoparasitoid wasps (see Mayhew and Blackburn 1999), and driven by different pressures on body-size evolution.

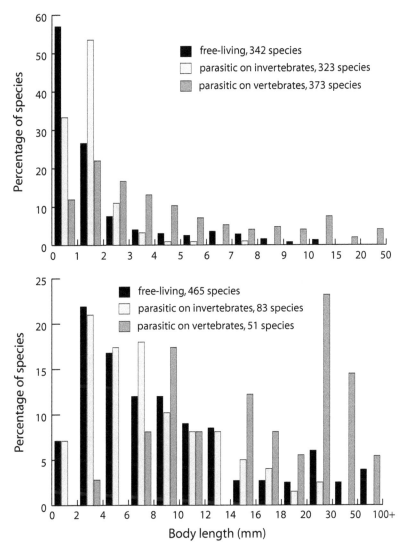

Figure 4.2 Frequency distribution of copepod (top) and isopod (bottom) body sizes with respect to mode of life and type of host. Terrestrial isopods are excluded, and only isopods that are obligate parasites of fish as adults (i.e., family Cymothoidae) are included in the vertebrate parasite category. Note the contraction of the scale on the right-hand side of the figure. (Data from Poulin 1995b, c)

There are three problems with the use of body-size distributions to infer evolutionary trends, however. First, the shape of the distribution may depend on whether or not a given taxon is polyphyletic. For example, among monogeneans, the majority of species belong to small, but not the smallest, size classes (fig. 4.3). This is because some small-bodied genera, for example, *Dactylogyrus* and *Gyrodactylus*, have diversified greatly compared to other genera (Poulin 1996c, 2002). There is, however, a second peak in the distribution, among intermediate size classes (fig. 4.3). These two peaks correspond roughly to the two major branches of the Monogenea, the Monopisthocotylea and the Polyopisthocotylea (Poulin 2002). The fact that these are separate clades has only become

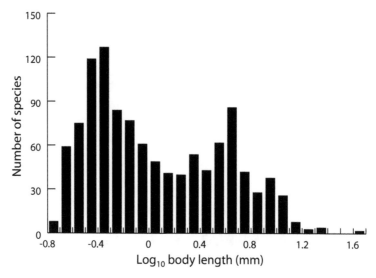

Figure 4.3 Frequency distribution of body sizes among 1,131 monogenean species parasitic on fish or other aquatic vertebrates. (From Poulin 2002)

clear following recent molecular phylogenetic studies (Justine 1998; Littlewood et al. 1999b). It is therefore possible to misinterpret body-size distributions when different clades are lumped because of insufficient phylogenetic information.

Second, if the diversification of taxa within the same size class is independent of body size, treating species as independent observations introduces a bias in the analysis. Among monogeneans, for instance, body-size distributions differ when generic values are used instead of species values, a procedure that gives less weight to a few very diverse genera (Poulin 1996c). Using species body-size distributions may thus lead to results confounded by phylogenetic influences (see Orme et al. 2002).

Third, the pattern observed may be an artifact of our incomplete knowledge of certain taxa. We can only examine the body-size distributions of known species, and not of all species. Many small species go unnoticed and therefore undescribed for long periods of time; as they are found and added to the body-size distribution of known species, the distribution becomes increasingly right skewed (Blackburn and Gaston 1994b). In some groups of parasites but not others, there is a negative relationship between species body size and date of description (Poulin 1996a, 2002; Poulin and Morand 2004). We may be left only with the smallest species to find and describe in some taxa, whereas much of the diversity of other taxa may have escaped us thus far.

Another much more rigorous method of contrasting body-size evolution in parasitic and free-living taxa consists of performing several independent comparisons between the mean body size of a parasitic taxon and that of its closest free-living sister group. These comparisons can then be analyzed statistically to see if they show a significant tendency for the parasitic lineages to be smaller or larger than their free-living relatives. This approach at least is free of phylogenetic influences because it treats evolutionary transitions between a free-living existence and parasitism as independent observations, rather than treating actual taxa as independent. Nematodes are a very suitable group to use in such an analysis. Parasitism has had a few independent origins among nematodes, and different types of hosts (plants, invertebrates, and vertebrates) have been colonized on separate occasions (Blaxter et al. 1998). A rough comparison of body sizes between free-living and parasitic nematodes shows clearly that parasitic nematodes are larger than their nonparasitic relatives (Morand and Sorci 1998).

A robust phylogenetic contrast approach has been used to examine the evolution of body size in parasitic crustaceans. In the phylogenetic history of copepods, parasitism on invertebrate hosts evolved on nine separate occasions from free-living ancestors (Poulin 1995b). Generally, the parasitic branch shows a larger body size than its sister branch, which remained free living; however, the average difference is not great enough to be significant. Parasitism of fish never evolved directly from a free-living ancestor among copepods; but on nine separate occasions, a lineage parasitic on invertebrates made the transition to parasitism on fish. Contrasts between these nine pairs of sister branches indicate that these transitions from invertebrates to fish hosts were followed by significant increases in body size. The average increase in size, >3 mm in body length, is huge for copepods (Poulin 1995b). These trends are also apparent in the distribution of copepod body sizes (fig. 4.2). The history of copepod parasites is thus one of increases in body size.

Body-size distributions of isopods also suggest little difference between free-living isopods and isopods parasitic on invertebrates, with fish parasitic isopods being considerably larger (fig. 4.2). This is a deceptive picture, however. There have been few transitions to parasitism in the phylogeny of isopods, but in all cases the branch that became parasitic displays a smaller body size than its free-living sister branch (Poulin 1995c). This counterintuitive result can be reconciled with fig. 4.2. The large-bodied Cymothoidae, parasitic on fish, evolved from morphologically similar ancestors that were only facultative parasites of fish. These in turn had evolved from free-living ancestors whose extant descendants are the Anuropidae, a family of giant isopods inhabiting oceanic depths. Thus while cymothoids are larger than most free-living isopods, they are most closely related to some of the largest of these free-living taxa, and have apparently evolved toward slightly smaller bodies than those of their ancestors. A recent phylogenetic hypothesis proposes that isopods parasitic on other crustaceans, such as the family Bopyridae, evolved from the Cymothoidae (Dreyer and Wagele 2001); if correct, then this host switch has been followed by a further reduction in body size.

Finally, the phylogenetic history of amphipods suggests seventeen independent transitions from a free-living existence to an obligate association with invertebrates. Although amphipod symbionts of invertebrates may not be true parasites, they face the same transmission problems and other pressures acting on parasites, and trends in the evolution of their body size may be instructive. In the majority of cases (thirteen out of seventeen), the amphipod lineage switching to symbiosis on invertebrates shows a smaller body size than its free-living sister group (Poulin and Hamilton 1995). The average decrease in body size for all seventeen contrasts, though, was small and not significant.

In other groups, parasitism has not evolved on many separate occasions but in only one or two lineages, providing limited opportunities for robust comparisons. Still, these

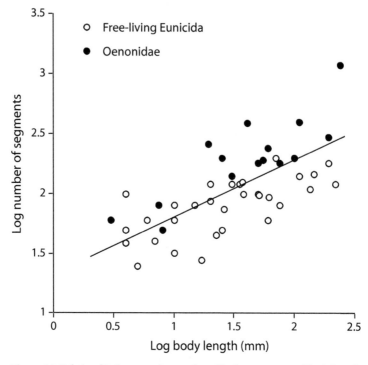

Figure 4.4 Relationship between the number of body segments and body length among polychaete species belonging to the clade Eunicida; the regression line computed across all species is shown. In members of the parasitic family Oenonidae, there are generally more segments for a given body length (most points above the line) than in the free-living species from other families (most points below the line). (From Poulin 2001a)

can be informative. In polychaetes, members of the family Oenonidae are endoparasites of other polychaetes, at least in the first part of their life; this family is the only polychaete family consisting of true parasites (Hernández-Alcántara and Solis-Weiss 1998). Morphologically, oenonids are very similar to their closest free-living relatives. Although they must fit in their polychaete hosts, they are often longer than their host (Hernández-Alcántara and Solis-Weiss 1998). So how do they compare with free-living polychaetes from closely related families? It turns out that they do not differ in body length from their nonparasitic relatives, but that their segmentation rate is significantly higher (fig. 4.4). Because the number of segments is a better indication of gamete production than body size in polychaetes, parasitic oenonids may have been selected to produce more segments than free-living polychaetes, rather than merely achieving longer bodies (Poulin 2001a). This finding still suggests a life history change following transition to parasitism.

The preceding discussion has focused on comparisons between parasitic lineages and their closest free-living relatives, as a way of assessing whether transitions to parasitism are generally accompanied by changes in body size. It is quite likely, however, that following the transition, parasite body size has evolved away from the ancestor's original size. Thus we may ask another question: how has parasite body size evolved since the transition to parasitism? Without fossil evidence, this is a difficult question to answer. However, using a phylogenetic approach, it is possible to compare the body size of basal and derived taxa, which is to say, ancestral taxa issued from the base of the phylogenetic tree and recent taxa originating from the branches of the tree. Consistent differences would indicate either an increase or a decrease in body size over evolutionary time, depending on the sign of the differences. Applied to digenean taxa, this approach indicates that there has been no consistent trend in the evolution of body size in these parasites (Poulin 2005b). In monogeneans, however, there is evidence for a gradual decrease in average body size over evolutionary time. In other words, recently evolved lineages tend to be smaller-bodied than ancestral lineages (Poulin 2005b). This weak pattern may be a consequence of the invasion of space-limited microhabitats, such as the gaps between gill lamellae, by the more recent monogenean taxa derived from ancestors that lived on fish skin.

Parasitism thus leads to either decreases or increases in body size, depending on which taxa are considered. The examples above illustrate how variable the evolution of parasite body size can be, and how much it depends on the size of the ancestor. There are no absolute rules governing body-size evolution, and each lineage may proceed down a different path. Similar studies of other parasite taxa would probably extend this conclusion to all parasites in general.

4.2.2 Correlates of Body Size

Following a transition to parasitism in a branch of a phylogenetic tree, and whatever general direction is taken by evolution with respect to body size, the lineage will undergo diversification and give rise to many species that do not all have identical body size. Various factors can act to shape body size, by placing pressures or constraints on its evolution (Poulin 1996b). These factors can originate from the host or from the external environment.

Host body size is one of the most obvious host-related factors that can affect the evolution of parasite body size. There are often differences in size between parasites of invertebrates and their relatives exploiting the larger vertebrates (figs. 4.1 and 4.2); these suggest that larger sizes can be attained in larger hosts. Within any host taxon, such as mammals for instance, space constraints faced by parasites may vary inversely with host body size. The volume of the lumen in the mammalian gut increases as we go from a mouse to an elephant. Large size is correlated with fecundity in most animals including parasites (Peters 1983); given the space, selection should favor larger-bodied parasites over their smaller conspecifics, and drive the evolution of body size toward the largest size that can be accommodated by a host without decreasing host (and parasite) survival. Larger hosts may also provide a greater supply of nutrients to parasites. Since life span correlates positively with body size, larger hosts are also likely to provide parasites with a more permanent habitat, which can favor individual parasites that delay maturity and reach larger sizes (Stearns 1992). These are some of the reasons why we may expect a relationship between host size and parasite body size.

One fact is often overlooked, however. Whereas large parasites can only live in large-bodied hosts, small parasites can exploit a wide range of host sizes. The variability in

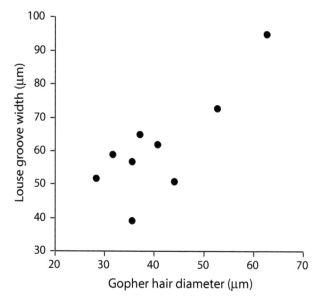

Figure 4.5 Relationship between the width of the groove on the head of chewing lice and the average diameter of body hairs from their pocket gopher host species. The groove is used by lice to grasp a hair on their host; it is thus not only essential for attachment to the host, but is also proportional to overall louse body size. Each point represents a different louse-gopher species combination. (Modified from Morand et al. 2000)

parasite body size may therefore increase along with mean parasite body size as we move from small to large hosts. This would weaken the relationship between host size and parasite size but would not eliminate it altogether if it exists.

In some parasite species, the body size of individual parasites in the population correlates with the body size of their host individual (see Poulin 1996b). There have been a few attempts to find such a relationship across species or higher taxa. Kirk (1991) found a positive relationship between the body size of flea species and that of their bird and mammal hosts. Harvey and Keymer (1991) found that, after controlling for phylogenetic influences, host size and parasite body size correlated positively among species of chewing lice infecting rodents and among species of pinworms parasitizing primates. Recently, Morand et al. (2000a) have not only confirmed Harvey and Keymer's (1991) results on chewing lice, but they have also provided a mechanistic explanation for the host-size-versus-parasite-size relationship. They have demonstrated that the width of the groove on a louse's head used to hold on to host hair covaries positively with the average diameter of host hair (fig. 4.5). In this case, the host size versus parasite size relationship is simply the by-product of both the fact that lice need to be larger to have wider grooves, and the correlation between host body size and hair diameter (Morand et al. 2000a). A similar phenomenon is observed among species of wing lice of the genus *Columbicola*: louse size is correlated with wing feather size, which is itself correlated with the overall size of the bird host (Johnson et al. 2005).

Harvey and Keymer's (1991) findings on nematodes have also been confirmed. Extending the analysis to oxyurid nematodes parasitic of both invertebrates and vertebrates,

Morand et al. (1996a) found a strong relationship between parasite body size and host size, independent of parasite or host phylogeny. Among nematodes parasitic in mammals, it is possible to generate a theoretical distribution of nematode body sizes using the distribution of mammalian body masses and an equation that predicts the optimal body size of a parasite for a given host size, based on a maximization of the parasite's net reproductive rate, R_0 (see chapter 7); when this simulated distribution is compared with the known distribution of nematode body sizes, the two are found to be very similar and have the same mean value (Morand and Poulin 2002). This strongly suggests that host body size is an important influence on the evolution of nematode body sizes.

Among acanthocephalan species, there is also a positive correlation between worm size and the mass of the vertebrate definitive host, although the relationship is very weak when phylogenetic influences are taken into account (Poulin et al. 2003a). Among digeneans, some of the largest species are found in very large host species. For instance, among didymozoid digeneans parasitic in teleost fishes, one of the largest (if not the largest) known species, over twelve meters in length, is found in one of the largest teleosts, the marine sunfish *Mola mola* (Noble 1967). Given the thousands of digenean and fish species, this particular combination seems unlikely to be a mere coincidence. Indeed, a comparative analysis across all digenean taxa gave contrasting results depending on the analytical method used to control for phylogenetic effects, but also provided some evidence for a positive relationship between vertebrate host mass and digenean adult body size (Poulin 1997b). Finally, host size also correlates positively with parasite size among species of rhizocephalans, a group of barnacles parasitic in other crustaceans (Poulin and Hamilton 1997).

The positive host size versus parasite size interspecific relationship is not universal, however. Some studies have failed to find any association between host size and parasite size (e.g., among copepods parasitic on fish; Poulin 1995b). In some cases, the relationship was not apparent across genera or families when phylogenetic effects were removed, but became clear in analyses across species within genera or genera within families when taxa were treated as independent observations. For instance, a strong relationship between host size and parasite size across genera within the diverse monogenean family Dactylogyridae failed to translate into a similar relationship among monogenean families (Poulin 1996c). Also, host size and parasite size are consistently and strongly correlated among species in the most speciose genera of ticks, but not across tick genera (Poulin 1998b).

Given the preceding results, it is probably fair to say that large host size facilitates the evolution of parasites toward large body sizes and the high fecundity that goes with it, but for some reason not all groups of parasites can take advantage of the benefits offered by large hosts. The influence of host body size on parasite life-history traits can be mediated by other host factors. For instance, larger hosts are longer-lived than small hosts, and place fewer temporal constraints on parasite growth (Harvey and Keymer 1991; Sorci et al. 1997; Morand and Sorci 1998). In addition, the mortality rate experienced by parasites in the host prior to their maturation is expected to be a pivotal determinant of maturation time: higher mortality should lead to reduced age and size at maturity (Morand and Sorci 1998). Since host immunity is a major cause of parasite mortality, variation in immune responses among host species could influence the age at maturity and size of their parasites.

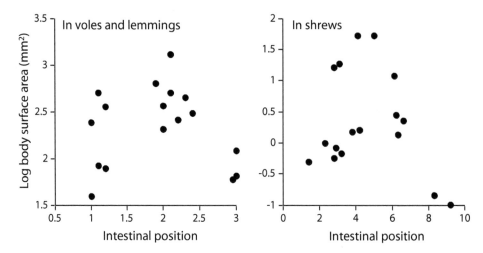

Figure 4.6 Cestode body size as a function of their position in the intestine of their mammalian host. Body size is taken as the surface area of the worm, or the product of mean length and width. Intestinal position is the median position of attachment of cestodes when the host intestine is divided into ten parts of equal length, or in the case of rodents (voles and lemmings), when the small intestine in divided into three parts. Each point represents a different cestode species. The humped distribution of body sizes indicates that the largest species occur in the middle intestine. (Modified from Haukisalmi et al. 1998)

Along those lines, Sorci et al. (2003) reported a negative correlation between nematode body size and the concentration of eosinophils in host blood. Eosinophils are white blood cells involved in fighting nematode infections, and nematode species exploiting host species with a strong eosinophil response mature at smaller sizes, thus increasing their chances of reaching their reproductive phase before being killed by the host. Similarly, Haukisalmi et al. (1998) have argued that the attachment position of cestodes in the gut of small mammals determines the optimal size a given cestode species should achieve via its impact on nutrient availability and mortality. Assuming higher mortality for species attaching in the anterior intestine, because of stronger peristalsis and the more rapid flow of ingesta, the model they developed predicted larger body sizes for cestode species attaching in the middle intestine. This prediction was supported by empirical evidence (fig. 4.6). These examples illustrate why host body size is not a universal determinant of parasite body size: many other host-derived factors are also involved.

Among the many other forces that intervene to determine parasite body size, some deserve mention. For example, schistosomes live in the blood vessels of their host, and their site of infection will place an upper limit on the size they can attain independently of the body size of the host. Some important factors may even come from the external environment, since environmental variables can shape life-history traits in free-living animals. Among free-living poikilothermic animals, for instance, temperature during development usually affects growth rates and body size, with most animals reaching smaller sizes at higher temperatures (Atkinson 1994). This effect has led to adaptive adjustments in life-history traits in free-living animals, with species living in cold environments differing in body size and other traits from their relatives in warmer habitats (Sibly and Calow 1986). The latitudinal gradient in body sizes observed within certain taxa, with

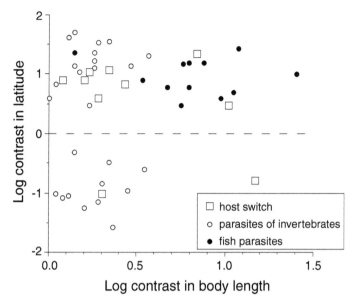

Figure 4.7 Relationship between latitude and body size in parasitic cope-pods. Each point represents a contrast between two sister branches in the copepod phylogeny. Values for the branch with smallest body size were sub-tracted from the branch with greater body size, giving contrasts in body size that are all positive. In most cases the contrast in latitude is also positive, in-dicating that copepod taxa with greater body size inhabit higher latitudes than their sister taxa. Host switches represent cases in which one taxa is par-asitic on invertebrates and its sister taxa is parasitic on fish. (Modified from Poulin 1995b).

body sizes increasing as one moves away from the equator, is known as Bergmann's rule (Blackburn et al. 1999). Internal parasites are only exposed to outside conditions for brief periods, and most if not all growth takes place in the host. External factors such as tem-perature may still have some impact on them if either these factors influence the physiol-ogy of the host with cascading effects on parasites, or the host does not buffer the para-site against external conditions. This latter scenario was not supported in comparisons of digenean taxa living in poikilothermic definitive hosts with their sister taxa living in homeothermic birds and mammals: host type had no detectable effect on digenean body size (Poulin 1997b).

Ectoparasites, on the other hand, are exposed to external conditions throughout their entire life and could show patterns of body size variation similar to those reported for free-living organisms. For instance, in parasitic crustaceans, body size tends to increase with increases in both latitude and water depth, two environmental correlates of water temperature (Poulin 1995b, c; Poulin and Hamilton 1995). In comparisons between sis-ter taxa, the lineage inhabiting higher latitudes or deeper water tends to be larger-bodied than its relatives in warmer waters, independently of other variables such as host type (fig. 4.7), a geographical trend that more or less conforms to Bergmann's rule. A similar pattern is observed among monogenean taxa (Poulin 1996c). In ticks, however, body size showed no correlation with average air temperature in analyses across species or higher taxa (Poulin 1998b). Again, these examples illustrate the many different responses of parasite body size to the same selective pressures, and reinforce the point that the evolution of body size in parasites does not follow any given route.

The preceding discussion deals exclusively with the body size of parasites in their definitive host; what about the body size of the various developmental stages of parasites in their intermediate hosts? Parker et al. (2003b) modeled the optimal growth strategies of larval helminths in their intermediate hosts, applicable for instance to the cystacanths of acanthocephalans, the metacercariae of digeneans, or the procercoids of cestodes. A central assumption of their model was that the optimal size of a larval helminth is determined by the trade-off between the benefits of being larger, and the costs imposed by a large size on the probability of transmission through adverse effects on host survival. The benefits of attaining a large larval size are substantial. A large initial size upon reaching the definitive host can mean shorter maturation time and greater fecundity. It can also mean a greater probability of establishment and initial survival, as seen in acanthocephalans (Steinauer and Nickol 2003). Parker et al. (2003b) considered two scenarios. First, in the resource-constraints scenario, larval parasites use all the available resources for growth regardless of how many other parasites are present in the host. Thus, if four helminth larvae share an intermediate host, they will on average be a quarter the size that a single larva would attain on its own. Second, in the life-history strategy scenario, larval parasites adjust their growth with regard to the number of conspecifics in the host, but in a way that maximizes their individual fitness in this competitive situation. In this scenario, with four helminth larvae sharing an intermediate host, each larva would try to appropriate more than a quarter of the resources available, and thus try to reach a size more than a quarter of that normally achieved by a single larva on its own. A review of the limited data available, all from studies of the growth of cestode procercoids in copepod hosts, suggests that both strategies are observed (Parker et al. 2003a). The models assume that all parasite larvae sharing a host are acting selfishly and are thus unrelated; as genetic relatedness increases among larvae, the predictions of the life-history scenario converge with those of the resource-constraints scenario (Parker et al. 2003a).

Some predictions of Parker et al.'s (2003b) model can be tested intraspecifically, but they are of little help in explaining the considerable interspecific variation in larval body size observed within many helminth taxa. For instance, in digeneans, metacercarial sizes vary among species across three orders of magnitudes (Poulin and Latham 2003); the same applies to cystacanth volume in acanthocephalans (Poulin et al. 2003a). Increases in size from these larval stages to the adult stage also vary substantially across species, from almost no change to more than hundredfold increases in digeneans, and from twofold to more than thousandfold increases in size in acanthocephalans (Poulin and Latham 2003; Poulin et al. 2003a). Neither the type of definitive host used, either ectotherm or endotherm, nor its size influences the relative growth of these parasites from the larval to the adult stage. There are, however, consistent correlations between the adult size of a species and its size at other stages of its development, from the egg to the final larval stage (Poulin and Latham 2003; Poulin et al. 2003a). Thus, a relatively large adult worm is always relatively large, whatever the stage in its life cycle. There may be no point in searching for general reasons why some parasite species have large larvae, other than to look at their adult size.

4.2.3 Sexual Size Dimorphism in Parasites

The discussion above applied to parasite body size in general. In many dioecious parasites, however, another factor creates differences in body size within species. That factor is sex, and its effect is expressed as differences, sometimes very large, between the body sizes of males and females. The causes behind these differences vary according to mating systems, the dependence of fecundity on size, and so forth (Shine 1989). Whatever their nature, asymmetric selective pressures on males and females can result in different evolutionary responses.

Because fecundity is related to body size and because the transmission of most parasites involves a series of unlikely events, we may expect selection to exert more pressure on female body size than male body size. This appears to be the case in most parasitic crustaceans, where males are often dwarves compared to females (Raibaut and Trilles 1993). In the isopod *Ichthyoxenus fushanensis* (family Cymothoidae), for example, a male and a female live paired within a sac in the body cavity of the fish host. The limited available space limits the total volume of both parasites, and thus does not allow both sexes to evolve toward larger sizes simultaneously; selection has favored females that are twice as large as males as a way to maximize clutch size (Tsai et al. 2001). In copepods, free-living lineages that switched to parasitism on invertebrates experienced small increases in both male and female body size, with little resulting change in sexual size dimorphism (fig. 4.8). When lineages parasitic on invertebrates switched to fish hosts, though, female body size increased substantially while male size did not change significantly. The result is a highly female-biased sexual-size dimorphism among copepods parasitic on fish (fig. 4.8). A possible explanation for the influence of host type on sexual-size dimorphism may be that fish are mobile and more difficult to infect than the generally sessile invertebrates serving as hosts to copepods. Selection could have favored high fecundity, and thus large body size, in female copepods parasitic on fish to a greater extent than in their sister taxa parasitic on invertebrates (Poulin 1995b, 1996d).

A female-biased size dimorphism also exists in oxyurid nematodes, but in contrast to copepods it is more pronounced in taxa exploiting invertebrates than in those parasitic in vertebrates (Morand and Hugot 1998). Competition for food among nematodes of both sexes may be intense in the digestive tract of small invertebrate hosts. In addition, sexual selection may favor early maturation in male parasitic nematodes (Morand and Hugot 1998). These factors may have acted in tandem to push the evolution of male oxyurids toward relatively small body sizes, especially in parasites of invertebrates. Sexual selection may have been an important driving force in the evolution of sexual-size dimorphism in all parasitic nematodes, not just in oxyurids, as well as in acanthocephalans (Poulin 1997c; Poulin and Morand 2000b). Among all kinds of parasitic nematode species, for instance, the size of males relative to females increases as the average sex ratio becomes less female

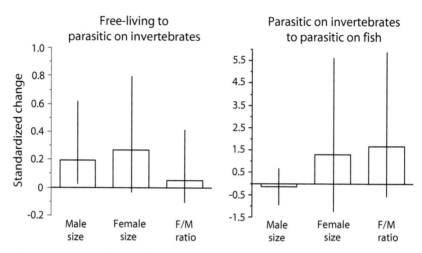

Figure 4.8 Mean (and range) change in body size and female-to-male size ratio for nine transitions from free living to parasitic on invertebrates, and nine transitions from parasitic on invertebrates to parasitic on fish, during the phylogenetic history of copepods. Change was estimated as the difference between sister taxa with different modes of life or different hosts, standardized for branch length. (Data from Poulin 1996d)

biased, an observation compatible with situations in which male-male competition for access to females intensifies with decreasing availability of mates (Poulin 1997c).

The situation may be different in other helminths. Schistosomes evolved from hermaphroditic ancestors and now have separate sexes, with males having larger body sizes than females. Females lodge themselves in a groove on the ventral surface of the male and the two form long-lasting (mostly) monogamous pairings. One hypothesis explaining male-biased sexual-size dimorphism in schistosomes is that a division of labor has been favored by selection (Basch 1990). Females have lost much of their body musculature and become highly specialized for reproduction. The large males provide all the muscle that the pair needs to move and feed. There is variability in sexual size dimorphism across schistosome genera: it ranges from moderate (e.g., genus *Schistosomatium*) to extreme (e.g., *Schistosoma*). The considerable variation in sexual-size dimorphism among schistosomes and the wealth of information available for this family make them an ideal model for tests of evolutionary hypotheses, such as the division-of-labor hypothesis, using comparative analyses.

The constraints imposed by phylogeny on sexual-size dimorphism are apparent when two unrelated groups under similar selective pressures are contrasted with each other. Nematomorphs, which emerge from insect hosts as free-living adult worms, display a male-biased sexual-size dimorphism (e.g., Poulin 1996e). In adult nematomorphs, competition among males for access to females in the external environment may be a key factor shaping body size of the two sexes. Adult nematomorphs form large clusters of dozens of individuals, where presumably a few males account for the majority of copulations. The situation is different in mermithid nematodes, which have independently evolved a life cycle strikingly similar to that of nematomorphs, but without mating within clusters. In mermithids, females are usually much larger than males (Poinar 1983), mirroring the pattern in other nematode taxa (Morand and Hugot 1998; Poulin 1997c). As evident from this and all preceding examples, parasite body size and sexual-size dimorphism evolved independently and often differently in different parasite lineages.

4.3 Parasite Age at Maturity

Adult body size is generally associated with age at maturity (Peters 1983). Small animals typically reach sexual maturity earlier than large ones. In parasitic animals, age at maturity can sometimes be difficult to quantify because of periods of arrested development in certain parasites (e.g., hypobiosis in certain nematode genera such as *Cooperia*, *Haemonchus*, and *Ostertagia*; Michel 1974); these can generate much intraspecific variability in actual age at maturity. Here, age at maturity of helminths is defined as equivalent to the prepatent period, the time elapsed between infection of the definitive host and the onset of egg production. Early maturity allows a parasite to begin producing eggs sooner, whereas delayed maturity allows it to reach a larger size and produce eggs at a higher rate (see next section). Age at first reproduction shows some plasticity among individuals of the same species. For instance, in the hermaphroditic cestode *Schistocephalus solidus*, solitary individuals delay their reproduction to increase their opportunity for outcrossing, whereas worms in pairs reproduce early (Schjørring 2004). The adjustment made by a single cestode in response to its immediate social-breeding environment allows it to minimize the number of lower-quality offspring produced by selfing. The discussion that follows focuses mainly on differences in age at maturity among species; these are more pronounced than intraspecific variation, and they can be examined using a comparative approach.

The optimal age at maturity of a parasite species is that which results in maximal lifetime reproductive success, and it should depend strongly on age-dependent mortality rates and the relationship between body size and fecundity (Stearns 1992; Roff 1992). The evolution of age at maturity in parasites has been studied most intensely in nematodes, a group for which good data are available. Across nematode species, maturation time, or the length of the prepatent period, determines adult female size, and female size is closely associated with fecundity (Skorping et al. 1991). Age at maturity is therefore a central feature of nematode life-history strategies. A recent model, based on a trade-off between increased fecundity associated with greater size and the risk of dying before attaining that size, predicts that nematodes should mature earlier when prematurational mortality is high (Gemmill et al. 1999; Read et al. 2000). There is a relatively good fit between the model's predictions and empirical data from a range of nematode species, with the model accounting for approximately 50 percent of the variation in observed ages at maturity (Gemmill et al. 1999). In contrast, the model's predictions were not supported by the results of an experiment in which host immune responses, the main source of prematurational mortality, were manipulated (Guinnee et al. 2003). Therefore other factors must influence age at maturity in nematodes. Adult, or postmaturational, mortality may also be important (Morand 1996b; Morand and Sorci 1998). Host mortality itself could also influence age at maturity in nematodes and other parasites. Surely, if host mortality rates are high, the chances of a parasite reaching maturity before host death would be greatly improved if its maturation time was decreased. However, the daily in-host mortality rates of nematodes, due to immune attack or resource competition, are an order of magnitude greater than daily mortality rates of vertebrates, and thus the contribution of host mortality to overall parasite mortality may be trivial in many cases (Read et al. 2000). Still, both theoretical and empirical evidence suggest that nematodes are selected to delay their maturity in long-lived hosts with low mortality rates (Morand and Poulin 2000). The optimal age at maturity in nematodes is therefore the maturation time that best meets the combined pressures from different mortality sources, acting mainly on immature parasites.

In apicomplexans such as malaria parasites, age at maturity can be viewed as either the time at which asexual multiplication begins, or the onset of gametocyte production. The timing of these events, and the rate of asexual multiplication and gametocyte production, emerge as fundamental life-history traits in lizard malaria: they correlate with other traits, and account for most of the variation in total life-history variation among strains (Eisen and Schall 2000). Inside the vertebrate host, asexual reproduction competes with the production of gametocytes, that is, it competes with sexual reproduction, and we would expect natural selection to maintain the balance between the two that maximizes the parasite's fitness. The physiological or immunological status of the host can influence the parasite's commitment to the production of gametocytes (Taylor and Read 1997; Dyer and Day 2000). Gametocyte production also shows some plasticity, and can be accelerated if the parasite detects that mosquito vectors are probing its vertebrate host (Billingsley et al. 2005; but see Shutler et al. 2005). However, strong selection on such an important trait is detectable. In lizard malaria, for instance, time to first production of gametocytes was the only life-history trait out of eleven studied traits not showing significant variation among genetic strains (Eisen and Schall 2000). No rigorous studies of age at maturity are available for parasites other than nematodes and apicomplexans, but from what these two groups have revealed it is clear that the timing of life-history events is of great importance for parasites, no doubt because of the dynamic and ephemeral nature of their habitats.

4.4 Egg Production in Parasites

The evolution of parasite body size is not the only aspect of parasite biology that has been plagued by misconceptions. Parasite fecundity is also commonly assumed to evolve in one direction only, toward higher egg output. Because of the massive losses suffered by infective stages during transmission, high fecundity is seen as a form of compensation ensuring that at least a few offspring will make it (Price 1974). But surely, selection should act to maximize fecundity in all organisms and not just in parasites, whatever mortality they incur at any stages in their life. Contrasts between resource availability in free-living and parasitic organisms provide a better explanation for the high fecundity of some parasites (Jennings and Calow 1975; Calow 1983). Because parasites do not experience shortages of food, they have the resources necessary to maintain a high rate of egg production. Parasites are also often viewed as perfect examples of *r*-selected organisms, possessing many characteristics of the *r* end of the *r-K* continuum of life-history strategies (Pianka 1970; Stearns 1992). For instance, they are generally seen as short-lived, early-maturing, small-bodied and highly prolific egg producers. Other strategies, such as the production of fewer offspring of better quality or the delay of maturity, are not considered viable options for parasites.

This section will demonstrate that the evolution of reproductive strategies in parasites has shown much more flexibility than generally believed, and produced a wide range of outcomes. Within species, a combination of environmental and host-related factors can influence egg output in parasites and produce substantial intraspecific variability (e.g., see review of egg-production patterns in monogeneans by Tinsley 2004). In this section, I will again focus on interspecific patterns. I will first examine some correlates of egg production, and then the trade-offs among various components of reproduction. The aim throughout will be to show how strategies other than the massive production of small eggs can sometimes be adaptive.

4.4.1 *Correlates of Fecundity*

Animals typically use one of two strategies of egg production. They either produce eggs more or less continuously, or they partition their reproductive effort into discrete clutches. Estimating lifetime fecundity from either a daily rate of egg production or from an average clutch size can be difficult. Rates of egg production can vary with age, and lifetime fecundity can only be quantified properly by repeated measurements on the same individuals during their whole reproductive life. Similarly, average clutch size does not necessarily reflect lifetime fecundity, as clutch size and the number of clutches in a lifetime are not correlated (Godfray 1987). The following discussion is based mostly on studies of either rates of egg production or clutch sizes; the conclusions are therefore limited to these components of fecundity and not to lifetime fecundity itself.

Were transitions to parasitism followed by increases in fecundity? This question can only be answered by comparing parasitic taxa with their free-living sister taxa. Copepods and isopods parasitic on fish are no doubt more fecund than their closest free-living relatives (Poulin 1995b, c), but a lack of data precludes any robust analysis. In flatworms (phylum Platyhelminthes), there is no difference in total reproductive output relative to body size between free-living taxa (turbellarians) and parasitic taxa (cestodes, monogeneans, digeneans). Reproductive output, here measured as estimated lifetime fecundity multiplied by average egg volume, increases with body size at the same rate, independently of mode of life (Trouvé and Morand 1998; Trouvé et al. 1998). Again, nematodes might be an ideal group to use in resolving this issue.

Once a transition to parasitism is completed, many variables will shape and constrain fecundity. In most animals, fecundity correlates with body size (Peters 1983). Parasites are no different, with egg production varying as a function of body size both within species and across taxa. In nematodes, for instance, larger species have a higher rate of egg production (Skorping et al. 1991; Morand 1996b). Long prepatency periods, which is to say, delayed maturity, results in longer adult lifespan and large body size in parasitic nematodes and may have been favored as a means of increasing rates of egg production and lifetime fecundity. Larger species of parasitic copepods, especially among those infecting fish, also produce larger clutches of eggs (Poulin 1995b). Digeneans may be exceptions to this trend (Loker 1983; Poulin 1997b). For example, egg production in mammalian schistosomes does not appear to be dependent on schistosome body size (fig. 4.9). In general, though, selection for higher fecundity may drive the evolution of body size in many parasites. One may hypothesize that high fecundity would be under stronger selection in some parasites than others. For instance, parasites with complex life cycles may be under greater pressure to evolve high fecundity than their relatives with simple life cycles, because each egg has such an infinitesimal probability of completing the cycle. Morand (1996a), however, compared rates of egg production of nematodes with simple and complex life cycles, and found no differences.

Pressures from the host or the environment may also push or constrain the evolution of egg production. In the phylogenetic history of parasitic copepods, switching from invertebrates to fish hosts tended to be associated with increases in clutch size (Poulin 1995b). At the same time, latitude affected egg production in copepods, creating a latitudinal gradient in copepod clutch sizes independent of body size (Poulin 1995b). As is the case for most of the trends reported in this chapter, further comparative analysis on other taxa are needed before any generalizations can be made.

4.4.2 Trade-offs and Strategies of Egg Production

Larger parasites tend to produce more eggs, but they can also produce larger eggs. A positive relationship exists between body size and egg size in taxa ranging from parasitic ascothoracidan crustaceans (Poulin and Hamilton 1997) to digeneans (Poulin 1997b). In nematodes, egg sizes show considerable variability both across (Skorping et al. 1991) and within (Yoshikawa et al. 1989) species but egg size does not appear to covary with any other life-history trait. In most other taxa, though, since both egg numbers and egg size are unlikely to be maximized simultaneously, selection may have favored different ways in which to partition reproductive effort. Indeed, trade-offs between egg numbers and egg size are observed among taxa of schistosomes (fig. 4.9) and parasitic copepods (fig. 4.10).

The strategy favored by selection in a given taxon can range from the production of huge quantities of tiny eggs to the production of few large eggs. Parasites will not all evolve toward the many-small-eggs end of the spectrum. Again, copepods provide a good illustration of what sort of factors can shape life history traits. Contrasts between lineages parasitic on invertebrates and their sister lineages parasitic on fish indicate that fish parasites favor the production of many small eggs whereas parasites of invertebrates opt for few large eggs (Poulin 1995b). Some copepods parasitic on invertebrates produce only one or two eggs per clutch, but these eggs are huge relative to the copepods' body size. Because of the greater difficulties associated with locating and infecting fish, selection

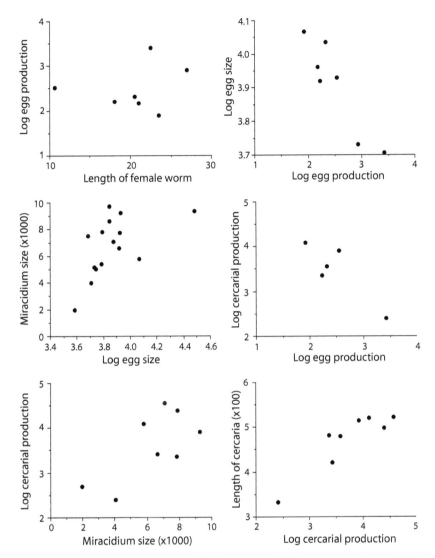

Figure 4.9 Relationships among various life-history traits of schistosome species parasitic in mammals; each point represents a different parasite species. Length of female worms is in mm, egg production is the number of eggs produced per female per day, egg size is the product of egg length (μm) and egg width (μm), miracidium size is the product of miracidium length (μm) and miracidium width (μm), cercarial production is the total number of cercariae produced per infected snail intermediate host, and cercarial length includes the tail and is in μm. (Data from Loker 1983)

has favored investments in offspring numbers rather than offspring size among copepods exploiting fish hosts (Poulin 1995b).

The probability of finding a host may also influence selection for offspring size in unionid mussels. The larvae, or glochidia, of these freshwater mussels are ectoparasitic on fish whereas adults are free living. Bauer (1994) reported a positive relationship across mussel species between glochidial size and the number of fish species that a mussel species can use as host. Bauer did not correct for phylogenetic effects, but the relationship

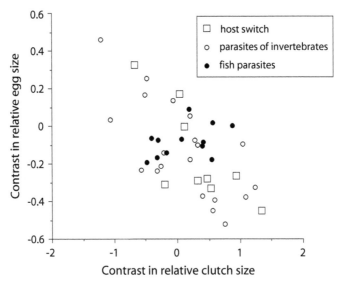

Figure 4.10 Relationship between egg size and clutch size in parasitic copepods. Each point represents a contrast between two sister branches in the copepod phylogeny. Both variables were corrected for parasite body size, which covaries with the number and size of eggs produced. The negative relationship suggests that there is a trade-off between the two components of egg production. Host switches represent cases in which one taxa is parasitic on invertebrates and its sister taxa is parasitic on fish. (Modified from Poulin 1995b).

he found holds even when independent contrasts derived from a molecular phylogeny of unionids (Lydeard et al. 1996) are used in the analysis (Poulin, unpublished). Bauer (1994) suggested that the evolution of large glochidial size preceded the expansion of the host range to many fish species. Large glochidia are already at an advanced stage of development when they first attach to a fish. The parasitic life of large glochidia is a short one and they detach from the fish before a specific immune response is activated, unlike species with small glochidia in which more time must be spent on the fish. Species with large glochidia may thus have faced little pressure to specialize on a particular host taxon and consequently they could have colonized a number of new host species. There is another explanation for the relationship, however, one that involves the probability of infecting a host. In this alternative scenario, low host specificity evolved before large glochidial size: in species that can utilize any one of a large number of host species, the probability of transmission should be higher, and investments in offspring size may have been favored at the expense of fecundity. This is similar to the copepod example mentioned above, where copepods parasitic on sessile or colonial invertebrate hosts, which experience a relatively high probability of transmission, invest in large offspring rather than many offspring.

Factors other than the probability of transmission can influence the trade-off between egg size and fecundity. Calow (1983) suggested that homeothermic hosts provide better growing conditions for endoparasites than poikilothermic hosts. He then hypothesized

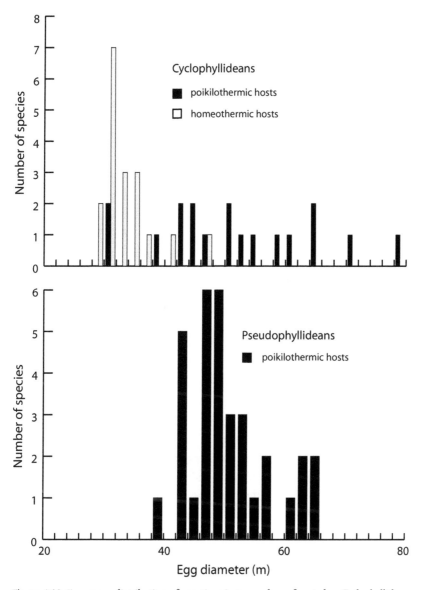

Figure 4.11 Frequency distribution of egg sizes in two orders of cestodes. Cyclophyllidean species developing in homeothermic hosts tend to have smaller eggs (average diameter, 32.3 μm) than either cyclophyllideans or pseudophyllideans using poikilothermic hosts (average diameter, 51.1 and 49.7 μm, respectively). (Modified from Calow 1983)

that parasites using homeothermic intermediate hosts should be selected to produce small eggs since their larvae can grow at higher rates in the intermediate host and make up for their small initial size. Comparisons of egg sizes between and within cestode families support this prediction (fig. 4.11). Cyclophyllidean cestode species developing in homeothermic hosts produce smaller eggs than cyclophyllidean and pseudophyllidean species developing in poikilotherms (Calow 1983). In digeneans, though, egg sizes appear independent of the nature of the host exploited (Poulin 1997b), so the growth potential of the host may be of limited importance in the egg size versus egg numbers trade-off.

Furthermore, the trade-off may be influenced by environmental variables. Rohde (1993) suggested that there may be a latitudinal gradient in reproductive strategies among monogenean species. For instance, the viviparous gyrodactylids, which produce few but large offspring, are rare in warm seas but extremely diversified in northern waters. Among free-living invertebrate taxa, there is often a trend for species at low latitudes to produce more and smaller offspring than their relatives at higher latitudes (Sibly and Calow 1986). Whether the apparent relationship in monogeneans is related to latitude or the product of some historical or phylogenetic accident remains to be determined rigorously.

The vast majority of studies on life-history strategies in parasites have focused on mean trait values, such as mean egg volume and how it covaries with other traits or with environmental variables. A life history trait is characterized not only by its average or typical value, however, but also by its variability, or the dispersion of individual values around the mean. Given the unpredictable nature of parasite transmission, there are good reasons to believe that the intraspecific variability in certain parasite life-history traits may also have been under selection. A recent individual-based model shows that when the probability of transmission is both low and highly variable, because of the unpredictable availability of hosts, the ideal strategy for a parasite is to produce offspring that differ in their infection strategies (Fenton and Hudson 2002). This would allow the parasite to hedge its bets and spread the risk of failure, increasing the chances that at least some of its offspring will successfully reach a host. In many parasites, egg volume may reflect not only the amount of energy invested in each offspring, but also the amount of resources available to each offspring for host finding. Intraspecific variation in egg volumes may thus mirror variability in the infection strategies of parasites. In digeneans, intraspecific variability in egg size varies according to whether eggs are released in aquatic or terrestrial environments, and also with latitude (Poulin and Hamilton 2000). The patterns observed suggest that greater variability in offspring sizes is favored in low-latitude terrestrial habitats, where the spatial heterogeneity of the external environment, and thus the probability of failed transmission, are likely higher than in temperate or aquatic habitats (Poulin and Hamilton 2000). The allocation of resources between quantity and quality of eggs is therefore not a simple trade-off; other factors need to be considered.

Parasites with complex life cycle in which there are two episodes of reproduction, one sexual and one asexual, may face another trade-off, one between egg production by adults in the definitive host and asexual multiplication of larval stages in the intermediate host. Mammalian schistosomes species provide a good example of this possibility (Loker 1983). Species producing eggs at a lower rate produce larger eggs, larger eggs produce larger miracidia, and larger miracidia generate greater numbers of cercariae in the snail intermediate host (fig. 4.9). The net result is a trade-off, or a negative relationship, between egg production and cercarial production. Schistosome species either produce many eggs as adults, or many cercariae as larvae, but not both. Energy invested at the adult stage in egg size limits fecundity but produces larger offspring better able to reproduce asexually. Surprisingly, there is a positive relationship between the number of cercariae produced and their size (fig. 4.9). This correlation may be an artifact of variation in the size of the intermediate host, as cercarial production is proportional to snail size (Loker 1983; Rondelaud and Barthe 1987).

Asexual reproduction by larval stages has also evolved in taeniid cestodes. Theoretically, it should be favored in highly predictable and stable environments, where the genetic heterogeneity associated with sexual reproduction is not essential. The occurrence of asexual proliferation by larval stages in taeniid species is not linked with any potential indicators of predictability, such as the nature of the host or host specificity (Moore 1981; Moore and Brooks 1987). However, as in schistosomes there may be a trade-off between sexual and asexual reproduction. Moore (1981) observed that taeniid species with an asexual phase are characterized by small, short-lived adults and those with no asexual phase are characterized by large, long-lived adults. Thus adult fecundity may be maximized at the expense of larval multiplication, and vice versa.

4.5 Conclusion

Life-history strategies are combinations of biological traits that have been favored by selection because they result in higher fitness than alternative combinations. Under certain conditions, more than one strategy can lead to high fitness, and evolution can proceed in different but equally adaptive directions. Contrary to what has often been said about parasites, their evolution has followed the same diversity of paths as that of free-living organisms. Far from evolving toward small body sizes, parasites often become larger than their free-living relatives and, presumably, their free-living ancestors. Similarly, instead of invariably progressing toward higher fecundity, several parasite lineages have opted for different ways of partitioning their reproductive investments, often favoring fewer but larger offspring.

Detailed comparative studies of other life-history traits are yet to be performed. Only in nematodes have entire life-history strategies been examined from an evolutionary viewpoint (Skorping et al. 1991; Morand 1996b; Morand and Sorci 1998; Read et al. 2000). Often data are unavailable, or derived from experimental infections of laboratory hosts and not representative of natural situations. As these hurdles are overcome and as more investigations of parasite life histories get published, we will no doubt confirm that parasite strategies are as varied as those of free-living animals, and not the inevitable outcome of rigid evolutionary rules applying only to parasites.

5 | Strategies of Host Exploitation

Most if not all definitions of *parasitism* involve the concept of harm, and restrict the use of the term *parasites* to symbionts that have a negative impact on the fitness of their host. In fact, the level of harm done to the host is often the main criterion used to categorize symbioses in which the host does not benefit. Symbioses fall along a continuum, from commensalism in which the host incurs no harm, to highly virulent parasitism in which host fitness is greatly reduced. This wide range of effects among contemporary symbioses has long been viewed as a series of snapshots of different associations at different stages in their evolution. Indeed, it was assumed that symbiotic associations progress through time from an initial period of adjustment, in which the symbiont causes harm to the host because it is not yet adapted to exploit it without side effects, to a final stage of peaceful coexistence achieved when the well-adapted symbiont becomes benign to its host. Whatever the initial level of virulence, evolution was thought to proceed in one direction only, toward reduced virulence, with commensalism as the end point of parasitism.

No other evolutionary paths were seen as possible, because a parasite population evolving to become more virulent would eventually drive its host population, and therefore itself, toward extinction. This argument is clearly flawed and based on misconceptions about the process of evolution. Natural selection has no foresight; it acts blindly by favoring the genotypes among the currently available ones that lead to the highest reproductive success under the present conditions. If these genotypes also cause host mortality and if an increase in their frequency in the parasite population could threaten the host population fifty generations down the line, they will still be favored by selection and spread over subsequent generations. The realization that parasite virulence could proceed toward both lower and higher levels has changed the way we perceive the evolution of parasitism (Anderson and May 1982; Ewald 1983, 1994; Ebert and Herre 1996).

The host is simply an ephemeral resource that the parasite uses to maximize its fitness, either by increasing its own probability of completing its life cycle or by increasing its production of offspring. The way the parasite exploits the host as a resource will depend on which of the possible strategies maximizes its fitness, and not on any concern for the welfare of the host. This chapter discusses how strategies of host exploitation have evolved. The first section addresses virulence and the factors determining whether it increases or decreases through evolutionary time. The next three sections examine three well-studied and more specific ways in which parasites exploit hosts, using the same evolutionary framework applied to virulence in order to determine when such strategies are adaptive.

5.1 The Evolution of Virulence

Although the notion of virulence is intuitively easy to grasp, there is no consensus on a rigorous definition of the term that specifies how virulence should be measured (Toft and Karter 1990; Poulin and Combes 1999; Schall 2002). In mathematical models, virulence is usually taken as the parasite-induced host mortality rate (e.g., Ebert and Herre 1996; Frank 1996; Day 2001; Gandon 2004). This is a quantitative definition of the term that allows its effect to be modeled. In practice, however, the parasite-induced mortality rate is not easy to estimate, especially if one wants data from a large number of host-parasite systems for a comparative analysis. In such studies, the definition of virulence ranges from subjective evaluations of the general level of harmfulness caused by a parasite (Ewald 1995) to more quantitative measures such as parasite-induced reductions in host lifetime reproductive success (Herre 1993). The latter measure is probably the most significant way of estimating virulence from a host-centered evolutionary point of view.

From the parasite's perspective, however, what happens to the host as a consequence of its exploitation by the parasite may be of no importance. Selection will favor an aggressive exploitation of the host only because it leads to improved parasite fitness, not because it causes reductions in host fitness. For instance, nematomorphs and mermithid nematodes almost invariably cause the death of their insect host when they are finished using it. The adult worms emerge from the host and have no further interaction with it. If the host were to recover and miraculously survive the infection, we might be tempted to view the worms as less virulent. But from the point of view of parasite evolution, selection would favor the same rate of exploitation of host resources since the subsequent death or survival of the host is of no consequence to parasite fitness. Our notion of virulence originates from medical science and is focused on impact on host fitness, whereas selection in parasites acts on rates of host exploitation, often irrespective of effects on host fitness (Poulin and Combes 1999).

Another reason why virulence measured as pathology can be misleading is that pathology is the result not only of parasite actions but also of host responses. Often, closely related host species incur different fitness losses from infection by the same parasite (Park 1948; Thomas et al. 1995; Jaenike 1996a). Should we conclude that the parasite uses different exploitation strategies in different suitable host species? Not necessarily. Some hosts may simply be better able to cope with infection than others, making pathology a poor indicator of parasite strategies. In the discussion that follows, this should not be forgotten. To avoid confusion, I will specify what is meant by virulence in the various models or empirical examples presented below.

5.1.1 The Theory

In some host-parasite systems, certain manifestations of virulence may benefit the parasite directly (Ebert 1999). For instance, diarrhea can facilitate the transmission of gut parasites, even if prolonged diarrhea can compromise host survival. Host castration and manipulation of host phenotype, both discussed later in this chapter, are also examples of direct benefits of virulence for the parasites.

In many cases, however, parasites only benefit indirectly from virulence, because virulence is an unavoidable side effect associated with some other fitness component of the parasite, such as fecundity or transmission rate (Ebert 1999). Thus, parasites face a trade-off between greater exploitation and higher reproduction on the one hand, and the time before the host dies on the other hand. Some parasites exploit their hosts' resources at a gentle rate, in a way that yields sustainable but not maximal returns. Other parasites exploit their hosts more aggressively and cause it harm, deriving maximum returns at the expense of long-term availability of resources. The trade-off between high reproductive rate and host survival may prevent the unchecked escalation of virulence, but will not prevent relatively high levels of virulence from being favored. For each parasite-host system there may be an optimal strategy of host exploitation that maximizes the parasite's lifetime fecundity (Combes 1997; Ebert 1999; Galvani 2003a). In fact, because local conditions exert strong selective pressures on parasite transmission, there may not be a species-wide optimal virulence, but instead each parasite population should evolve toward a virulence level optimal for the local conditions in its host population (Dybdahl and Storfer 2003).

The trade-off between the parasite's reproduction rate and host longevity can be analyzed using mathematical models. In standard epidemiological models (Anderson and May 1982, 1991), the fitness of a parasite is described by its lifetime reproductive success, R_0. Despite the limitations of this parameter for modeling the evolution of virulence (Garnick 1992b; Ebert and Herre 1996), it provides a good approximation of parasite fitness and is already a central feature of epidemiological theory. There are several mathematical expressions for R_0, but it is generally obtained as follows:

$$R_0 = \frac{\beta(N)}{\mu + a + v}$$

where β is the rate at which an infected host transmits the parasite to susceptible hosts (β is dependent on host density, N), μ is the mortality rate of parasite-free hosts, a is virulence or the parasite-induced host mortality rate, and v is the host recovery rate. If transmission is strictly horizontal, that is, among host individuals of the same generation rather than from parents to offspring, strains of parasites with the highest R_0 are usually favored. In the above equation, R_0 is highest when a equals zero. However, there is often a linkage between traits that constrain their simultaneous evolution, resulting in genetic

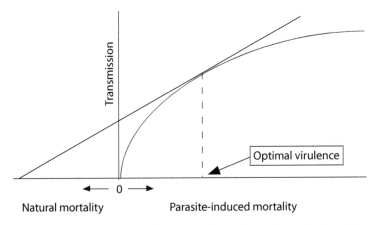

Figure 5.1 Optimal virulence of a parasite derived from the application of the marginal value theorem to the functional relationship between transmission rate, β, and the parasite-induced mortality rate, a. Parasite fitness is maximized at the point where the tangent touches the curve. (Modified from Ebert and Herre 1996)

correlations and trade-offs. A genetic linkage between β and a, for instance, can maximize R_0 at values of a different from zero (Ebert and Herre 1996). The optimal level of virulence, the level that maximizes R_0, can be obtained using a simple rate-maximizing approach (fig. 5.1). It is the level below which improvements in the transmission rate are still possible without being annulled by the death of the host, and above which any increase in transmission rate is offset by high host mortality.

The most interesting aspect of models of virulence in the context of this chapter is their ability to make predictions about the course of the evolution of virulence in different conditions. For instance, a higher natural mortality rate of hosts due to causes other than parasites, or μ in the above equation, should select for evolutionary increases in parasite virulence (fig. 5.2a). Similarly, optimal virulence should be lower in parasites that achieve relatively large increases in transmission rate, β, from small increases in virulence, a, than in parasites that obtain more modest returns from increased virulence (fig. 5.2b). Recent mathematical models have highlighted the importance of a wide range of host-related, parasite-related, or external factors on the evolution of virulence (see Schall 2002; Galvani 2003a). Some of these factors are discussed below.

It is possible to model various components of β, the parasite transmission rate, to see how they can affect the evolution of virulence. In directly transmitted parasites with simple life cycles, one of the main determinants of the probability of transmission apart from host density will be the frequency of contacts or interactions between individual hosts. In other words, the probability of transmission depends on the frequency of opportunities for transmission. If the host's social or sexual behavior promotes contacts among individuals, transmission may be facilitated and lead to the selection of more virulent parasites (Møller 1996). In spatially structured or fragmented host populations, parasite transmission will often occur only among individuals within a locality. Connectedness among localities, however, whatever its nature, can allow the transmission of parasites over greater distances and increase the number of host individuals they can reach. Boots and Sasaki

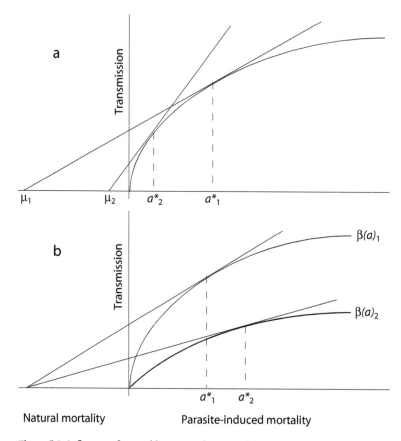

Figure 5.2 Influence of natural host mortality, μ, and the shape of the functional relationship between transmission rate, β, and parasite-induced host mortality, a, on the optimal level of virulence, a^*. A decrease in natural host mortality favors a decrease in parasite virulence (a), and a lower rate of returns in terms of transmission efficiency as a function of parasite-induced host mortality selects for increased virulence (b).

(1999) used a spatial model to predict that parasites with the ability to infect distant host individuals as well as local ones should evolve toward higher virulence.

In sexually transmitted parasites, however, host specificity is often high and the absence of alternate hosts means that there are no refuges allowing the persistence of a virulent parasite when its host population becomes really small (Lockhart et al. 1996). Low virulence should thus be favored in sexually transmitted parasites even if the host is promiscuous. Strict host monogamy, the other extreme, can have similar effects: theoretical models suggest that when opportunities for transmission and contacts between the same individuals are frequent, maximum β and R_0 are achieved at low levels of virulence (Lipsitch et al. 1995). Therefore, host behavior and population structure will be important factors driving the evolution of parasite virulence.

Other features of host biology can modulate the evolution of parasite virulence. For instance, in host populations where resistant individuals cannot be infected by a parasite, selection should favor less virulent parasites. In contrast, in host populations where resistance does not fully protect a host against infection but instead allows it to fight the parasite with immune responses and reduce the within-host replication or growth rate of the parasite, selection should favor higher virulence (Gandon and Michalakis 2000). Because virulence is commonly seen as a parasite strategy, but is measured in terms of host fitness, it cannot be assumed that the host is a passive participant. Its optimal investment in resistance will in turn evolve in response to parasite virulence. This creates a dynamic system in which optimal levels of parasite virulence will continuously track the changing conditions.

The above host-related factors may obscure any trade-offs between maximal transmission rates and the long-term survival of the host. In fact, certain parasite transmission modes can result in a total uncoupling of host health and parasite fitness. If the transmission route becomes independent of host health and survival, and if more propagules can be produced at high levels of virulence, evolution should proceed quickly toward high virulence (Ewald 1983, 1994). For instance, if transmission is achieved by vectors, selection should favor high virulence in the vertebrate host: strains producing more propagules at the expense of host health would achieve a greater transmission success than more benign strains. If the parasite is transmitted from host to host during interactions among individual hosts, however, selection may favor low virulence: a healthy and socially active host is essential for transmission in these parasites, and strains killing their host rapidly would not spread. The timing of transmission matters, too: following infection, any time lag between the onset of parasite transmission and the first symptoms of infection will select for higher virulence (Day 2001, 2003). If host death is necessary for parasite transmission, as in many protozoan parasites that multiply in their invertebrate host and are only released upon the latter's death, then the optimal time at which the parasite should kill the host will be determined by its within-host rate of replication and other variables affecting its fitness (Ebert and Weisser 1997). Finally, the evolution of virulence in parasites with transmission routes involving more than one host species, that is, multihost life cycles, although more complex to predict, will also depend on the link between host fitness and parasite transmission (Gandon 2004). Clearly, this link and the trade-off associated with it are of fundamental importance to the evolution of virulence.

The effects of other variables acting to uncouple host health and transmission have also been examined in mathematical models. Long-lived parasite propagules, for example, can under certain conditions favor an evolutionary increase in virulence, since parasites whose propagules are able to survive long periods outside the host can benefit from a rapid and aggressive exploitation of the host (Bonhoeffer et al. 1996). Many more models have recently been published; they are reviewed by Anderson (1995), Ebert and Herre (1996), Frank (1996), Ebert (1999), Schall (2002) and Galvani (2003a).

The above discussion applies to horizontally transmitted parasites. In parasites that are also transmitted vertically, from parents to offspring, the optimal level of virulence should vary inversely with the proportion of the total transmission that is vertical (Mangin et al. 1995; Ebert and Herre 1996). A parasite that is always transmitted vertically and that reduces the lifetime fecundity of its host may go extinct; selection should favor more benign strains that do not jeopardize their opportunities for transmission. In fact, vertical transmission can serve to align the interests of the parasites with those of the host, and is seen as a factor that may promote not only a decrease in virulence, but also the evolution of true mutualism (Herre et al. 1999). As transmission becomes increasingly horizontal, increases in virulence have a decreasing impact on the overall frequency of transmission opportunities and can become adaptive. Under certain conditions, strictly horizontally transmitted parasites can face selection pressures identical to those acting on vertically transmitted parasites. For instance, if the prevalence of infection by a horizontally transmitted parasite is approximately 100 percent and is stable through time, the transmission is essentially vertical: the only uninfected and susceptible hosts available are the next generation. In this scenario, highly virulent strains of parasites would compromise the next generation of hosts and could drive their host population, and themselves, to extinction very quickly. All else being equal, we should expect a continuum between horizontally transmitted, virulent parasites and vertically transmitted, benign parasites (Lipsitch et al. 1996).

One other important factor might influence the evolution of virulence in any given parasite. Different parasite species do not exist isolated from one another; often, strains or related species of different virulence compete within the same host population. How should this influence the evolution of virulence? All else being equal, if two different parasite strains or species co-occur frequently in the same host individual, within-host competition between them for the same resources should select for higher levels of virulence (May and Nowak 1995; van Baalen and Sabelis 1995; Read and Taylor 2001; Gandon and Michalakis 2002). What good would it do one parasite to restrain itself in order to keep its host alive if that host is being harmed by another parasite? Escalating virulence appears to be the only solution to *within-host selection* acting on different parasites. If, on the other hand, the two different parasite species coexist in the same host population but rarely co-occur in the same host individual (because they both occur at low prevalence), then lower levels of virulence will be favored. In this case, *between-host selection* favors the parasites that maximize their total reproductive output, which depends in part on keeping the host alive (fig. 5.3). Thus the optimal virulence of parasite strains or species in a community may rise and fall over time, along with the fluctuating prevalences of these parasites. Other scenarios are possible, too. For instance, in stochastic models, chance events can lead to the extinction of the most virulent species, such that the average virulence of frequently co-occurring and competing strains does not necessarily increase over time (Leung and Forbes 1998). Also, in models where the effects of the parasites on their host are always sublethal, such as reduced growth, multiple infections by many parasites generally lead to lower virulence (Schjørring and Koella 2003). The important take-home message of these models is that the evolution of virulence in a parasite species cannot be considered independently from that in other co-occurring parasites.

5.1.2 Empirical Tests

Mathematical models are useful conceptual tools with which to generate testable predictions. The parameters used in models of the evolution of parasite virulence, however, have proven difficult to estimate in practice. The fitness, transmission rate, and virulence of parasites (R_0, β, and a in the models) are often impossible to measure in nature and

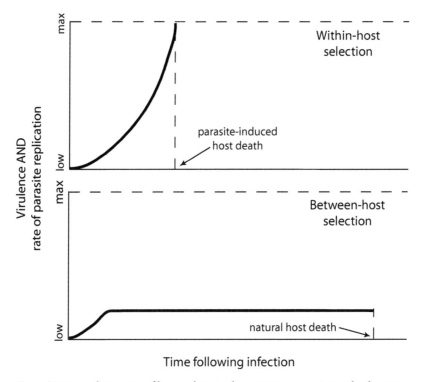

Figure 5.3 Optimal strategies of host exploitation by a parasite competing with other parasites. *Within-host selection,* occurring when the different parasites co-occur in the same host individuals, favors high rates of host exploitation and consequently high virulence. A virulent parasite may kill its host quickly, but it achieves a greater production of propagules in this short time than any competing strain or species within the host. In contrast, *between-host selection* favors lower rates of host exploitation and lower virulence. When competing parasites occur in different host individuals, parasites that keep their host alive for a longer time achieve greater lifetime reproductive output than virulent parasites, because reproductive output is a function of both host exploitation rate and the duration of infection.

can only be quantified approximately in the laboratory. Not surprisingly, then, there have been relatively few empirical tests of some of the predictions made by the models.

The two assumptions underpinning most theoretical models have received some empirical support. First, a trade-off between the rate at which the parasite exploits the host and host survival is expected to constrain the evolution of high virulence. Serial passage experiments, in which parasites are transferred repeatedly from one individual host to another by an experimenter instead of via natural transmission routes, provide very consistent results: virulence, defined as a reduction in host fitness, increases very rapidly with the number of passages to new hosts (see reviews in Ebert 1998, 2000). In these experiments, there is no cost of virulence for the parasite, because the experimenter ensures transmission regardless of the level of virulence. With transmission uncoupled from host health, virulence can increase unconstrained.

The second assumption is related to the first one: higher virulence is assumed to lead to greater replication of the parasites. Serial-passage experiments confirm that the rate of

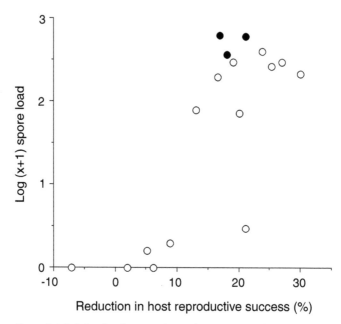

Figure 5.4 Relationship between the virulence (measured as a reduction in host reproductive success) and the spore production of the microsporidian parasite *Pleistophora intestinalis* infecting the planktonic crustacean *Daphnia magna*. Across the seventeen combinations of hosts and parasites of either sympatric (filled circles) or allopatric (open circles) origin, higher virulence leads to greater production of parasite spores. (Modified from Ebert 1994)

parasite reproduction or multiplication within the host correlates positively with virulence, across many types of parasites (Ebert 1998, 2000). The relationship is also commonly observed in comparisons of different laboratory strains of several parasite species (e.g., Turner et al. 1995). Supporting evidence from natural systems has proven more difficult to obtain. Ebert's (1994) study of the virulence of a microsporidian protozoan parasite of the crustacean *Daphnia magna* is the only study to demonstrate that high virulence in the field translates into higher production of parasite propagules. Ebert (1994) infected hosts with parasites either of local origin or from distant localities. The various combinations of hosts and parasites produced different levels of parasite virulence, with high virulence associated with higher levels of spore production by the parasite (fig. 5.4). Finally, a comparison of a few species of strongyle nematodes indicates that the more virulent species are indeed those achieving greater transmission potential, based on egg production and larval survival, except in species where damages to the host result mostly from larval worm migration following infection (Medica and Sukhdeo 2001). More evidence of this kind would be welcome if virulence is to be considered as an adaptive parasite strategy of host exploitation.

Most empirical research to date has focused on the influence of a parasite's transmission mode on the evolution of its virulence. Ewald (1983, 1994) has investigated the virulence of protozoans, bacteria, and viruses causing diseases in humans. He defined virulence as the probability of an infection resulting in the death of the host. Comparisons among pathogens that differ in mode of transmission provide support for the predictions of theoretical models. Pathogens that frequently use water systems for transport are more virulent than those strictly dependent on host-to-host contacts. Vector-borne pathogens, which do not need to spare the host to achieve transmission, are typically much more virulent than those transmitted directly from host-to-host during contact. Also, pathogens capable of surviving long periods in the external environment are much more virulent than those that only survive briefly outside their human host (Walther and Ewald 2004). Medical science provides a wealth of information on human diseases and their consequences; comparative studies such as those of Ewald will be more difficult to perform with parasites of nonhuman animals, but are needed if we are to accept the general validity of the models' predictions.

A series of experiments on the rodent malaria parasite *Plasmodium chabaudi* provides insights into the evolution of virulence in vector-transmitted parasites. In these studies, virulence was measured as either the amount of red blood cells destroyed by the parasite or the amount of weight lost by the rodent host, both of which correlate with host mortality (Mackinnon and Read 2004). Across different clonal lines of *P. chabaudi*, the most virulent ones also achieved the highest asexual multiplication rates within rodents and the highest transmission rates of sexual forms to mosquito vectors (Mackinnon and Read 1999a, 2004). In serial-passage experiments, where parasites are transferred from rodent to rodent by a syringe rather than via a mosquito, both virulence and the rate at which the parasite multiplies and produces sexual-transmission forms increased rapidly (Mackinnon and Read 1999b). This increase was observed even under conditions expected to impose a strong selection for low virulence. Within-host selection for high multiplication rates and thus high virulence appears to be very strong in this parasite. This becomes apparent when two competing clonal lines are used in mixed infections: they typically achieve higher rates of multiplication and transmission to mosquitoes, and higher virulence to their rodent host, than in single-clone infections (Taylor et al. 1997, 1998). The most virulent strain consistently outcompetes the least virulent one in mixed-strain infections in the rodent, and achieves a higher transmission rate to mosquitoes (de Roode et al. 2005). The virulence of malaria cannot increase indefinitely, even if it correlates with the parasite's transmission success to mosquitoes. So what constrains it? One possibility is that parasites that are more virulent to the rodent host also cause vector mortality, which would limit their transmission success back to other rodents. Different clonal lines of *P. chabaudi* indeed vary in their harmful effects on mosquito vectors. However, although mixed-clone infections tend to have greater negative impacts on mosquito fitness, virulence in the rodent host is generally not correlated with virulence in the vector (Ferguson and Read 2002; Ferguson et al. 2003). In rodent malaria and similar parasites, the vector is responsible not only for transmission to new vertebrate hosts, but also for larger-scale dispersal among host patches, and thus selection should favor parasites that are relatively benign to their vectors (see Elliot et al. 2003). The lack of a clear genetic correlation between virulence for the rodent host and that for the vector does not provide a clear limiting factor acting to constrain the evolution of virulence in *P. chabaudi*. In this system and with other vector-borne parasites, virulence in the vertebrate host may be capped by factors such as the diversity of host genotypes and/or the availability of resources within hosts such as red blood cells (Ferguson et al. 2003).

In parasites with complex life cycles, as in vector-borne parasites, we might expect parasites to specialize on certain hosts as resource bases and on other hosts in the cycle as agents of transport or dispersal in space and time. Virulence could then evolve to be high in one host but lower in the next one in the cycle. Ewald (1995) gathered information on helminths with life cycles involving predation as a means of transmission, and contrasted the virulence of helminths in their definitive and intermediate hosts. He found that helminths generally have severe effects on their intermediate hosts but are usually benign to their definitive hosts. This pattern was interpreted as evidence that evolution favored the use of intermediate hosts as resource bases and definitive hosts as vehicles for dispersal, with the debilitating effect of helminths on intermediate hosts serving to increase transmission success via predation. Ewald (1995) argues that when definitive hosts are not infected through predation, high virulence in intermediate hosts is not beneficial but instead costly; selection would instead favor the use of the definitive host as a resource base for the production of propagules. Indeed, helminths such as schistosomes that are not transmitted to the definitive host by predation tend to be virulent in their definitive host (Ewald 1995). This virulence results from inflammatory responses around eggs lodged in host tissues rather than from an aggressive exploitative strategy of the parasite, however. In addition, schistosomes are also virulent to their intermediate hosts. In *Schistosoma mansoni*, the level of virulence in the mouse definitive host correlates positively with rates of egg (or miracidial) production in that host, whereas virulence in the snail intermediate host correlates negatively with rates of cercarial production (Davies et al. 2001; Gower and Webster 2004). Strains of *S. mansoni* that achieve high replication in the definitive host tend to achieve low rates of replication in the intermediate host, and vice versa (fig. 5.5). Thus, the parasite would indeed benefit from low virulence in its intermediate host and higher levels of virulence in its definitive host, in accordance with Ewald's (1995) arguments. When co-occurring in the same snail, *S. mansoni* strains with low virulence for the snail host outcompete those with high virulence, because of the negative association between virulence and replication rate in the snail (Gower and Webster 2005). A high frequency of multiple infections in snails could promote the evolution of less virulent parasites. In any event, the genetic trade-off in parasite success between the two hosts (fig. 5.5) may favor strains showing intermediate levels of virulence in both hosts.

Although the trends pointed out by Ewald (1995) are interesting, they do not present the complete picture. In his assessment of the effects of helminths on their hosts, virulence was measured qualitatively rather than quantitatively, and the use of intermediate hosts as resource bases by helminths transmitted via predation is debatable since they frequently do not grow inside the intermediate host. When they do, they can be quite virulent, such as digenean larvae castrating their snail intermediate hosts (see section 5.2). Often the only detrimental effect that a parasite has on its intermediate host is a small

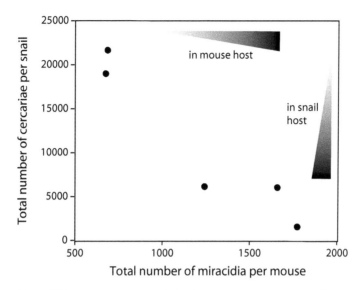

Figure 5.5 Negative genetic correlation between parasite success in the intermediate and definitive hosts, across five substrains of the digenean *Schistosoma mansoni*. Cercarial production in snail intermediate hosts and miracidial production in mouse definitive hosts were assessed under standard conditions for all five substrains; each point represents mean values for each substrain, averaged across numerous infected hosts. The shaded gradients indicate how virulence covaries with parasite success in both hosts. (Data from Davies et al. 2001)

change in its behavior without any apparent harm; taken out of context, this may appear trivial, but in nature it can translate into a higher risk of mortality via predation by definitive hosts (see section 5.3). Therefore a more detailed comparative analysis is necessary to confirm Ewald's suggestion.

Among parasites with direct or one-host life cycles, the mode of transmission also varies extensively in ways that can modulate the evolution of virulence. For instance, a comparative analysis of parasites from many taxa revealed that sexually transmitted parasites are generally less virulent, that is, less likely to induce host death, and more persistent than their relatives using other routes of transmission (Lockhart et al. 1996). This is possibly a consequence of their narrow host specificity: any virulent strain of a sexually transmitted parasite causing a decline in its host population cannot take refuge in an alternate host species and would go extinct along with its host.

The influence of horizontal versus vertical transmission on the evolution of virulence has received particular attention. In an elegant laboratory study in which the probability of horizontal transmission could be controlled, selection favored the most benign strains of a parasitic bacteriophage when transmission among bacteria hosts was restricted to vertical (Bull et al. 1991). When the conditions were changed to allow the horizontal spread of the bacteriophage, the more virulent strains became predominant. Experimental evidence of this nature is difficult if not impossible to obtain for host-parasite systems with longer generation times, so we must turn to comparisons among parasites that differ in how they are transmitted. Among arthropods ectoparasitic on birds, for instance, high virulence is restricted to species that rely mostly on horizontal transmission, and low virulence is characteristic of vertically transmitted species (Clayton and Tompkins 1994).

Parasites with both horizontal and vertical transmission stages can also shed light on the relationship between the mode of transmission and the level of virulence. Agnew and Koella (1997) found that in the microsporidian parasite *Edhazardia aedis*, horizontally transmitted spores are associated with higher virulence in the mosquito host than vertically transmitted spores. Selection on virulence acts in opposite directions in the two transmission modes: host death is required for horizontal transmission, since spores are only released into the environment following the rupture of the host's cuticle, whereas virulence should be minimized for vertical transmission. The measures of virulence used in this study, the degree of fluctuating asymmetry in host wing length and the size of host blood meals (Agnew and Koella 1997; Koella and Agnew 1997), are far from ideal but the results are interesting nonetheless. In a subsequent experiment, Koella and Agnew (1999) have shown that the parasites adopt the transmission mode, and associated level of virulence, that best fits the genetically determined life-history strategy of their host. In rapidly developing mosquito larvae, *E. aedis* is benign and opts for vertical transmission, whereas in mosquitoes with a prolonged larval development, the parasite is highly virulent and transmits horizontally (Koella and Agnew 1999). Thus, in parasites with both vertical and horizontal transmission, host life-history features can favor one mode over the other, and therefore influence the evolution of the parasite's virulence (see also Ebert and Mangin 1997; Koella 2000; Vizoso and Ebert 2005).

By far the most remarkable and most widely cited comparative evidence for a link between parasite virulence and opportunities of horizontal transmission comes from a study of nematodes and their fig-wasp hosts in Panama (Herre 1993, 1995). One or more gravid female fig wasps simultaneously enter a fig flower (that eventually ripens to become the fig fruit), lay their eggs, and die. Their bodies can be counted to determine the number of foundress wasps per fig fruit. As the fruit ripens, the wasp offspring mature and mate inside the fig before the winged females leave to begin the cycle anew. In broods founded by a single female wasp, a count of the offspring provides a direct measure of the female's lifetime reproductive success. Each species of wasp is host to a specific species of parasitic nematode of the genus *Parasitodiplogaster*. Infective nematode larvae crawl onto newly hatched female fig wasps within figs, enter their bodies to consume them, and are carried by them to other figs. Adult nematodes emerge from the bodies of dead, infected female wasps; they mate and lay eggs in the same fig in which the host wasp has laid her eggs. The nematode eggs hatch synchronously with the next generation of wasps to repeat the cycle.

The fig wasp–nematode system allows both virulence (the reproductive success of infected female wasps relative to that of uninfected females) and opportunities for horizontal transmission (the percentage of multiple foundress broods) to be quantified for each host-parasite species pair. Herre (1993) found that the more virulent nematode species parasitize the wasp species providing the most opportunities for horizontal transmission

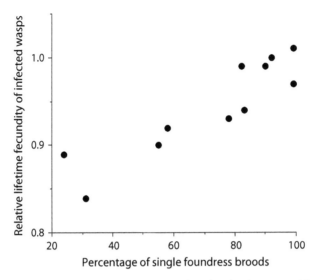

Figure 5.6 Relationship between nematode-induced reductions in lifetime fecundity and the percentage of single foundress broods among eleven species of Panamanian fig wasps. Virulence of the nematode species infecting each species of wasps can be estimated as the lifetime reproductive success of infected female wasps relative to uninfected conspecifics. A higher incidence of broods founded by multiple female wasps provides increased opportunities for nematode transmission: the more virulent nematode species are associated with wasp species presenting greater opportunities for horizontal transmission. (Modified from Herre 1993)

(fig. 5.6). In wasp species where nematode transmission can take place almost only from parent to offspring, the success of the parasite is tightly linked to that of the host, and selection has favored low levels of virulence. In wasp species offering good opportunities for horizontal transmission, there is a decoupling between host and parasite fitness, and selection favors the most virulent parasite strains (Herre 1995). Unfortunately, whereas the lifetime fitness of hosts could be quantified, that of parasites was not. Demonstrating that highly virulent nematode species are more fecund, for instance, than less virulent species would nicely complete the story.

Herre (1993, 1995) suggests that the association between nematodes and fig wasps is an ancient one, which began as a commensal, phoretic relationship in which the nematode was simply using the wasp for transport. The history of these associations is thus one of evolutionary increases in virulence. But how frequently are host-parasite associations evolving toward greater parasite virulence? This is a difficult question to answer, as it depends on the initial conditions in these various associations. Conventional wisdom, based on a "lack of adaptations" between host and parasite, asserts that novel associations are characterized by high virulence and that over time parasites evolve to become less virulent. Recent evidence, however, tends to refute this notion, and indicates that parasites in novel hosts are often less virulent or less infectious than in their original host (e.g., Ballabeni and Ward 1993; Ebert 1994; Dufva 1996; Ebert and Herre 1996).

In contrast, some new host-parasite associations are characterized by high parasite virulence. One explanation may be that initially hosts have no defenses against a new parasite but may evolve some over time. For instance, the mite *Varroa jacobsoni* is a parasite of the Asian honeybee *Apis cerana*. The mite is a brood parasite, sucking haemolymph from drone (male) pupae; it never reaches a high abundance, and does not threaten the bee colony (Oldroyd 1999). Over the past decades, the mite has been introduced to other parts of the world, and as a result it is now also parasitizing the western honeybee, *Apis mellifera*. The mite's virulence on its new host is dramatically higher than on its original host. In *A. mellifera*, the mite can also feed and reproduce in worker cells, and its population typically increases exponentially until the host colony dies (Oldroyd 1999). Unlike other *Apis* species, *A. mellifera* has no natural mite parasites, and was apparently not equipped to handle this new enemy. Another example comes from the striped bass, *Morone americana*. Introduced one hundred years ago from the east coast of North America to the west coast, this fish was at first suffering severe pathology from infection by larvae of the west coast fish cestode *Lacistorhynchus dollfusi* (Sakanari and Moser 1990). Today, striped bass from west-coast populations are less debilitated and more resistant to cestode infections than fish from east-coast populations. Since the bass is only one of many host species used by the cestode in the west, this is unlikely to reflect a decrease in parasite virulence: it is an evolved host response. This example illustrates how difficult it is to use host pathology as a measure of virulence if virulence is seen as a parasite strategy.

An important and consistent prediction of many mathematical models has been that when a parasite species co-occurs with others, its rate of host exploitation and associated virulence should increase because of within-host selection among these competing parasites (e.g., May and Nowak 1995; van Baalen and Sabelis 1995; Gandon and Michalakis 2002). The results of serial-passage experiments support this prediction (see review in Ebert 2000). In mixed infections of two parasite strains, both the virulence of each strain and their rates of replication are typically higher than in single-strain infections (e.g., Taylor et al. 1998; Davies et al. 2002; de Roode et al. 2005), although there are exceptions to this trend (see Imhoof and Schmid-Hempel 1998). When parasites in mixed infections display higher virulence than in pure infections, this enhanced virulence appears to be facultative, that is, it seems to be a strategy adopted only under conditions of competition for host resources (Taylor et al. 1998; Davies et al. 2002). We might expect higher virulence levels to become fixed only in situations where mixed infections are frequent generation after generation. Since in most cases they are not, it seems that between-host selection has maintained the average virulence of parasites well below the levels that intense within-host selection would favor.

5.2 Parasitic Castration and Host Gigantism

Many models of parasite virulence usually assume that achieving a high transmission rate by a ruthless exploitation of the host can only be realized at the expense of host survival. Some parasites, however, have become specialized to exploit the reproductive organs or reproductive effort of the host. This can be achieved directly, by feeding on the gonads of the host, or indirectly by either diverting energy away from gonad development or by the secretion of "castrating" hormones (Baudoin 1975; Coustau et al. 1991; Schallig et al. 1991). For example, the cestode *Hymenolepis diminuta* secretes a modulator molecule that directly suppresses vitellogenesis in its insect intermediate host (Webb and Hurd 1999). The ensuing reduction in the host's reproductive output may benefit the cestode either if it makes more nutrients available to the parasite, or if it results in the insect host living longer, and thus being exposed for longer to predation by the parasite's definitive host.

Many parasitic taxa include castrators that channel energy away from host reproduction toward their own or the host's growth; the better known are digeneans in their mollusc host and crustaceans such as rhizocephalans and epicaridean isopods in their crustacean hosts (Kuris 1974; Baudoin 1975). Castration does not have to be complete. In fact, "castrated" hosts are often capable of some gamete production, and the term *castration* is used loosely. Whether complete or not, host castration may be the ideal strategy of host exploitation: by attacking nonvital organs, castrators do not reduce the host life span, and they can obtain a high transmission rate without trading off longevity. Obrebski (1975) modeled the benefits of castration, and found that castration is particularly advantageous to the parasite if the natural death rate of the host (μ in the models of section 5.1.1) is high. He predicted that castration would be more likely to evolve in parasite taxa exploiting hosts with short life spans, a prediction that remains untested. Jaenike (1996a) also modeled the evolution of parasitic castration and concluded that it should be favored under a wide range of ecological conditions.

Advantages other than a normal host life span can be associated with the exploitation of host gonads. Rhizocephalans usurp the space normally occupied by the host gonads and eggs, and obtain the protection and ventilation usually provided by the host to its own offspring (Baudoin 1975; Høeg 1995). Also, castrated hosts often divert energy toward somatic growth instead of reproduction, and often grow to larger sizes than their unparasitized conspecifics. The increased growth of castrated hosts is not caused by the secretion of growth factors by the parasite, as in the well-known and noncastrating cestode *Spirometra mansonoides* (Phares 1996); rather, it is an indirect consequence of castration. This phenomenon is common in molluscs infected with digeneans. It can benefit the parasites since not only their reproductive output is typically proportional to host size, but also host survival is likely to be improved by investments in growth. Digeneans are also capable of targeting specific organs within the snail host; for instance they can stimulate the growth of the host's digestive gland in which they live (Théron et al. 1992). In addition, infection by digeneans results not only in greater size per se, but often also in altered shell shape in the mollusc host (Krist 2000; McCarthy et al. 2004; Hay et al. 2005). Periwinkles infected by the digenean *Microphallus piriformes*, for example, have a higher length-to-width ratio than uninfected individuals, resulting, on average, in a 12 percent increase in the volume available to the parasite in the shell spire (McCarthy et al. 2004). The host gigantism often linked with castration is therefore frequently seen as a parasite adaptation, whereby the host phenotype is controlled by parasite genes (Baudoin 1975; Dawkins 1982).

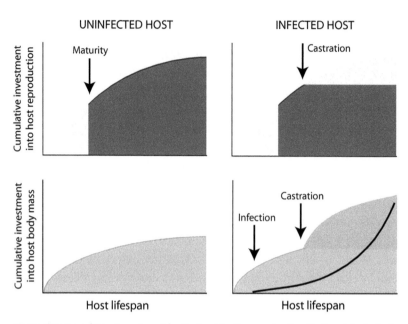

Figure 5.7 Cumulative investment by the host in its reproduction (top) and in its body mass (bottom), over its life span, for both uninfected hosts and hosts infected by a parasitic castrator. Upon castration, host resources are no longer invested into reproduction and are instead channeled into body growth. The curve represents the resource requirements of slow-growing and long-lived parasites. (Modified from Ebert et al. 2004)

A recent hypothesis proposes that host gigantism resulting from castration provides delayed benefits for long-lived parasites in hosts with indeterminate growth, such as digenean larval stages in snails (Ebert et al. 2004). The temporal-storage hypothesis argues that host resources liberated by castration are invested, or stored, in host body mass for later use by the parasite (fig. 5.7). Because any fitness returns associated with host castration are only achieved late in the infection, this hypothesis predicts that the parasite should induce little or no host mortality. Some of its predictions fit with what is known about many parasitic castrators, and resource storage may therefore represent a general influence in the evolution of host castration as a parasite adaptation.

As often the case when the label adaptation is applied to phenomena as complex as gigantism, other explanations are possible and just as valid (see Hurd 2001). Gigantism could be a side effect of the destruction of the gonads, or it could be an adaptive response of infected hosts. Minchella (1985) argues that gigantism of molluscs harboring larval digeneans could be a host strategy to prolong survival in the hope of outlasting the infection. One argument in favor of gigantism being a host adaptation comes from the observation that often molluscs that were exposed to digenean miracidia but did not become infected also increase their growth rate (e.g., Sorensen and Minchella 1998). In infected molluscs not completely castrated, normal host reproduction could resume once the parasite has died. Because gigantism is not the rule, and decreases in growth rates are also observed in molluscs parasitized by larval digeneans (e.g., Minchella 1985; Mouritsen and Jensen 1994; Taskinen 1998), it may be easier to reconcile the occurrence of gigantism

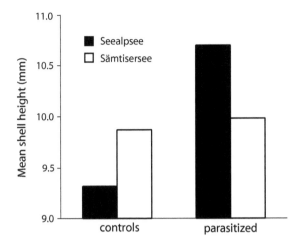

Figure 5.8 Effect of the castrating digenean parasite *Diplostomum phoxini* on the growth of its snail intermediate host, *Lymnaea peregra*. Several weeks following a laboratory infection, parasite-induced gigantism was only observed in snails from the Seealpsee Lake, where the parasite occurs naturally, and not in snails from the Sämtisersee Lake from which the parasite is absent. (Modified from Ballabeni 1995)

with life-history patterns of the mollusc hosts if gigantism is viewed as a host adaptation rather than a parasite adaptation. For instance, Sousa (1983) suggested that gigantism would only be favored as a host adaptation in short-lived semelparous mollusc species. Using different arguments, Gorbushin and Levakin (1999) predicted that molluscs with intermediate life spans were most likely to show gigantism following infection. Sorensen and Minchella (2001) reviewed the published evidence in order to match the occurrence of gigantism with life-history traits among mollusc species infected with digeneans. Their literature survey tends to support Gorbushin and Levakin's (1999) prediction. Short-lived snails have been selected to invest very little in repair mechanisms, and the damage they incur from infection may be too severe to allow subsequent increases in growth rates. At the other extreme, long-lived snails invest heavily in repair mechanisms, and the high cost they pay for long-term survival following infection will restrict any gigantism.

An experimental study attempted to distinguish between gigantism as a parasite strategy of host exploitation, and gigantism as a host adaptation to compensate for the effects of parasitism (Ballabeni 1995). In the laboratory, the digenean *Diplostomum phoxini* only induced gigantism in snails from a population naturally infected with this parasite, but not in snails from another, uninfected population (fig. 5.8). This result strongly suggests that gigantism is a host adaptation that has only appeared in the host population exposed to parasites over evolutionary time.

Therefore, gigantism following infection by a parasitic castrator may be an adaptive response of infected hosts to the adaptive parasite strategy of castration. However, the main benefit for parasites associated with host castration, prolonged host survival, is also not always apparent. Recent studies report contrasting results regarding mollusc survival following infection by digeneans (see Huxham et al. 1993; Jokela et al. 1993; Mouritsen and Jensen 1994; Ballabeni 1995; Sorensen and Minchella 1998), and the adaptiveness of castration may not be as general as often believed.

5.3 Manipulation of Host Behavior by Parasites

Host exploitation by parasites can take more subtle forms than the ones mentioned thus far in this chapter. Under a wide range of natural conditions, many parasites cause no measurable changes in host growth, reproduction, or survival. They can nevertheless alter host biology in other ways. Larvae of the nematode *Trichinella spiralis*, for instance, invade skeletal muscle cells in their mammalian hosts, and then proceed to induce profound structural alterations in the cell, turning it into a nurse cell in which a larva can reside for the life span of the host (Despommier 1990, 1993). Plant-parasitic nematodes achieve something very similar in their plant host (Niebel et al. 1994). These nematodes reprogram their host's genomic expression, providing clear examples of the parasite genotype being expressed in the host's phenotype. Other intracellular parasites are capable of delaying or even preventing apoptosis, or programmed cell death, of the host cells they inhabit, and promote it in immune cells that attack them (Lüder et al. 2001; James and Green 2004). More generally, many parasites can suppress various components of the host's immune response as a means of prolonging their exploitation of the host (Wikel 1999; Rigaud and Moret 2003; Boëte et al. 2004). These are all instances of alterations in host phenotype that are much more insidious than gross pathology or increased mortality; these also represent very specific host exploitation and manipulation strategies on the part of the parasite.

The best understood of these subtle and specific effects on host phenotype are parasite-induced modifications in host behavior. Parasites that manipulate the behavior of their host may be seen as virulent, because often behavioral changes both maximize parasite transmission and increase the probability of host death. In parasites transmitted from the intermediate host to the definitive host by predation, behavioral manipulation of the intermediate host can increase its susceptibility to predation by the next host. This means all-or-nothing fitness costs for the intermediate host. As a result of the parasite's actions, the host has a probability q of being captured by a predator and a probability $1 - q$ of suffering no fitness cost at all. Seen from the host perspective, when q is small such parasites may appear nonvirulent. However, they have been selected to do all they can to get ingested by their next host. Their success is absolute or nil, never partial. Because of the probabilistic nature of their transmission, their virulence, as traditionally measured by effects on host fitness, will often appear minimal. This is yet another situation in which a parasite-centered measure of host exploitation would be preferable to actual virulence in order to investigate parasite evolution. The following sections explore the main questions surrounding the evolution of parasite manipulation of host behavior.

5.3.1 Adaptive Manipulation?

Parasite-induced alterations in host behavior or coloration have been reported in a wide range of protozoan and metazoan parasites, many of which have complex life cycles (Holmes and Bethel 1972; Moore and Gotelli 1990; Poulin 1995e; Moore 1995, 2002; Poulin and Thomas 1999; Barber et al. 2000). The simplest explanation for these changes is that they are nonadaptive, coincidental side effects of parasitic infections. The very complex nature of some parasite-induced behavioral changes, however, often suggests that they could benefit either the host or the parasite, and therefore be adaptations. The best-documented examples consist of apparent increases in the vulnerability of intermediate hosts to predatory definitive hosts. Alterations of host behavior that appear adaptive have also been reported in systems where transmission takes a different route. Their nature varies greatly. Vector-borne parasites can render their vertebrate host more susceptible to vector feeding (Day and Edman 1983) and affect the feeding behavior of their vector host (Moore 1993). Fungi parasitic in insects can make their host climb trees and die perched in what seems to be an optimal position for the dispersal of fungal spores by wind (Maitland 1994). Some digeneans make their snail intermediate host change microhabitat and move to ideal sites for the release of cercariae (Curtis 1987; Lowenberger and Rau 1994). Nematomorphs and mermithid nematodes make their arthropod hosts enter water or seek water-saturated habitats, where the adult worms must emerge (Vance 1996; Poulin and Latham 2002; Thomas et al. 2002b). Strepsipterans, an unusual group of endoparasitic insects, cause their wasp hosts to desert their colony and aggregate on vegetation; this makes it easier for male parasites to find and mate with females, which remain within wasps for their entire life (Hughes et al. 2004). Usurpation of host behavior leading to increased transmission of parasites can thus take many forms and is common in many parasite taxa. It is also observed in insect parasitoids (Fritz 1982; Brodeur and McNeil 1989; Brodeur and Vet 1994; Eberhard 2000). But, are these truly parasite adaptations? This section will discuss several lines of evidence that can provide answers to this question.

Alterations in host behavior following parasitic infection are often exactly what we would expect to see if the host were to start acting in a way that benefits the parasite. They can appear to be adaptations rather than mere pathological side effects because their purposive nature fits our a priori expectations. Consider the changes induced by the protozoan *Toxoplasma gondii* in its rodent intermediate host, and how they may influence the parasite's transmission to a cat, its definitive host. Normally, rats show an innate aversion to cat odors; however, following infection by *T. gondii*, their aversion is turned into an attraction (Berdoy et al. 2000). Infected rats do not respond to neutral odors such as their own or those of a rabbit, but only to cat odors. Isn't this exactly what one might have predicted as a means to increase parasite transmission? In the absence of demonstrated fitness benefits, this line of evidence can be convincing. Complex traits that suggest a purposive design are less likely to evolve by chance and are therefore more likely to be the adaptive products of selection (Poulin 1995e). For a further example, take the well-known digenean *Dicrocoelium dendriticum*, which must be transmitted from an ant to a grazing sheep by ingestion (Carney 1969; Wickler 1976). *Dicrocoelium* causes infected ants to climb to the tip of grass blades and stay there waiting for sheep. Another digenean, *Leucochloridium* spp., alters the shape, size, and coloration of the tentacles of its snail intermediate host and causes them to pulsate in response to light (Wesenburg-Lund 1931). This seems an excellent way to get the attention of bird definitive hosts. Fitness benefits for the parasites have not been demonstrated in either of these two digeneans. But given the close fit between the presumed function of these behavioral changes and the parasites' life cycles, one can reasonably assume that such complex alterations of host behavior and coloration are extremely unlikely to arise by chance. In cases like this, changes in host behavior after infection may be perfect examples of the extended phenotypic effects of genes, which reach out of their own bodies to influence other organisms (Dawkins 1982, 1990). Note, however, that the argument can also be misused: a strong correspondence between the phenotypic expression of a host behavior modification and the design that an engineer might specify for the function of transmission is seen by some as evidence of divine creation (Smith 1984).

This is why stronger evidence of adaptation is necessary. The fact remains that the adaptiveness of most parasite-induced changes in host behavior is yet to be demonstrated convincingly (Moore and Gotelli 1990; Poulin 1995e; Moore 2002). The most important criterion that must be met by any behavioral change to be labeled a parasite adaptation is that it lead to improved parasite transmission or fitness. Definitive hosts may have little to gain by actively avoiding infected prey; infected prey with altered behavior are easier to capture, and typically the adult parasite causes little pathology in the definitive host (Lafferty 1992; Kuris 1997; Aeby 2002). Despite these arguments, empirical evidence of fitness benefits is necessary. In systems involving a parasite species capable of altering host behavior, only a handful of studies have shown that infected intermediate hosts are more susceptible than uninfected individuals to predation by their definitive hosts under field conditions (fig. 5.9). Results of laboratory studies usually (but not always) show similar trends (Poulin 1995e). Thomas et al. (1998a) reviewed all studies of behavior-altering parasites published before 1998 that examined predation on infected and uninfected intermediate hosts. They found that predation rates were higher on infected prey in all fourteen studies. The increase in predation rate caused by parasite infection ranged from less than 10 percent to over 50 percent, with an average increase of about 30 percent across all studies (Thomas et al. 1998a). Since their review was published, a laboratory study on the cestode *Hymenolepis diminuta* in its beetle intermediate host, has been the first to show quite conclusively that, at least in this system, parasitism does not increase predation by definitive hosts, although it changes the behavior of intermediate hosts (Webster et al. 2000). Conversely, there are examples of parasites increasing the susceptibility of their hosts to predation in systems where the parasite is not transmitted by predation (e.g., Hudson et al. 1992). Overall, though, the evidence generally indicates that parasites manipulating the behavior of their intermediate host achieve a higher transmission rate to their definitive host; still, only a fraction of known parasite-induced changes in host behavior have been tested for fitness effects.

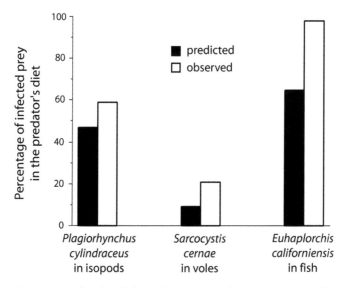

Figure 5.9 Predicted and observed percentage of parasitized prey in the diet of definitive hosts of three parasites transmitted by predation. Predicted values are the prevalences of parasites in the populations of available prey, and observed values are from actual prey captures under natural conditions. The acanthocephalan *Plagiorhynchus cylindraceus* passes from its terrestrial isopod intermediate host to starlings, the protozoan *Sarcocystis cernae* goes from voles to kestrels, and the digenean *Euhaplorchis californiensis* is transmitted from fish to piscivorous birds. (Data from Moore 1983; Hoogenboom and Dijkstra 1987; Lafferty and Morris 1996)

What if the parasite-induced alteration in host behavior can benefit *both* the parasite *and* the host? Consider the acanthocephalan *Moniliformis moniliformis* in its cockroach intermediate hosts. The parasite induces behavioral changes that result in cockroaches spending more time in microhabitats where they are more exposed to predation by rats, the parasite's definitive host (Moore et al. 1994; Moore and Gotelli 1996). However, these altered behaviors could also place the cockroaches in areas exposed to sunlight, where their body temperature would be elevated. In the laboratory, the development of the acanthocephalan is suppressed in cockroaches maintained at high temperatures (Moore and Freehling 2002). Which antagonist, the host or the parasite, really benefits from these behavioral modifications? Until these potential benefits are assessed in natural conditions, this question will remain unanswered.

Since quantitative estimates of fitness benefits associated with host manipulation are only rarely available, perhaps information on the adaptiveness of the manipulation can be obtained from the nature of the behavioral change itself. The first examples that found their way to the scientific literature were the spectacular cases still used today as textbook illustrations of the phenomenon. Over the years, however, a number of new cases have been documented, and as a whole they suggest that host manipulation by parasites usually involves much more subtle alterations whose effects are less obvious (Poulin 2000). At present, the majority of known parasite-induced behavioral changes are simple increases or decreases in the proportion of time that infected hosts perform

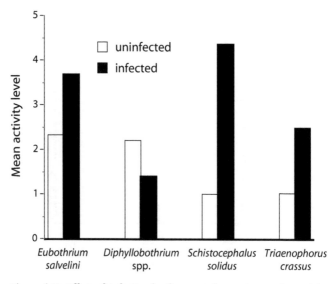

Figure 5.10 Effect of infection by four cestode species on the activity of freshwater planktonic copepods (*Cyclops* spp.). The activity of copepods harboring *Eubothrium salvelini* was measured as distance traveled (×10 mm), that of copepods harboring *Diphyllobothrium* spp. as hops per second, that of copepods infected with *Schistocephalus solidus* as the number of visits to different depth zones, and that of copepods with *Triaenophorus crassus* as the number of swimming bursts (×10). In all cases behavioral measurements were taken after the onset of infectivity to the definitive host. Infections with *E. salvelini*, *Diphyllobothrium* spp., and *T. crassus* resulted in greater susceptibility of the copepod host to predation; infections with *Schistocephalus solidus* had no effect on risk of predation. (Data from Poulin et al. 1992; Pasternak et al. 1995; Urdal et al. 1995; Pulkkinen et al. 2000)

certain behaviors with respect to uninfected hosts (Poulin 1995e). Such simple alterations are more likely to be accidents of chance or pathological side effects than the complex changes in host behavior induced by *Dicrocoelium* or *Leucochloridium*, and caution is needed when evaluating their adaptiveness. For instance, small changes in host activity caused by parasites can go in opposite directions and have different effects on transmission, even among closely related host-parasite systems (fig. 5.10). Increases in activity are just as likely to be observed as decreases in activity, and either change may affect transmission success. Intraspecific variability also exists in manipulative ability among parasite strains (Minchella et al. 1994), and among various combinations of host and parasite strains (Yan et al. 1994). In some cases this variability may simply reflect the mistaken lumping of distinct parasite species under a common name (see Perrot-Minnot 2004); this cannot explain most cases of intraspecific variation, however. Thus, if not all individual parasites in one species can induce similar changes in host behavior, how can we conclude with confidence that host manipulation is an adaptive species character of the parasite?

Two features of parasite-induced behavioral changes can shed further light on their adaptiveness. First, the timing of their appearance following infection can be informative.

If they are adaptive, we would expect the timing of parasite-induced changes in host be-havior to be synchronized with the parasite's developmental schedule inside the interme-diate host. At the very least, if their effect on parasite fitness cannot be verified directly, simple changes in host behavior should coincide with the onset of parasite infectivity to the next host (Bethel and Holmes 1974; Hurd and Fogo 1991; Poulin et al. 1992; Pulkki-nen et al. 2000). Alternatively, the strength of host manipulation should be strongly de-pendent upon parasite size in species with extensive growth in the intermediate host, such as plerocercoids of pseudophyllidean cestodes (Brown et al. 2001b). When changes in host behavior clearly precede infectivity (e.g., Shirakashi and Goater 2005), the most parsimonious explanation is that they are not adaptive.

Second, we would expect adaptive manipulation of host behavior to show a certain de-gree of specificity. It should increase predation rates on the intermediate host, but only by definitive hosts, and not by animals that are not suitable definitive hosts. The freshwater snail *Potamopyrgus antipodarum*, when infected by the digenean *Microphallus* sp., dis-plays an altered behavior that enhances predation by waterfowl definitive hosts while re-ducing predation by fish (Levri 1998). Infected snails stay on top of rocks only in the morning when waterfowl are foraging, and move under rocks for the rest of the day. Not only is this behavioral change aimed specifically at suitable definitive hosts and not at all predators, but it is also not a simple response of the snail host to increased energy de-mands (Levri 1999). This therefore appears to be a specific manipulation rather than some nonadaptive consequence of infection. It is crucial to demonstrate that alterations in host behavior are not simple host responses aimed at compensating for infection (Hechtel et al. 1993) or manifestations of general host debilitation (Robb and Reid 1996). For instance, the burrowing depth and crawling behavior of the bivalve *Macoma balthica* change following infection by the digenean *Parvatrema affinis*; this had long been ac-cepted as an adaptive manipulation by the parasite serving to increase rates of predation on the bivalve by bird definitive hosts of the parasite (Swennen 1969; Hulscher 1973). However, recent studies indicate that starvation and reduced energy stores can elicit the same behavioral changes in the bivalve host (Mouritsen 1997; Edelaar et al. 2003), and the behavioral alteration induced by the digenean now appears to be a general host re-sponse to nutritional stress rather than a specific parasite manipulation.

One indirect way of demonstrating that parasite-induced behavioral changes are par-asite adaptations would be to identify the mechanism causing the changes (Moore 2002; Thomas et al. 2005). The presence of a specialized structure or organ in the parasite re-sponsible for the manipulation of host behavior would not be a fortuitous product of nondirectional evolution. In some cases, host manipulation may simply result from the precise location of parasites in organs such as the host brain, which play fundamental roles in behavior (e.g., Barber and Crompton 1997). Often, though, alterations of host behavior appear to result from active interference with host neurochemistry involving the production of neuroactive substances by the parasite (Helluy and Holmes 1990; Holmes and Zohar 1990; Hurd 1990; Thompson and Kavaliers 1994; Øverli et al. 2001; Biron et al. 2005). We generally still know very little about the mechanistic processes un-derpinning parasite-induced alterations in host behavior (see Moore 2002; Thomas et al.

2005), but research in this area could shed much light on the adaptive nature of host manipulation.

Adaptations can also be recognized at the macroevolutionary scale. If different parasite lineages with similar life cycles have independently evolved the ability to cause identical alterations in host behavior, then surely this ability is adaptive. Comparative analyses of parasite-induced behavioral changes can be powerful tools for understanding the evolution of host exploitation strategies. To date, only one study has examined host behavioral changes in a phylogenetic context. Moore and Gotelli (1996) showed that the ability of the acanthocephalan *Moniliformis moniliformis* to alter the behavior of several cockroach species showed no concordance with a cockroach phylogeny derived from morphological data. In other words, behavioral alterations caused by the same parasite are not legacies from an ancestor but are derived traits that have arisen repeatedly and independently in different host lineages. Similarly, the acanthocephalan *Pomphorhynchus laevis* induces more pronounced behavioral changes in the amphipod species that has served as its historical intermediate host than in a recent invasive amphipod that can also serve as intermediate host (Bauer et al. 2000). These findings suggest that alterations of host behavior can be the adaptive outcome of coevolutionary history with certain host species. The truly interesting test, however, would be to compare the ability of different parasite species with identical life cycles to manipulate the behavior of the same host. Convergence of traits in different organisms under similar selective pressures can be instructive, but so can differences in these traits if they relate to differences in selective pressures. Superficially similar host manipulations induced by phylogenetically distinct parasites may have fundamentally distinct mechanistic bases. As mentioned earlier, both nematomorphs and mermithid nematodes make their arthropod hosts seek open water or water-saturated habitats, where the adult worms must emerge (Vance 1996; Poulin and Latham 2002; Thomas et al. 2002b). In spite of the marked analogies between these manipulations, they seem to result from unrelated physiological mechanisms: alterations in brain chemistry by nematomorphs (Thomas et al. 2003; Biron et al. 2005) and changes in host haemolymph osmolality by mermithids (Williams et al. 2004). Within parasite taxa, the ability of different species to alter host behavior varies greatly when measured as the magnitude of the induced changes (Poulin 1994b). In some taxa, host manipulation seems widespread and well developed (e.g., acanthocephalans; Moore 1984), whereas in other taxa it is more variable (see fig. 5.10). These latter taxa offer ideal opportunities for further comparative analyses of parasite-induced alterations in host behavior.

This section has summarized several lines of evidence that can be used to demonstrate that parasite-induced alterations in host behavior represent an adaptive manipulation of the host by the parasite. These include the complexity and purposiveness of the behavioral change, whether it confers fitness benefits to the parasite, its timing and specificity, the type of mechanisms that cause it, and whether similar host manipulations have evolved independently in other parasite lineages with similar life cycles. In the end, it is the agreement, or consilience, among these different lines of evidence that provides the strongest overall evidence for adaptation. Some documented examples come through all these hurdles with flying colors, whereas others fail one or more tests. Do changes in host behavior induced by parasites represent true parasite adaptations? There is no general answer; this question must be addressed on a case-by-case basis.

5.3.2 Evolution of Host Manipulation

From a theoretical perspective, variability in the ability to manipulate host behavior is expected both among and within parasite species, because the net benefits of manipulation must not be the same for all parasites. One reason for this is that manipulation may be costly, and to be favored by selection, the fitness benefits must outweight the costs incurred. Some parasites may produce changes in host behavior without incurring any costs, simply by being in the right organ by chance. For most parasites, however, two types of costs of manipulation may have to be paid. First, most alterations of host behavior appear to result from active interference with host neurochemistry that involves the secretion and release of substances by the parasite (Helluy and Holmes 1990; Hurd 1990; Thompson and Kavaliers 1994; Maynard et al. 1996; Øverli et al. 2001; Thomas et al. 2005). The development of specialized glands or tissues for the production of these chemicals must be costly.

Second, in addition to any physiological costs associated with inducing the manipulation, parasites that manipulate their host may face a higher probability of mortality (Poulin et al. 2005). This is a probabilistic cost, only paid in the event that the parasite dies, but it is a cost nonetheless. Ideally, to assess this cost, one would need to compare the fitness cost paid by a manipulative parasite with the cost, if any, incurred by a conspecific parasite that benefits from manipulation without itself inducing it. Examples of manipulative digenean species where such a contrast is possible allow the cost to be evaluated. For instance, metacercariae of *Microphallus papillorobustus* that encyst in the cerebral region of the amphipod second intermediate host induce a strong positive phototaxis and aberrant evasive responses in the host. This manipulation of host behavior results in infected amphipods being more susceptible to predation by aquatic birds, which serve as definitive hosts for the parasite. However, not all metacercariae of *M. papillorobustus* encyst in the cerebral region of amphipods, some also encyst in the abdomen. Amphipods are capable of mounting an immune response against invading parasites, involving both encapsulation and melanization, and they use this cellular defense reaction against metacercariae. Thomas et al. (2000) have found that approximately 17 percent of cerebral metacercariae of *M. papillorobustus* are killed by encapsulation and melanization, whereas less than 1 percent of abdominal metacercariae suffer this fate. Three other digenean species, belonging to the same family (Microphallidae) as *M. papillorobustus*, parasitize the same amphipod; they all encyst in the amphipod's abdomen, and none of them is attacked by the host immune system (Thomas et al., 2000). The host's defenses are apparently targeted specifically at those metacercariae most likely to cause it harm. The end result is that in this system, manipulative parasites incur a much greater probability of death from immune attack than their conspecifics opting not to manipulate and encysting in the abdomen of the host.

Another example involves the two echinostome digeneans *Curtuteria australis* and *Acanthoparyphium* sp., which live in sympatry on New Zealand intertidal mudflats and have the same life cycle. After their release from snail first intermediate hosts, cercariae of both species penetrate a bivalve second intermediate host, the cockle *Austrovenus stutchburyi*, through its inhalant siphon; they encyst as metacercariae in the foot of cockles and await predation by oystercatchers, their common definitive host. As metacercariae accumulate in the foot of a cockle, they replace host muscle tissue and stunt further foot growth. The outcome is that heavily parasitized cockles lose their ability to burrow, and are left stranded on the sediment surface (Thomas and Poulin 1998). Two independent field experiments have shown that manipulated cockles are about five to seven times more likely to be eaten by bird definitive hosts than healthy, buried cockles (Thomas and Poulin 1998; Mouritsen 2002b). Both digenean species appear to contribute equally to this manipulation: their relative abundances are roughly equal, and they both tend to encyst near the tip of the foot (Babirat et al. 2004), where their debilitating effect on the host's burrowing ability is most intense (Mouritsen 2002b). Importantly, however, about 25 percent of metacercariae of both species are found in the middle of the foot, and about 10 percent near the base of the foot (Babirat et al. 2004). The benefits of host manipulation are shared by all metacercariae: although only those near the tip of the foot impair the burrowing ability of cockles, once an oystercatcher feeds on a cockle, it ingests all parasites along with host tissues. However, an opportunistic predatory fish also picks on surface-stranded cockles, eating exclusively the tip of the foot of surface-stranded cockles trying in vain to burrow. This predator, a labrid fish, is not a suitable definitive host for the two digeneans: any metacercaria ending up in a fish dies. Up to one third of all cockles show signs that part of their foot has been cropped by the fish (Mouritsen and Poulin 2003). The frequency of foot cropping is particularly high among manipulated cockles lying on the sediment surface, and field estimates indicate that close to one-fifth of metacercariae are thus lost to fish predation (Mouritsen and Poulin 2003). All of those were encysted near the tip of the foot of cockles, not at its base. Therefore, in this system, all metacercariae in a cockle have the same probability of reaching a bird definitive host, but those that are responsible for manipulating host behavior face a greater risk of mortality.

Thus, in some systems, changes in host behavior that benefit the parasite may be fortuitous and cost free; in general, however, various costs of manipulation exist even if they are difficult to quantify (Poulin 1994a). From an energetic perspective, any energy invested in host manipulation will not be available for growth, reproduction, or fighting the host's immune system. These trade-offs, combined with the risks of parasite mortality associated with manipulation, mean that selection will not always favor large investments in host manipulation. Investments in manipulation, or manipulation effort (ME), will tend toward an optimal value at which parasite fitness is maximized. Under some ecological conditions, low values of ME will be favored, and the associated changes in host behavior may sometimes be very small.

It is possible to make general predictions about the optimal ME, or ME*, expected in different systems (Poulin 1994a; Brown 1999). Even with no investment in host manipulation (ME = 0), the transmission success of the parasite is unlikely to be nil. For instance, some infected intermediate hosts would still be eaten by predatory definitive hosts even in the absence of manipulation by the parasites. Typically, prevalence of infection is

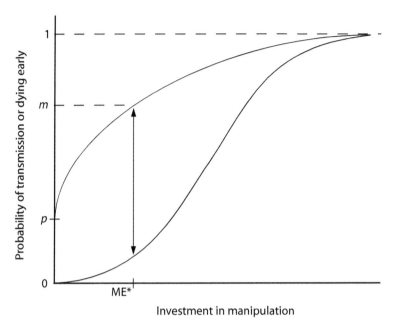

Figure 5.11 Probability of parasite transmission and probability of the parasite dying early as a function of investment in the manipulation of the intermediate host (i.e., manipulative effort, ME). Without any investment in manipulation, the parasite has a passive transmission rate (p) that is greater than zero but less than one; increasing investment in manipulation yields higher transmission probabilities but with diminishing returns (top curve). At the same time, the cost of manipulation, or the probability of dying early, increases with the level of investment, following a sigmoidal function in this hypothetical example (bottom curve). The gross benefits of manipulation equal $m - p$, or the difference between the realized transmission rate and the passive transmission rate. The optimal investment in manipulation (ME*) is the level at which the net gain (benefits minus costs) in transmission probability is maximized. (From Poulin 1994a)

extremely low among invertebrates serving as intermediate hosts for helminths (see review in Marcogliese 1995). However, because definitive hosts ingest large numbers of prey over time, chances are that random prey selection will result in some infected intermediate hosts being captured. Investments in manipulation (ME > 0) will only increase the probability of transmission above the passive transmission rate, or p (fig. 5.11). The gross benefits of manipulation correspond to the difference between the transmission rate achieved through manipulation and the passive transmission rate. As ME gets higher, the rate of increase in the probability of transmission is likely to follow a law of diminishing returns (fig. 5.11): small investments yield greater returns per investment unit than large investments. The costs associated with manipulation are also likely to increase with increases in ME. The shape of the cost function will vary, from roughly linear if the costs are mainly due to production of neuroactive substances, to an all-or-nothing step function if mortality of manipulative individuals is much higher than that of nonmanipulative individuals. The investment favored by selection, ME*, will depend on p, and on the exact shape of the transmission and cost curves in figure 5.11.

These variables will in turn be influenced by the ecological conditions prevailing in the host-parasite system. Consider the mean infrapopulation size, or the mean number of conspecific parasites sharing a host individual, and its effect on ME* (Poulin 1994a; Brown 1999). In systems where infrapopulations are small, for instance, an average of a single parasite per host, the task of modifying host behavior cannot be shared and selection may favor a high ME*. In contrast, if a parasite is likely to share its host with several conspecifics, the costs of manipulation can be shared: if the impact of manipulation by several individual parasites is cumulative, then small ME* values can achieve the same transmission rate obtained by the single parasite doing all the work on its own. Sharing the costs of manipulation with many conspecifics is no guarantee of greater individual fitness, however: larger infrapopulation sizes in intermediate hosts create crowded conditions and lead to reductions in growth and subsequent fecundity of individual parasites. This has been documented in helminth species known to manipulate their intermediate hosts (e.g., Dezfuli et al. 2001a; Brown et al. 2003). In addition, larval helminths at different stages of development do not all benefit from host manipulation. For example, stages of the acanthocephalan *Acanthocephalus dirus* that are infective to the definitive host induce marked coloration changes in their isopod intermediate hosts, whereas noninfective stages only have a modest effect on host color (Sparkes et al. 2004). When infective and noninfective stages co-occur in the same intermediate host, it is the manipulation typical of the infective stage that is expressed; any resulting increase in predation by the definitive host could only harm the noninfective stages.

To keep things simple, let's ignore any costs or risks associated with host sharing and see how infrapopulation size can influence ME*. Because of the aggregated distribution of parasites among their hosts (chapter 6), the mean infrapopulation size is meaningless, as it does not represent the infrapopulation size experienced by the majority of parasites. Selection cannot, therefore, favor a fixed ME* based on infrapopulation size as it is too variable. One solution may be to favor higher ME* values than those expected strictly from a consideration of mean infrapopulation size. Parasites finding themselves crowded with several conspecifics in their host would pay a higher cost than necessary, but this strategy would guarantee a decent chance of transmission to parasites ending up alone in their host (Poulin 1994a). Another, better strategy would be to develop a plastic ME rather than a fixed ME*. This would allow a parasite to adjust its investment in manipulation to the conditions it happens to encounter rather than follow an inflexible investment plan.

In populations where there is a more or less fixed ME*, what is preventing some cheating parasite from paying a lower cost (ME < ME*) while still obtaining all the benefits of manipulation by relying on the action of its conspecific (Brown 1999)? Cheating genes could spread quickly through a parasite population when they first appear. Eventually, however, the need for some honest, manipulative parasites to do the work would place limits on the spread of cheating genes. The ME* strategy would return as the most profitable strategy (Poulin 1994a).

In many systems, entire parasite infrapopulations may consist of related individuals. For example, most if not all cercariae of *Dicrocoelium dendriticum* ingested by an ant are thought to be derived from the same sporocyst inside the snail serving as first intermediate host. In other words, all cercariae come from the same egg layed by a single adult parasite (Wickler 1976; Wilson 1977). In the ant, only one larva attaches to the ant's subesophageal ganglion and induces a behavioral change in the host. This parasite dies and is not transmitted when the ant is ingested by the sheep definitive host (another good example of the cost of manipulation). Nevertheless, this individual benefits in terms of inclusive fitness, since its sacrifice helps its identical kin complete their life cycle. In this case, selection has apparently favored one individual paying the full cost of manipulation rather than the sharing of the cost among relatives. Because of the asexual multiplication of many parasite larvae, infrapopulations of close relatives may be common in intermediate hosts, and various methods of paying the costs of manipulation may have been favored over the all-or-nothing strategy of *D. dendriticum*. In theoretical models, genetic relatedness among co-occurring parasites is a key parameter influencing ME* (Brown 1999). Unfortunately, there exists at present practically no hard data on the genetic structure of infrapopulations in manipulative parasite species.

Many factors other than infrapopulation size or composition can affect ME* (Poulin 1994a). Transmission by predation from an intermediate host to a definitive host is an unlikely event in most cases. To counter the odds, parasites can use several strategies. They can produce more eggs at the adult stage, or they can increase their longevity in the intermediate host in order to wait for the unlikely predation event (see chapter 2). If there are trade-offs between ME and either adult fecundity or longevity in the intermediate or definitive host, lower values of ME* may be expected. Kuris (1997) argued that host manipulation is only one of many interrelated life-history traits. For instance, parasites that have a long life span and grow to large sizes in their definitive host should not manipulate their intermediate host, because in such parasites an enhanced transmission to the definitive host would impose serious fitness costs to this host. Host manipulation is not expected to evolve if the definitive host incurs a cost associated with infection; if it does, it will evolve ways to discriminate between manipulated and nonmanipulated prey, and the manipulation will fail (Lafferty 1992; Kuris 1997). Two helminths that have identical life cycles, use the same host species, and live in the same Norwegian lake provide a perfect example. The nematode *Cystidicola farionis* is long-lived in its fish definitive host, and causes no change in the susceptibility of its amphipod intermediate host to predation by fish; in contrast, the cestode *Cyathocephalus truncatus*, which has a short life span and matures quickly in fish hosts, causes an eightfold increase in amphipod susceptibility to fish predation (Knudsen et al. 2001). Different strategies, involving different combinations of manipulation, fecundity, larval and adult longevity, and so on, may all result in similar overall fitness. Which of these strategies actually evolves depends on the genetic variability of the different traits and on various other constraints. The important point is that we cannot expect manipulation of host behavior to be the rule in systems where such manipulations may appear adaptive. Just as with high virulence, host manipulation is only one of many strategies that may be selected. Perhaps the only approach we have to investigate the evolution of host manipulation is to relate either its occurrence or the degree of its expression (ME*) to some of the many life-history and ecological variables to which it may be linked (Poulin 1994a). Comparative analyses across species, once again, are the only way to obtain general answers.

5.3.3 Host Manipulation in a Multispecies Context

Parasite species rarely occur on their own in a host population. As a consequence, most larval helminths, whether or not they are capable of manipulating host behavior, will often share their intermediate host with other parasite species. These other species may have identical life cycles, or they may have completely different cycles. This means that when different parasite species share an intermediate host, their interests may or may not be the same. If they have the same definitive host, then they could cooperate to reach their common final destination. In contrast, if they have different definitive hosts, there will be a conflict between them regarding what should happen to their shared intermediate host. When one of these parasites is capable of host manipulation, a range of strategies may be favored in the other co-occurring species. Lafferty (1999) and Lafferty et al. (2000) have summarized the many possible scenarios, and what follows is based on their ideas.

When considering what influence a manipulator parasite can have on the evolution of transmission strategies in other sympatric parasites, it is useful to consider intermediate hosts as vehicles and definitive hosts as destinations. With the manipulator parasite acting as a pilot steering the intermediate host toward the manipulator's definitive host, the host-as-vehicle analogy allows us to illustrate the various scenarios of co-occurrence between a manipulator and other parasites with the same or different definitive hosts, and with or without the ability to manipulate the phenotype of intermediate hosts (fig. 5.12). The simplest scenario is one in which a nonmanipulating parasite shares both intermediate and definitive hosts with a manipulator. When infecting a manipulated intermediate host, this "lucky passenger" gets a cost-free ride toward its definitive host (scenario b in fig. 5.12). Assuming it infects intermediate hosts at random, the passenger's luck depends on the prevalence of the manipulator. If the manipulator is extremely common in the intermediate host population, the passenger is practically always lucky; but if the prevalence of the manipulator is low to moderate, selection may favor the evolution of mechanisms for locating and infecting only manipulated intermediate hosts (Thomas et al. 1998a). Of course, for this to happen, the transmission and fitness benefits obtained by actively seeking manipulated hosts must outweight the costs of this active search. Because manipulated hosts sometimes occupy a microhabitat distinct from that of uninfected hosts, discriminating between them may be easy, and the transition from lucky passenger to "hitchhiker" may be straightforward (from scenario [b] to [c] in fig. 5.12).

The ideal demonstration of hitchhiking would include both a positive association between a nonmanipulating parasite and a manipulator among intermediate hosts in the field, and experimental evidence that the larvae of the nonmanipulating parasite actively seek and infect manipulated hosts. The best-documented example is provided by the digenean *Maritrema subdolum*, which is positively associated with another digenean,

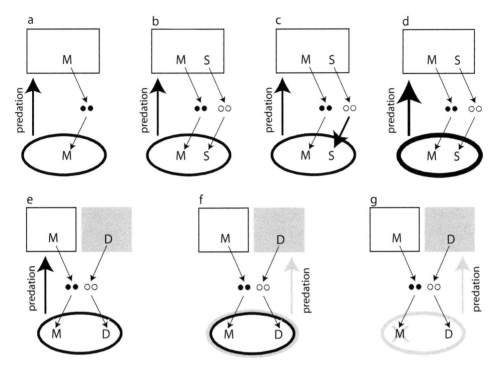

Figure 5.12 Schematic representation of some possible strategies available to a parasite species sharing an intermediate host with a manipulative species, using the host-as-vehicle analogy. As shown in (a), the manipulative parasite M alters the phenotype of its intermediate host (bold outline) and increases its probability of transmission to the definitive host (bold arrow). In some cases, a second parasite S may have the same life cycle as M, with both species having the same destination, i.e., the same definitive host. In (b), S has no active strategy and is merely an occasional lucky passenger benefiting from M's efforts. In (c), S preferentially infects intermediate hosts already harboring M (bold arrow from S's eggs), thus hitchhiking a free ride to the definitive host. In (d), S is also a manipulator, and the two copilots join forces to achieve higher transmission rates to the definitive host (thicker outline of the intermediate host and thicker arrow). The parasite sharing an intermediate host with M may have a different definitive host, however, creating a conflict of interest between the two parasites regarding their destination. In (e), D is an occasional unlucky passenger, taken to an unsuitable definitive host frequently when it shares the intermediate host with M. In (f), D is also a manipulator: it manages to hijack the intermediate host by altering its phenotype in a way that makes it more susceptible to predation by D's definitive host (gray outline and gray arrow); here D nullifies M's own manipulative efforts. Finally, in (g), D just kills M, takes over the intermediate host, and manipulates it without interference.

Microphallus papillorobustus, among amphipod intermediate hosts (Thomas et al. 1997). *M. subdolum* is not a manipulator, but *M. papillorobustus* infections cause amphipods to swim closer to the water surface where they are more susceptible to predation by bird definitive hosts. The cercarial larval stage of *M. subdolum* actively swims toward the water surface where they are more likely to infect amphipods already manipulated by the other digenean (Thomas et al. 1997). Interestingly, two other digenean species with the same intermediate and definitive hosts as *M. papillorobustus* show no association with the manipulator, and are thus merely lucky passengers (Thomas et al. 1998b). Evolution toward hitchhiking is therefore not a universal outcome of co-occurrence with a manipulator. In fact, in geographical areas where the prevalences in amphipods of both *M. subdolum* and

M. papillorobustus are high, there is no positive association between the two species in the field (Kostadinova and Mavrodieva 2005), indicating that active hitchhiking is only beneficial when the prevalence of the manipulator is low. There are a few other examples of positive associations between a nonmanipulator and a manipulator, but it is not known in these cases whether the larvae of the presumed hitchhiker actively seek manipulated hosts. These include larval helminth parasites in fish intermediate hosts (Lafferty and Morris 1996), and a digenean sharing a fish host with a protozoan causing the liquefaction of the host's muscles and thus facilitating the release of both parasites' propagules (Pampoulie and Morand 2002).

A parasite sharing intermediate and definitive hosts with a manipulator may itself be capable of host manipulation (scenario [d] in fig. 5.12). Each manipulator may affect a different feature of host phenotype, both leading to increased transmission rate to the definitive host. If the effects of both "copilots" on transmission success are additive, selection could favor synergism between the two so that they would evolve mechanisms to find and associate with one another in intermediate hosts. There is empirical evidence that such strategies may have evolved in several parasite communities. First, the manipulative acanthocephalans *Pomphorhynchus laevis* and *Acanthocephalus clavula*, which both mature in fish hosts, are positively associated in their amphipod intermediate hosts, and neither species shows any association with a third acanthocephalan species maturing in bird hosts (Dezfuli et al. 2000). Second, two echinostome digeneans with the same life cycle, *Curtuteria australis* and *Acanthoparyphium* sp., encyst as metacercariae in the foot of cockles to await predation by their bird definitive host. Metacercariae accumulating in the foot of a cockle impair its ability to burrow, and cockles thus manipulated are more vulnerable to bird predation than healthy, buried cockles (Thomas and Poulin 1998; Mouritsen 2002b). Both digenean species appear to contribute equally to this manipulation: their relative abundances are roughly equal, and they both tend to encyst near the tip of the foot (Babirat et al. 2004), where their debilitating effect on the host's burrowing ability is most intense (Mouritsen 2002b). Finally, recent evidence suggests that digeneans and acanthocephalans sharing the same crab intermediate hosts and the same bird definitive hosts have a greater effect on the crab's brain chemistry, and thus presumably on its behavior, when they co-occur than on their own (Poulin et al. 2003b).

Conflicts of interest arise when a parasite species shares an intermediate host with a manipulator parasite but has a different definitive host. If the first parasite species is not capable of host manipulation (scenario [e] in fig. 5.12), then it is an "unlucky passenger" that is likely to take a no-return detour toward the wrong definitive host each time it co-infects an intermediate host with the manipulator. If the manipulator is rare, this is not a major concern. If the manipulator is moderately to highly prevalent, however, selection may favor the evolution of mechanisms for the unlucky passenger to avoid manipulated intermediate hosts and infect preferentially uninfected hosts.

Perhaps the most interesting scenario of all involves two manipulator species with the same intermediate host but different definitive hosts. The host features modified by these two manipulators are likely to be different since they are targeted at different definitive hosts. Depending on many factors, such as the mechanisms of manipulation used by the two parasites or which one arrives first in the intermediate host, one of them is likely to be stronger and it is mainly its effects on host phenotype that are expressed. Because the intermediate host can only be ingested by one predator, there has to be a winner. One of the two manipulators may overpower the other without killing it, thus "hijacking" the intermediate host (scenario [f] in fig. 5.12). This situation could lead to arms races between the two manipulators since the costs of losing are so high.

For instance, the acanthocephalans *Pomphorhynchus laevis* and *Polymorphus minutus* both use the same amphipod as intermediate hosts, but the former matures in fish whereas the latter matures in birds. Both induce different behavioral alterations in their amphipod host, each aimed at a different type of predatory definitive host. When they co-occur in the same amphipod, however, it is mostly the behavioral manipulation of *P. laevis* that is expressed (Cézilly et al. 2000). This suggests that *P. laevis* is the winner of this conflict, hijacking the shared intermediate host. Under these circumstances, *P. minutus* would benefit by avoiding amphipods infected by *P. laevis*; in the field, however, the two species are randomly associated (Outreman et al. 2002). Given that acanthocephalans infect their intermediate hosts when the latter feed on parasite eggs, it is not surprising that one parasite cannot discriminate against hosts already harboring a conflicting species.

There are other solutions to the conflict between two manipulators with different definitive hosts. For instance, one of the two manipulators could evolve a "shoot the other pilot" strategy (scenario [g] in fig. 5.12). This may have happened in a conflict between two cestodes using the same beetle intermediate hosts. Both can manipulate beetle behavior, *Hymenolepis diminuta* in order to reach a rat definitive host, and *Raillietina cesticillus* in order to reach a chicken definitive host. In experimental infections, the presence of *R. cesticillus* larvae in a beetle prevents the establishment of *H. diminuta* larvae (Gordon and Whitfield 1985). A similar situation exists in a system involving two acanthocephalan species sharing the same amphipod intermediate host but having different fish definitive hosts. The presence of *Pomphorhynchus bulbocolli* in amphipods interferes with the establishment and development of larvae of *Leptorhynchoides thecatus* (Barger and Nickol 1999). In a very similar system, the presence of one acanthocephalan species in an intermediate host negatively affects the growth of other acanthocephalans (Dezfuli et al. 2001a). Considering that the ability to manipulate intermediate hosts is almost universal among acanthocephalans (Moore 1984), these examples also suggest that in a situation of conflicting interests, one manipulator either eliminates or harms the other one.

Although all these scenarios and similar ones (e.g., the sabotage hypothesis: Thomas et al. 2002c) are appealing, the fact remains that they generally lack solid empirical support. To some extent, this is simply because few investigators have looked for these strategies in natural systems. Given the costs and benefits of manipulation discussed in the previous section, it would not be surprising to find that parasites have come up with several alternative transmission strategies when another co-occurring species is inducing a manipulation and already paying all the costs.

5.4 Manipulation of Host Sex Ratio by Parasites

Among parasites transmitted vertically, several protozoans (microsporidians) use host gametes to get from one host generation to the next (Canning 1982; Dunn et al. 1993, 2001). This transmission mode is also common in viruses and bacteria (Werren et al. 1988; Hurst 1993), notably in the *Wolbachia* bacteria (Bouchon et al. 1998; Bourtzis and O'Neill 1998). These parasites inhabit the cytoplasm of host gametes and are passed on to host offspring during reproduction. Because of the difference in size and in the amount of cytoplasm between ova and sperm, the parasites are generally transmitted only from infected female hosts to their offspring. In male hosts, if horizontal transmission does not occur, the parasites are at an evolutionary dead end. Of the potential strategies of host exploitation open to maternally transmitted parasites, biasing host investment in favor of female offspring appears to be a common option (Hurst 1993).

Parasites capable of increasing the proportion of host offspring that are of the transmitting (female) sex and thus not dead ends are referred to as sex-ratio distorters. They manipulate the sex ratio of host offspring in either one of three ways (Dunn et al. 1995, 2001). First, they can kill the male offspring of their host, such that more resources are diverted toward female offspring that serve as routes of transmission. The release of infective stages from dead hosts for horizontal transmission can be another benefit obtained from killing male offspring (Dunn et al. 2001). Second, they can distort the primary sex ratio of the host by altering gene expression in the host and converting genotypic males into genotypic females. Third, they can feminize the male offspring of the host, turning genotypic males into functional phenotypic females. By turning male hosts into females capable of transmitting the infection, the parasite increases its base for transmission to the next generation of hosts. Feminization can be achieved by suppressing the development or functioning of the male host's androgenic gland during or after embryogenesis (Rodgers-Gray et al. 2004), or by other physiological manipulations. Whatever the mechanism, feminization can produce strongly female-biased sex ratios among the offspring of infected hosts (fig. 5.13).

The effect of a feminizing parasite on host sex ratio will depend on the host's intrinsic primary sex ratio, the parasite's transmission efficiency from the mother host to her offspring, and the parasite's feminizing influence (fig. 5.14). These last two parameters, as well as parasite prevalence, can vary considerably among host populations (e.g., Dunn and Hatcher 1997), but where they are high, the parasite's influence can be very pronounced. Whereas natural selection acting on the parasite should favor increases in the parasite's transmission and feminizing efficiency (t and f in fig. 5.14), it may at the same time favor hosts producing more offspring of the rarer sex (Hatcher and Dunn 1995). Despite the evolution of compensatory sex ratios in the host population, increases in the transmission and feminizing efficiency of the parasite will be favored as long as they do not affect other components of the parasite transmission success, such as host fecundity. When transmission and feminization attain their maximum value ($t = 1$ and $f = 1$ in fig. 5.14), the parasite may drive its host population, and hence its own population, extinct because of a lack of

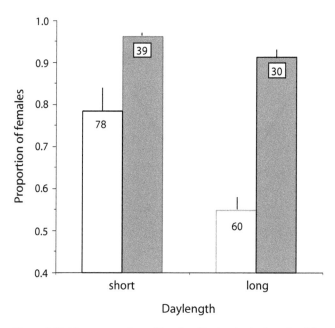

Figure 5.13 Mean proportion of female offspring in the broods of the female amphipod *Gammarus duebeni* either uninfected (open bars) or infected by the transovarially transmitted microsporidian *Octosporea effeminans* (shaded bars). Standard errors and numbers of broods in each treatment are also shown. Female amphipods were raised under different day-length conditions, because photoperiod is an environmental factor known to influence sex determination in *G. duebeni*. Whatever the photoperiod, parasitized amphipods produced many more female-biased broods than uninfected ones, with more than 90 percent of offspring of parasitized mothers becoming females due to the feminizing effect of the protozoan. (Modified from Dunn et al. 1993)

male hosts. In fact, even at values of t and f less than unity, it is theoretically possible for the parasite to drive its host population to extinction, for instance if the host population is of small size. Thus parasitic sex-ratio distortion is a risky strategy for host exploitation and parasite transmission: traits that maximize transmission also jeopardize the host population. The relatively short list of parasites known to distort host sex ratio may reflect a lack of research effort. It may also be a consequence of the risks of adopting this strategy, the highly successful taxa having gone extinct without leaving a trace.

The sex-specific nature of virulence in these vertically transmitted sex-ratio distorters is worth emphasizing. Microsporidians that are transmitted horizontally are generally highly virulent (Dunn et al. 2001). They produce huge numbers of spores to compensate for the low probability of individual spores being ingested by another host; the demands that this strategy places on the host typically lead to marked reductions in host fitness and survival. In contrast, microsporidians that are transmitted vertically, or transovarially, cause little or no pathogenicity in female hosts responsible for their transmission (Dunn et al. 2001). The story is different when the parasites are in males: infected male hosts are either killed or feminized. Thus virulence in these parasites is host sex specific (Bandi

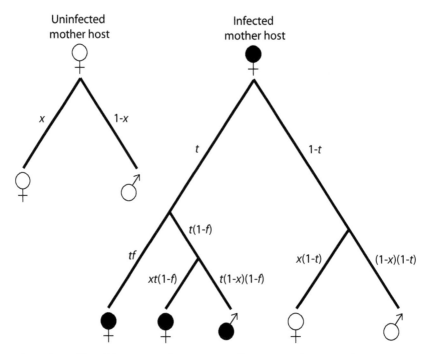

Figure 5.14 Effect of feminization by a transovarially transmitted, sex-ratio-distorting parasite on the proportion of male and female offspring produced by uninfected and infected female hosts. Infected hosts are indicated by black-filled symbols. The proportion of female offspring in the brood of uninfected female hosts equals x, which is usually approximately 0.5
in uninfected populations. Broods of infected female hosts are likely to have female-biased
sex ratios, determined by t, the proportion of the brood that inherit the parasite, and f, the
proportion of infected offspring that are feminized by the parasite. (Modified from Hatcher
and Dunn 1995)

et al. 2001). Because the fitness of female hosts and the transmission of the parasite are
so tightly coupled, selection should also favor beneficial effects of the parasite on female
host fecundity. Indeed, Haine et al. (2004) have found that two different species of microsporidians have positive effects on female reproduction of their amphipod hosts, causing them to start breeding earlier in the reproductive season. This once again illustrates
that parasites have evolved very sophisticated strategies of host exploitation: they are not
merely agents of pathology and disease.

5.5 Conclusion

The past few years have witnessed much progress in our understanding of parasite evolution. It is now generally accepted that parasites do not invariably evolve to become
harmless to their host. A diversity of strategies for the exploitation of the host have flour-
ished among the many taxa of parasites, such that parasites cause anywhere from unde-
tectable to drastic changes in host fitness. With the data currently available and the rapid

advances in the resolution of parasite phylogenies, we can soon start using comparative studies to elucidate the causes behind the variation in patterns of virulence, castration, and sex ratio or behavior manipulation.

Yet, we may need to rectify our approach to these problems. Currently, we investigate the evolution of parasite strategies by investigating their effects on hosts. To quantify how quickly parasites exploit hosts as food resources, we quantify the pathology incurred by the host rather than the actual exploitation rates and how they relate to parasite fitness (see Poulin and Combes 1999). To examine how parasites manipulate their hosts, we measure changes in host behavior instead of quantifying how much energy is invested in manipulation by parasites or how it translates into fitness gains for that parasite. What happens to the host is more a reflection of how well the host can cope with a parasite than of what the parasite does. A final example shows how focusing on the effects of parasites on hosts can be misleading when we are really interested in parasite adaptations. Following infection with metacercariae of the digenean *Podocotyloides stenometra*, coral polyps become distended, causing coral colonies to suffer reduced growth (Aeby 1992, 2002). Butterfly fish, the parasite's definitive host, feed preferentially on parasitized polyps, which appear pink and differ in coloration from healthy polyps. This then looks like a case of host manipulation by the parasite, in which the parasite benefits and the host suffers from lower growth and higher predation risk. However, predation enhances the regenerative capabilities of the coral colony, such that the net effect of parasitism on the coral colony's fitness may be positive (Aeby 1992). Focusing exclusively on fitness effects on the host, or on the changes in the susceptibility of hosts to predation, would give us a biased picture of what strategy is really employed by the parasite. Clearly, studying parasites through their impact on hosts is not the best way to proceed.

6 | Parasite Aggregation: Causes and Consequences

The previous chapters have focused on properties of individual parasites, somewhat variable among members of the same species but shared by all. The next level of organization is the population, consisting of conspecific parasites interacting and coexisting in time and space. This and the next chapter step up to the parasite population level, in particular the distribution of parasite individuals, their dynamics and their persistence. Because of their complex life cycles, different members of parasite populations occupy completely distinct habitats. For instance, infective larval stages may be in the external environment, immature stages in intermediate hosts, and adults in definitive hosts. This fragmentation of the parasite population complicates its study unless the different stages are examined separately. This chapter looks at the distribution of parasites in space; it will discuss mainly parasitic stages in hosts as opposed to free-living infective stages, as parasitic stages have received the most attention from parasitologists.

Free-living animals are not evenly distributed across their geographical range; the abundance of a species varies in space in response to heterogeneity in the suitability of the habitat or the availability of resources. The abundance of a species is usually highest in the parts of the range where conditions are near optimal and lower elsewhere, resulting in an uneven distribution of abundance across the geographical range (Hengeveld and Haeck 1982; Hengeveld 1992). On smaller scales, individuals also show a patchy distribution, reflecting the patchy distribution of resources. To some extent, the habitat of parasites differs from that of most free-living animals: it is not spatially continuous but consists of discrete cells or islands. Hosts represent patches of suitable habitat in an otherwise inhospitable environment. Parasites are not uniformly distributed among these patches, so that some patches contain many more parasites than the average, and others harbor fewer. Typically, parasites are aggregated among the available hosts, such that most host individuals harbor no or few parasites and few hosts harbor many parasites. The heterogeneity in numbers of parasites per host is usually so pronounced that, on its own, the average number of parasites per host is often a meaningless descriptor of infection level (see Rózsa et al. 2000). Parasitologists often use the word *overdispersion* to describe the uneven distribution of parasites among hosts. The word is confusing: to many biologists it suggests a wide dispersion rather than clumping. Here the term *aggregation* is preferred, and will be used as synonymous with *overdispersion*: it is intuitively easier for a nonparasitologist to form a mental picture of the distribution of parasites when aggregation is mentioned.

The parasite individuals of the same species inside the same host individual form an infrapopulation. Infrapopulations are discrete subsets of the entire parasite population. In terms of infrapopulations, parasite aggregation means that most infrapopulations will be small and only a few infrapopulations will be large (fig. 6.1a). Reformulating this with respect to parasite individuals is more complicated: the total number of parasites belonging to the many small infrapopulations may be roughly equal to the total number of parasites belonging to the few large infrapopulations (fig. 6.1b). From an individual point of view, there may be two equally likely types of infrapopulations, small ones and large ones, in which an individual parasite may end up. This will have ecological and evolutionary consequences. These consequences, as well as the causes of aggregation, will be discussed later; first the chapter begins with some remarks regarding the measurement of parasite aggregation.

6.1 Measuring Parasite Aggregation

The distribution of parasites is not static through time. It is the product of processes that are not constant in time and therefore the distribution of parasites is a dynamic phenomenon. No single measure can capture the distribution. An estimate of aggregation obtained from a sample of hosts and parasites is merely a snapshot of an intricate and ever-changing distribution (e.g., Boag et al. 2001). Whatever measure one chooses to quantify aggregation, the dynamic nature of the distribution should never be overlooked.

6.1.1 Indices of Aggregation

There are many ways of quantifying parasite aggregation (Wilson et al. 2002). The simplest and most commonly used measure of parasite aggregation is the ratio of the variance to the mean number of parasites per host. If parasites are distributed at random among their hosts according to a Poisson distribution, the mean number of parasites per host will equal the variance. A variance-to-mean ratio greater than unity indicates a departure from randomness and a tendency toward aggregation, and aggregation levels increase as the value of the ratio increases. This index of aggregation is easy to compute and to understand, which explains its popularity. In addition, simple statistical tests exist to assess the significance of the deviation between the observed variance-to-mean ratio and unity, the value associated with a random distribution (e.g., Sun and Hughes 1994). Using this index, one finds that parasite aggregation is very much the rule among all kinds of host-parasite associations (Dobson and Merenlender 1991; Shaw and Dobson 1995).

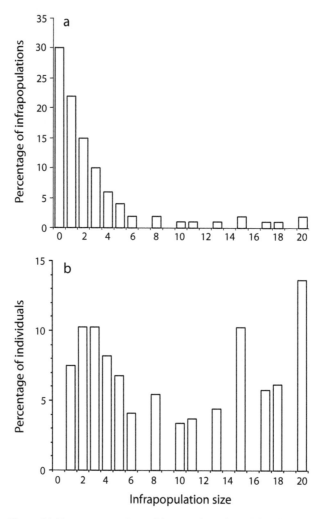

Figure 6.1 Two representations of the same hypothetical distribution of parasites among hosts. From a host-centered perspective, we can focus on the frequency distribution of numbers of parasites per host, or infrapopulation sizes (a). In addition, from the perspective of parasite individuals, we can instead look at the frequency distribution of individuals among infrapopulations of different sizes (b). Less than 25 percent of hosts harbor more parasites than the average number of parasites per host (3.0, including uninfected hosts). About a third of infrapopulations are larger than the average infrapopulation size (4.2, excluding uninfected hosts). However, almost two-thirds of the parasites occur in infrapopulations larger than the average infrapopulation.

However, the variance-to-mean ratio is not without problems. There has been much confusion regarding the inclusion of uninfected hosts in the computations of the mean and variance. Technically they should be included but often they are not. A more important issue is the perspective of aggregation that this measure gives us. The ratio of the variance to the mean number of parasites per host quantifies the variability in intensity of infection among hosts, or the variability in infrapopulation sizes including empty infrapopulations. This host-centered view may not be appropriate when we are interested in parasite individuals. Put differently, the variance-to-mean ratio characterizes the distribution in figure 6.1a but not that in figure 6.1b although these are two representations of the same parasite distribution.

Ever since Crofton (1971) suggested it, fitting the negative binomial distribution to observed parasite distributions has also become common practice. Just as the Poisson distribution exemplifies a perfectly random distribution, the negative binomial is the statistical representation of aggregation. Its use is not always straightforward (see Grafen and Woolhouse 1993), but in general if the observed parasite distribution does not depart from the frequencies predicted by the negative binomial, then the parasites are considered to be aggregated. As a rule, the negative binomial distribution fits observed parasite distributions very nicely (Shaw et al. 1998; but see Gaba et al. 2005). The fit is generally so good that any discrepancy is often taken as evidence of parasite-induced host mortality: heavily infected hosts exist but are missing from samples because of their high mortality (Gordon and Rau 1982; Adjei et al. 1986; Rousset et al. 1996). In various stochastic and statistical models, the negative binomial distribution and its relatives have proven very useful to simulate realistic patterns of parasite aggregation (e.g., Grenfell et al. 1995; Elston et al. 2001; but see Gaba et al. 2005). Consequently, the parameter k of the negative binomial can serve as a simple index of aggregation: as k tends toward zero, aggregation increases, but at values of k of eight or more, the negative binomial converges with the Poisson series (Southwood 1978), and the parasites are randomly distributed. Using k as a measure of aggregation, one finds aggregation is by far the predominant pattern across natural host-parasite systems (Anderson 1982; Shaw and Dobson 1995).

Like the variance-to-mean ratio, however, the use of k as an index of aggregation is not without problems. The parameter is not as easy to compute as the variance-to-mean ratio, and good approximations can only be obtained with the maximum-likelihood method of Bliss and Fisher (1953); other mathematical shortcuts generally provide unreliable estimates. The parameter k is also not very sensitive to the tail of the distribution, that is, to the heavily infected hosts, and does not change much as their numbers vary (Scott 1987). Finally, comparisons of aggregation levels between different samples using k as an index may be totally unreliable because k is highly dependent on the mean number of parasites per host, which is likely to vary among samples (Scott 1987). Thus k is far from an ideal index of aggregation.

Another index of aggregation is the patchiness index of Lloyd (1967). This measure has remained mostly ignored by parasitologists despite its interesting properties. It equals the mean number of conspecifics sharing a host with a parasite, calculated across all parasite individuals in a sample, divided by the mean number of parasites per host. Or, it equals the mean of the distribution in figure 6.1b, minus one, divided by the mean of the distribution in figure 6.1a. Lloyd's patchiness index relates to k as well, and can be computed as $1 + 1/k$. It is meant to represent the average aggregation experienced by individual parasites, and measures how much an individual is "crowded," on average, compared to what it would experience had the parasite population been randomly distributed.

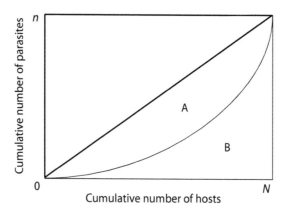

Figure 6.2 The relative discrepancy, *D*, between the observed distribution of parasites among hosts (curve) and their hypothetical uniform distribution (straight line) can be quantified as the ratio of area A to area (A + B). The cumulative number of parasite individuals is plotted against the cumulative number of host individuals, with hosts ranked from the least to the most heavily infected prior to being summed up. (From Poulin 1993)

In contrast to the variance-to-mean ratio and the parameter k, the patchiness index gives more weight to individual parasites and less to infrapopulations (see Wilson et al. 2002).

The final measure discussed here, the index of discrepancy or D, quantifies aggregation as the departure between the observed parasite distribution and the hypothetical distribution in which all hosts are used equally and all parasites are in infrapopulations of the same size (Poulin 1993). It is computed as follows:

$$D = 1 - \frac{2\sum_{i=1}^{N}\left(\sum_{j=1}^{i} x_j\right)}{\bar{x}N(N+1)}$$

where x is the number of parasites in host j (after hosts are ranked from least, i.e., $j = 1$, to most heavily infected) and N is the number of hosts in the sample. The index is represented graphically in figure 6.2. Plotting the cumulative number of parasites as a function of the cumulative number of hosts, we always get a concave curve, the more concave the curve the greater the degree of aggregation. The index of discrepancy is simply the relative departure between this curve and the straight line that would be obtained if the parasite distribution were perfectly uniform (fig. 6.2). In contrast to previous indices, the index of discrepancy has a finite range of values. The minimum value is zero, when the curve falls on the straight line and when there is no aggregation. The maximum value is one, when all parasites are in the same infrapopulation, that is, when aggregation is at its theoretical maximum. This limited range of values facilitates comparisons between distributions that vary in prevalence or mean number of parasites per host. The index is a measure of relative discrepancy, in which each distribution is compared to its own theoretical uniform counterpart. This property makes it reliable for comparisons of aggregation levels among samples.

There are other ways of quantifying aggregation, some of which have more specific uses. For example, Boulinier et al. (1996) have described a method for measuring aggregation at different hierarchical levels when the host population is spatially structured. The aggregation of parasites can be partitioned between two levels, for instance among individual bird hosts within a nest and among nests. This approach can be useful to understand the exact mechanisms generating aggregated distributions, especially if aggregation occurs at one level but not at another.

6.1.2 Problems with the Measurement of Aggregation

Whatever the index of aggregation one chooses, it is likely to covary with other infection parameters. Typically, both prevalence and the mean number of parasites per host are positively correlated; both measures also correlate negatively with aggregation as measured by k or D (Anderson 1982; Poulin 1993). Using the variance-to-mean ratio as a measure of aggregation gives different results (Dobson and Merenlender 1991; Poulin 1993; Shaw and Dobson 1995). Some of these relationships are no doubt simple statistical properties of an aggregated distribution and should not be interpreted as biological phenomena.

One variable that affects equally all measures of aggregation is host sample size (Ludwig and Reynolds 1988; Poulin 1993; Wilson et al. 2002). True population aggregation levels are always underestimated when computed on small samples, producing a positive relationship between sample size and aggregation levels. This is purely a statistical artifact. This is clearly seen by generating a hypothetical population of hosts harboring an aggregated parasite population, and drawing random samples of different sizes from this host population (fig. 6.3). At small sample sizes, estimates of both aggregation level and mean number of parasites per host have a high probability of underestimating (and a low probability of overestimating) true population values. This is an inevitable consequence of parasite aggregation: heavily infected hosts are rare and unlikely to be included in small samples. Small samples are therefore not representative of the population. The accuracy of estimates of aggregation and mean number of parasites per host improves with further sampling, such that sample values increase asymptotically toward real population values as host sample size increases (fig. 6.3). Similar results are obtained independently of the measure used to quantify aggregation or mean number of parasites per host (Gregory and Woolhouse 1993; Fulford 1994; Poulin 1996f). The measurement of prevalence is also plagued by the influence of host sample size (Gregory and Blackburn 1991). In the sampling of hypothetical populations (fig. 6.3), the effect of sample size on estimates of prevalence is not as marked as its effect on the measurement of other parameters (Gregory and Woolhouse 1993; Poulin 1996f). The calculation of prevalence is affected not by the absence of heavily infected hosts but by the proportion of infected hosts, and this should not vary substantially with sample size.

The main point of the above discussion is that host samples of small sizes are inappropriate for the estimation of aggregation. The more aggregated the parasite population,

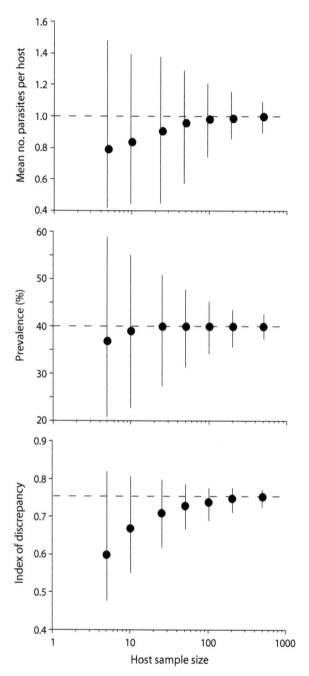

Figure 6.3 Relationship between host sample size and the mean number of parasites per host, the prevalence of infection, and the index of discrepancy, *D*. Data were obtained using bootstrap sampling of a hypothetical population of one thousand hosts harboring one thousand parasites. Broken lines represent the true population values. For each host sample size, the mean and range of twenty-five simulated samples are presented. (From Poulin 1996f)

the more aggregation will be underestimated by relying on small host samples. Mathematical corrections for small sample sizes cannot be used, as the true aggregation level is unknown. Therefore, whatever index is used, efforts should be made to obtain large, representative samples of hosts for computations of aggregation measures. This may be particularly relevant for studies of dynamics of parasite populations over time, in which host samples are taken at monthly intervals; because the availability of hosts or the ease with which they can be captured will vary with the seasons, the dynamic picture of aggregation thus obtained may often be a poor reflection of reality because of variable sample sizes.

6.2 Natural Patterns of Aggregation

No matter which index of aggregation one chooses, the results are the same: natural parasite populations are much more aggregated among their hosts than if parasite individuals were distributed randomly. Using the variance-to-mean ratio, Shaw and Dobson (1995) found that all but one of 269 populations of metazoan parasites had a variance-to-mean ratio significantly greater than unity, and thus a distribution departing from a Poisson distribution. Most had ratios smaller than fifty, but several showed values well above one hundred (fig. 6.4). The same pattern emerges when the parameter k of the negative binomial distribution is used: apart from a handful of cases, the vast majority of populations have k values approaching zero, and thus indicating strong aggregation (fig. 6.4). Shaw and Dobson's (1995) extensive survey of aggregation patterns covered only parasites in vertebrate hosts. There are no similarly exhaustive surveys of host-parasite systems involving invertebrate hosts, but the available evidence also suggests that aggregation is the norm. For instance, Rolff (2000) compiled data on the distribution of parasitic mites on aquatic insects, and found variance-to-mean ratios that were generally slightly lower than those from Shaw and Dobson's (1995) data set, but still greater than unity.

Some authors have argued that the aggregated distribution of parasites among their hosts generally follows a common quantitative pattern, with about 20 percent of the hosts typically harbouring 80 percent of the parasites. Several host-parasite systems appear to conform to this pattern (Woolhouse et al. 1997; Perkins et al. 2003). For example, in a wild mouse population, approximately 20 percent of host individuals, corresponding mostly to the oldest males, harbor between 75 and 90 percent of both ticks and intestinal nematodes (Perkins et al. 2003; Ferrari et al. 2004). This so-called twenty-eighty rule, however, does not apply to the majority of systems in Shaw and Dobson's (1995) data set, and it therefore does not represent a widespread pattern. Aggregation levels are not fixed; they vary among parasite populations along with other interrelated infection parameters, in particular prevalence. Clearly, the percentage of the host population used by parasites is a component of what we measure as aggregation. The higher the prevalence of infection, the greater the dispersion of parasites among available hosts, and thus the lower the aggregation; the inverse relationship between prevalence and aggregation is indeed well documented (Anderson 1982; Poulin 1993; Eppert et al. 2002). For this and other reasons,

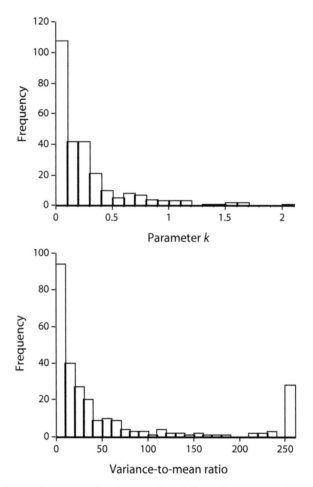

Figure 6.4 Frequency distributions of two indices of aggregation, the parameter k of the negative binomial distribution and the ratio of variance to mean number of parasites per host, for 269 natural populations of metazoan parasites in their vertebrate host. Values of k greater than 2 and values of the variance-to-mean ratio greater than 250 are pooled. (Data from Shaw and Dobson 1995)

we cannot expect the twenty-eighty rule or any other constant aggregation pattern to apply generally across host-parasite systems.

There are some constraints acting on aggregation levels, though: for a given abundance, that is, a given mean number of parasites per host, only a certain range of aggregation levels are likely to be observed. This becomes apparent when one plots log-transformed variance against log-transformed mean number of parasites per host (fig. 6.5). Not surprisingly, the relationship is linear. What is remarkable, however, is the excellent fit of the regression line to the points: about 87 percent of the variation in variance is explained by the mean. Shaw and Dobson (1995) suggested that the degree of parasite aggregation may be constrained for any given mean abundance (see also Wilson et al. 2002). A trade-off between high host (and thus parasite) mortality at very high levels of

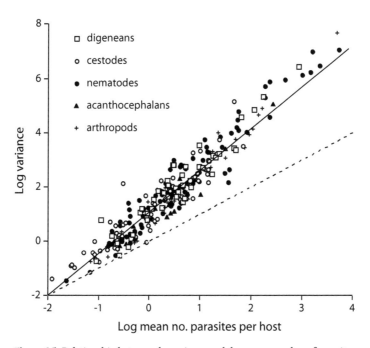

Figure 6.5 Relationship between the variance and the mean number of parasites per host across 269 natural populations of metazoan parasites in their vertebrate hosts. The solid line is the fitted regression line (log variance = 1.098 + 1.551•log mean; $r^2 = 0.87$), and the broken line represents the 1:1 relationship expected if the variance is equal to the mean, i.e., for a Poisson distribution with a variance-to-mean ratio of unity. (Data from Shaw and Dobson 1995)

aggregation, and reduced mating opportunities for parasites at very low levels of aggregation, could limit the observed levels of aggregation to a relatively narrow range of intermediate values.

Finally, although ubiquitous, parasite aggregation is not universal; there are some striking exceptions to the general pattern. Various ecological forces, such as intensity-dependent parasite mortality and parasite-induced host mortality, act to decrease aggregation (Anderson and Gordon 1982). These processes prevent the formation of stable infrapopulations consisting of many parasites: such infrapopulations may occur temporarily but are invariably reduced in size or removed altogether from the population at large. Systems in which both or either intensity-dependent parasite mortality and/or host mortality are believed to act strongly and in which parasites are uniformly distributed among their hosts have been reported (Zervos 1988a; Adlard and Lester 1994; Donnelly and Reynolds 1994). Often, these involve parasites that are almost as large as their hosts and that should probably be called parasitoids, as they almost invariably kill their host. Nematomorph worms in their insect hosts provide a well-known example of a large, lethal parasite that almost always occurs singly in its host. The same is true of polychaetes parasitic in other polychaetes: each infected host harbors only one parasite (see Uebelacker 1978). Parasite populations that are evenly distributed rather than aggregated among their hosts also occur in more typical taxa of parasites. For instance, Burn (1980) found that all fish hosts in a sample were infected with exactly one pair of digeneans, of the genus *Deretrema*, in their gallbladder, except for one fish harboring a single worm. The copepod *Leposphilus labrei*, parasitic on wrasse off the Irish coast, provides another example: 1,922 of 1,924 infected fish examined harbored a single copepod, and the other two harbored two copepods (Donnelly and Reynolds 1994). Some form of infrapopulation control, perhaps host mediated, prevents the existence of larger and stable infrapopulations in these parasites. Systems with such strict infrapopulation regulation are exceptions, however, but they illustrate that aggregation is not a universal feature of parasite populations.

6.3 Causes of Aggregation

It may not be universal, but the widespread occurrence of parasite aggregation suggests that similar processes may be acting to generate the same pattern in different host-parasite systems. Many biological forces operate to shape the distribution of parasites among hosts, some tending to increase aggregation and others tending to produce a more even distribution (Crofton 1971; Anderson and Gordon 1982). At any point in time, the degree of aggregation is determined by the balance of these forces. Clearly, in most parasite species, the forces promoting aggregation are stronger than the processes acting to dampen aggregation.

The biological forces shaping aggregation include stochastic demographic mechanisms in both the host and parasite populations (Anderson and Gordon 1982). For instance, simple biological processes, such as the direct reproduction of parasites within hosts, can increase aggregation. The vast majority of helminth and arthropod parasites, however, do not multiply within a host: their eggs or infective stages leave the host and infect other hosts. Anderson and Gordon (1982) used Monte Carlo simulations to investigate the interactions among the processes affecting the rates of gain or loss of parasites by host individuals. They showed that heterogeneity in these rates among hosts was the major cause of aggregation when direct reproduction of parasites in hosts did not occur. Even if acquisition of parasites by hosts is a random process, so that the distribution of acquisition-rate values follows a Poisson distribution, the resulting variance-to-mean ratio of parasite infrapopulation sizes will be greater than unity. Under these conditions, parasite aggregation only becomes less pronounced when offset by processes such as a highly density-dependent parasite mortality rate (Anderson and Gordon 1982).

Heterogeneity among hosts in rates of parasite acquisition can have at least two main origins. First, it may be due to heterogeneity in exposure to parasites among hosts. Hosts that are equally susceptible to infection may acquire different numbers of parasites simply because of the patchy distribution of parasites in space and time. For example, Keymer and Anderson (1979) showed that in experimental situations, aggregated distributions of infective stages in space accentuate the aggregated distribution of parasites in the host population (fig. 6.6). Empirical evidence and modeling studies suggest that the spatial distribution of infective stages is often sufficiently clumped to generate aggregated

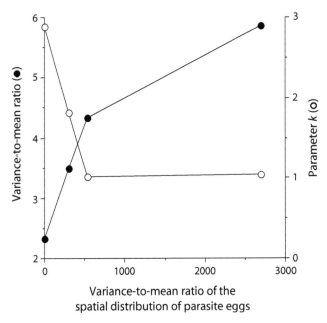

Figure 6.6 Relationship between the aggregation of parasites among their hosts (measured using both the variance-to-mean ratio and the parameter k of the negative binomial distribution) and the spatial distribution of parasite eggs. Data are from experiments in which equal numbers of beetles, *Tribolium confusum*, were exposed to identical numbers of eggs of the cestode *Hymenolepis diminuta*, for which it acts as an intermediate host; only the dispersion of eggs varied between treatments. (Modified from Keymer and Anderson 1979)

distributions of parasites among hosts, without the need to invoke any other mechanism (e.g., Harvey et al. 1999; Bohan 2000; Hansen et al. 2004). Indeed, computer simulations indicate that in randomly distributed host populations, where all individuals are identical and equally susceptible to infection, unequal distances between individual hosts and parasite infective stages inevitably lead to an aggregated parasite distribution (Leung 1998). From the perspective of the hosts, small differences in microhabitat use between individuals can therefore result in substantial differences in exposure to parasite infective stages, and to aggregated distributions of parasites among host individuals (e.g., Jaenike 1994; Wetzel and Shreve 2003). Heterogeneity among hosts in exposure to parasites has a temporal as well as a spatial component. Janovy and Kutish (1988) modeled the effect of the temporal dispersion of parasite infective stages on the distribution of the parasites in their host. They found that when infective stages are continuously available, parasites are not particularly aggregated in their hosts. However, when the temporal dispersion of infective stages is highly heterogeneous, that is, when infective stages are released in discrete waves, the parasites become highly aggregated in their hosts. The results of these simulations are in agreement with empirical measurements (Janovy and Kutish 1988) and suggest that temporal as well as spatial heterogeneity in rates of exposure are enough to produce aggregated distributions.

The second cause of variability in rates of parasite acquisition results from heterogeneity among hosts in susceptibility to infection (see Wilson et al. 2002). Some hosts are better than others at preventing the initial establishment of parasites, or at eliminating them at a later stage following establishment. Other hosts may provide better habitats for parasites, regardless of immunity. Thus, in a natural host population, we would expect various subsets of hosts, based on sex, age, or size, to harbor different numbers of parasites; sampling randomly across these subsets would provide a heterogeneous group of hosts with an aggregated parasite distribution. Differences in susceptibility to infection among individual hosts extend beyond the effects of age or sex, however. In experimental infections, hosts of the same age, size, and sex, given identical doses of infective stages are often found harboring different numbers of parasites. In such cases, variability in susceptibility to parasites reflects either or both differences in the ability of the host's defenses, including the immune system, to fight parasitic infection, or differences in the quality of the hosts as habitats for parasites. Individual differences among hosts in immunity are often all that is needed to generate aggregated parasite distributions (Galvani 2003a). There is considerable evidence that these differences have a genetic basis (e.g., Schad and Anderson 1985; Quinnell 2003). Most experiments with different genetic strains of hosts demonstrate between-strains differences in susceptibility to infection (Wakelin 1978, 1985). Other experiments with natural populations often identify genetic variability in susceptibility among individual hosts. For example, Lysne and Skorping (2002) found that among caged fish exposed to natural infection by parasitic copepods, there were two distinct groups of fish, one inherently susceptible to repeated infection, and the other apparently resistant to infection. In addition, resistance to parasite infections is known to break down in hybrid zones between related host species, due to the mixing of genotypes (Sage et al. 1986; Moulia et al. 1993). Finally, it is possible to breed for parasite resistance in domestic animals (Albers and Grey 1986; Woolaston and Baker 1996). All these lines of evidence indicate that some host individuals may be genetically predisposed to acquire large numbers of parasites relative to their conspecifics even given equal exposure to infective stages.

Susceptibility to parasites often has a genetic basis, and host genetics may be sufficient to produce an aggregated parasite distribution in the absence of all other ecological variables. Wassom et al. (1986) found that the immune response that expels the cestode *Hymenolepis citelli* from the small intestine of its definitive rodent host is under genetic control. Resistance is conferred by a dominant allele at a single gene locus. Patent infections with the cestode occur only in individuals that are homozygous for the recessive allele: these hosts are incapable of rejecting the cestode before it matures, and may accumulate several individual worms. Wassom et al. (1986) suggested that the aggregated distribution of *H. citelli* in natural host populations is due primarily to the high frequency of resistant host genotypes. The simple one-locus genetic resistance in this host-parasite system may not be representative of the majority of systems. Indeed, in a related system (same cestode in a congeneric host), the genetically based expulsion of the parasite plays only a minor role in generating an aggregated parasite distribution (Munger et al. 1989). Immunity may be generally unimportant in the field as a causative agent of parasite aggregation (see Quinnell et al. 1990) in contrast with variability in other host defense mechanisms.

Three further points must be made about heterogeneity among hosts in susceptibility to parasites and its role in parasite aggregation. First, differences in susceptibility need not involve immune or defense responses, but can apply to other host traits. For example, differences in dietary specialization among fish hosts are clearly sufficient to explain the aggregated distributions of nematodes and cestodes in the arctic charr, *Salvelinus alpinus* (Knudsen et al. 2004). Fish that prefer the intermediate hosts of one parasite will end up harboring large infrapopulations of this parasite, whereas fish with other diets will harbor only a few worms. Second, heterogeneity in host susceptibility may also be manifested via preferences for certain hosts over others, in parasite species capable of active host choice such as those with mobile infective stages (e.g., Valera et al. 2004). Third, heterogeneity in susceptibility may not be entirely innate, and some heterogeneity may be acquired. The best example is acquired immune resistance, which is a function of both innate mechanisms and environmental influences such as exposure to parasites at a young age. Often heterogeneity in susceptibility and thus parasite aggregation will be the product of interactions between innate and acquired resistance in the host population (Tanguay and Scott 1992). Acquired heterogeneity in susceptibility can take other forms, too. In many ectoparasitic arthropods, pheromones released by established parasites are believed to attract new parasites to infected hosts (e.g., Blower and Roughgarden 1989; Norval et al. 1989). In others, parasite-induced changes in host behavior seem to make infected hosts more easily detected by parasites and more vulnerable to further infections (Poulin et al. 1991). In these systems, early exposure to and infection by parasites could turn some hosts into "magnets" attracting further parasites at a higher rate than uninfected conspecifics. Acting in isolation, this process could easily and rapidly generate aggregated parasite distributions.

Obviously, parasite aggregation will usually result from a combination of factors. Still, which of the two main factors identified above, heterogeneity in exposure and heterogeneity in susceptibility, generally plays the biggest role? Many studies in natural or semi-natural populations can identify individual differences among hosts as a major determinant of parasite aggregation, but cannot elucidate the precise nature of those differences (e.g., Haukisalmi and Henttonen 1999). This kind of discrimination requires an experimental approach, and to date only a couple of studies have adopted a suitable experimental design to distinguish between exposure and susceptibility. The first is that of Karvonen et al. (2004a), who evaluated the relative importance of exposure and susceptibility as determinants of the aggregation of metacercariae of the digenean *Diplostomum spathaceum* among individual rainbow trout, *Oncorhynchus mykiss*. They found that the fish were capable of developing acquired resistance following an initial exposure to the parasites. Despite the heterogeneity among fish created by acquired immunity, fish kept in cages in the littoral zone of a lake and all experiencing similar levels of exposure to cercariae did not harbor aggregated parasite distributions (fig. 6.7). In contrast to wild-caught fish, the relationship between variance and mean number of parasites per host in caged fish did not depart significantly from that expected from a Poisson distribution.

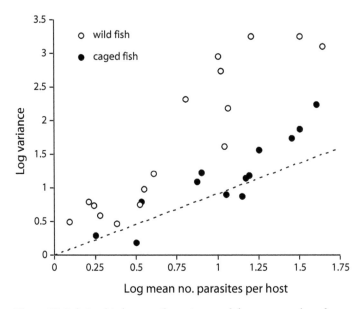

Figure 6.7 Relationship between the variance and the mean number of parasites per host across wild populations and caged groups of rainbow trout, *Oncorhynchus mykiss*, parasitized by metacercariae of the digenean *Diplostomum spathaceum*. The broken line represents the 1:1 relationship expected if the variance is equal to the mean, i.e., for a Poisson distribution with a variance-to-mean ratio of unity. (Data from Karvonen et al. 2004a)

Karvonen et al. (2004a) argue that aggregation of *D. spathaceum* in wild fish must result mainly from variability in exposure rather than heterogeneity in susceptibility among fish. Wild fish can visit several areas of the lake and thus experience a range of exposures to parasites, compared to caged fish restricted to the same area in the littoral zone.

Working with the same host species but with the ectoparasitic crustacean *Argulus coregoni*, Bandilla et al. (2005) found that there were no apparent differences in susceptibility among individual fish, and that fish did not develop acquired resistance following an initial infection by *A. coregoni*. When fish were exposed to identical numbers of infective stages, but for different periods of time, they showed different levels of parasite aggregation (fig. 6.8). As exposure time increased, aggregation levels decreased. Allowing more time for infection essentially reduced the variability in exposure, since it gave time for all free-swimming parasites to encounter a host. Thus, Bandilla et al. (2005) also conclude that differences among fish in exposure, rather than differences in susceptibility, may be the key factor generating aggregated parasite distributions. Whether or not these conclusions apply broadly to other types of hosts and parasites remains to be seen.

All the mechanisms influencing parasite distributions discussed above will vary in strength among host-parasite systems. There may be properties of hosts or parasites that affect the degree of aggregation observed in any host-parasite association. It is important to remember that when variance is plotted against mean number of parasites per host on a log scale, only about 13 percent of the variation in variance remains unexplained (see fig. 6.5); the influence, if any, of parasite or host characteristics on aggregation levels

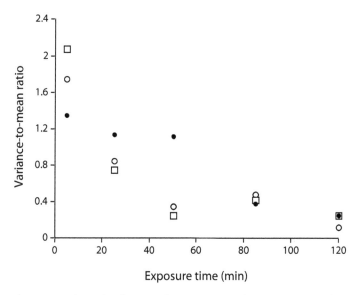

Figure 6.8 Relationship between the aggregation of parasites among their hosts, measured as the variance-to-mean ratio in numbers of parasites per host, and the time of exposure to infective stages. Data are from experiments in which rainbow trout, *Oncorhynchus mykiss*, were exposed to identical numbers of newly hatched branchiuran ectoparasites (*Argulus coregoni*), but for different periods of time. The different symbols represent three different treatments. (Data from Bandilla et al. 2005)

must therefore be rather limited. Nevertheless, in comparative analyses of published values of parasite aggregation, there is a tendency for lower levels of aggregation to be observed in intermediate than in definitive hosts, and for ectoparasites to be more aggregated than endoparasites (Dobson and Merenlender 1991). Also, in systems involving vertebrate hosts, parasites that enter the host passively when ingested accidentally, such as trichostrongylid and oxyurid nematodes, tend to be much more aggregated than parasites such as acanthocephalans that enter the host when the latter ingests an infected intermediate host (Shaw and Dobson 1995). Finally, among species of gastrointestinal nematodes parasitic in mammals, aggregation tends to be more pronounced in small-bodied nematode species than in large ones (fig. 6.9). One explanation for this trend is that for a given host size, space restrictions and intensity-dependent interactions among parasites should be more severe in large-bodied parasites than in small ones, thus preventing the accumulation of numerous large worms in the same individual hosts (Poulin and Morand 2000c). Parasite traits can therefore affect distribution patterns, and explain some of the interspecific variation in aggregation levels.

Among parasites with similar body sizes, life cycles, and modes of transmission, the spatial distribution of hosts may explain variation in levels of aggregation. In experimental situations, mobile infective larvae achieve a higher prevalence of infection and/or lower levels of aggregation when target hosts are clumped than when hosts are randomly or evenly distributed (Blower and Roughgarden 1989; McCarthy 1990). Along the same lines, Rózsa et al. (1996) and Rékási et al. (1997) found that avian lice species parasitizing

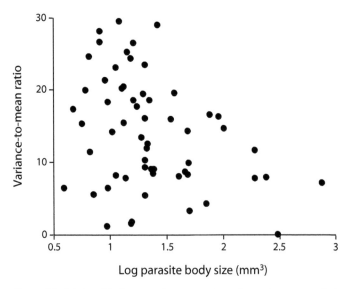

Figure 6.9 Relationship between the aggregation of parasites among their hosts, measured as the variance-to-mean ratio in numbers of parasites per host, and the body size of the parasites, across fifty-nine species of parasitic nematodes. Each point represents a different parasite population, all of which belong to different species (From Poulin and Morand 2000c)

colonial host species are less aggregated than related lice species parasitizing territorial hosts. Therefore, aggregation of hosts may weaken the aggregation of some parasites by facilitating transmission and minimizing differences among host individuals in levels of exposure to infective stages. Many other host characteristics can influence parasite aggregation. For example, Møler (1996) suggested that parasite aggregation should vary positively with the intensity of sexual selection in the host species. However, this relationship as well as the effects of other host and parasite traits, remain to be tested empirically.

Finally, parasites of any given species generally share a host population with other parasite species. Results of mathematical models indicate that if a parasite population experiences strong competition from other parasite species in the same host population, it is likely to display greater aggregation among host individuals (Bottomley et al. 2005). This follows from the fact that competing parasite species generate additional heterogeneity among hosts: parasites of a given species can only accumulate in hosts not already harboring competing species. Parasite aggregation should therefore be considered within the context of the whole parasite community, as the distributions of different parasite species are not mutually independent.

6.4 Consequences of Aggregation

Although there are exceptions, aggregation of individuals among hosts is the rule in metazoan parasite populations. Clearly this pattern of distribution can influence the evolutionary biology of parasites. We may expect the frequency of interactions between conspecific parasites to be determined by the level of aggregation. Different individual parasites may experience highly different levels of competition for resources, depending on the size of the infrapopulation in which they occur. Many parasites will be in small infrapopulations and will not suffer from competition, whereas many other parasites will belong to fewer but larger infrapopulations and will experience intense competition for food or space (fig. 6.1). The frequency of encounter between mates is also likely to be highly uneven among parasites because of aggregation, as are the individual reproductive success of parasites and the fate of entire infrapopulations. Chapter 7 will consider the impact of aggregation on the dynamics and regulation of parasite populations; this section, 6.4, will discuss some of the consequences of parasite aggregation on three selected aspects of parasite evolutionary ecology.

6.4.1 Effective Population Size and Genetic Diversity

The most obvious effect of parasite aggregation on parasite biology is that the intensity of intraspecific competition for space or nutrients will not be equal for all individuals in a population but will instead be proportional to infrapopulation size. Other living conditions in the host can also vary among infrapopulations as a consequence of the variability in infrapopulation size. For these and other reasons, parasites in infrapopulations of large sizes experience a "crowding effect" (Read 1951): the more parasites per host, the smaller the parasites on average. This phenomenon has been particularly well documented in cestodes (see Heins et al. 2002 and Fichet-Calvet et al. 2003 for recent examples). Nevertheless, it applies widely to other types of parasites: both growth and fecundity show pronounced density dependence in many helminth taxa (Keymer et al. 1983; Keymer and Slater 1987; Jones et al. 1989; Quinnell et al. 1990; Shostak and Scott 1993). The shape of the relationship between parasite growth or fecundity and infrapopulation size determines the strength of the density dependence; but for most realistic functions, we can expect that severe aggregation will lower average parasite fitness (Jaenike 1996c).

Although average fitness is lower in large infrapopulations, the variance in fitness is typically greater (Dobson 1986). At the level of both the various infrapopulations and the entire population, this may result in gross inequalities in size or fecundity among individual parasites (Dobson 1986). The adult parasite population in the definitive hosts may consist of several small individuals with low fecundity and a few large individuals with high fecundity, all sexually mature and producing eggs at a rate proportional to their size. This phenotypic plasticity in growth and development may have been favored over evolutionary time because the aggregated distribution of parasites prevents most individuals from reaching their potential adult size. It is not yet clear whether the minority of parasites that attain large sizes are individuals with superior genotypes, or whether adult size results solely from chance events following infection such as arriving first and securing a good attachment site.

Dobson (1986) proposed the use of Lorenz curves and Gini coefficients to measure the degree of inequality in size or fecundity among conspecific parasites in a population (fig. 6.10). The approach was also the inspiration behind the index of discrepancy used to

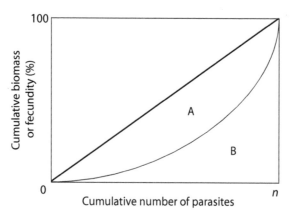

Figure 6.10 The Gini coefficient used to quantify inequalities among parasites in size or egg production is simply the relative discrepancy between the observed distribution of biomass or fecundity among parasites (curve) and their hypothetical uniform distribution (straight line). It can be quantified as the ratio of area A to area (A + B). The cumulative biomass or fecundity is plotted against the cumulative number of parasite individuals, with parasites ranked from the smallest or least fecund to the largest or most fecund prior to being summed up. (From Dobson 1986)

quantify aggregation levels (see section 6.1). Dobson (1986) suggested that Gini coefficients computed using the distribution of total biomass among individuals could indicate to what extent a large fraction of the reproduction in the population is undertaken by only a few individuals. Later studies in host-parasite systems where measures of parasite fecundity can be obtained directly have shown that inequalities in egg production are more pronounced than inequalities in body size (Shostak and Dick 1987; Szalai and Dick 1989). Typically, in highly aggregated parasite populations, only a few worms may be responsible for most of the total egg output, such that genetic contributions to the next generation are highly unequal among individuals.

In population genetics, it is the effective population size, or the number of individuals contributing genes to the next generation, and not the actual population size that determines the relative amount of genetic drift and the relative power of natural selection (Hartl and Clark 1997). In aggregated parasite populations experiencing inequalities in reproduction, the effective population size may be closer to the number of infected hosts than it is to the total number of parasites because only very few parasites per host achieve high egg outputs, regardless of infrapopulation sizes (Dobson 1986; Criscione and Blouin 2005). The smaller the effective population size, the greater the likelihood that the allele frequencies and the genetic heterogeneity characteristic of a parasite population at a point in time will not be transmitted to the next generation. At small effective population sizes, the spread of beneficial alleles may be slower, and random changes in allele frequencies caused by stochastic events become more likely (Hartl and Clark 1997; Nadler 1995). The negative impact of aggregation on effective population size certainly applies to hermaphroditic parasites such as cestodes and most digeneans. In dioecious

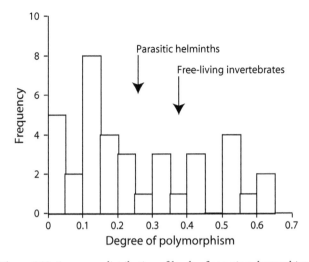

Figure 6.11 Frequency distribution of levels of genetic polymorphism among thirty-seven parasite populations, representing twenty-four helminth species. Polymorphism is measured as the proportion of polymorphic loci among those surveyed. The mean for the helminth populations is shown, along with that for free-living invertebrates in general (Data from table 16.2 in Bush et al. 2001)

parasite taxa, this effect can be further enhanced by skewed sex ratios, which are characteristic of many helminth taxa. There is some evidence, though, that parasite sex ratios approach unity at high intensities of infection (see section 6.4.2); if this is the case, the net effect of aggregation on effective population sizes in dioecious parasites could instead be positive (Criscione and Blouin 2005).

Does aggregation have any impact on levels of genetic variability in parasite populations? Studies of genetic variability within and among parasite populations have grown in number recently thanks to developments in molecular techniques (Nadler 1995; Criscione et al. 2005). In an early survey of the available evidence, Nadler (1990) found that levels of genetic variation in populations of helminth parasites seemed a little lower than those reported for free-living invertebrates. The level of heterozygosity, or the average proportion of heterozygous loci per individual, was 0.07 for parasites and 0.11 for free-living invertebrates (0.10 if insects are excluded; Nevo 1978). However, when genetic variation was measured as the proportion of polymorphic loci among the total number of loci surveyed, parasite populations appeared much less variable than those of free-living invertebrates (fig. 6.11). Even when including the few allozyme studies published after Nadler's (1990) survey, the average level of polymorphism among parasite populations is 0.253 whereas that of free-living invertebrates is 0.375 (about the same if insects are excluded; Nevo 1978). Studies based on mitochondrial DNA data have tended to redress the balance, and it now appears that parasitic helminth populations probably display levels of genetic diversity at least as high as those of free-living animals (Blouin et al. 1992; Bush et al. 2001). Any suggestion that they are lower is most likely due to the fact that we have only obtained estimates of variability for a handful of parasite species, and thus that our sample is not representative.

The very low levels of genetic variation observed in some helminth populations may be due to the cumulative reduction in genetic variability resulting from a small effective population size maintained over several generations (Bullini et al. 1986; Nadler 1995). In contrast, the effects of parasite aggregation and the associated inequalities in reproduction on effective population size and the genetic variation within parasite populations may be minimal for certain parasites. For example, directly transmitted nematodes such as *Ostertagia* and *Haemonchus* often form infrapopulations of several thousand individuals in their hosts. Their small size, relative to that of their hosts, limits the effects of crowding on their reproductive output, and inequalities in fecundity are not likely. Typically, these parasites show high levels of genetic variation within populations (Blouin et al. 1992, 1995). In contrast, *Ascaris suum*, a nematode with a similar life cycle and also parasitic in livestock, displays high levels of within-population homozygosity and high inbreeding coefficients (Nadler et al. 1995). The low genetic variation among *A. suum* individuals is most probably due to their much larger body size, and crowding may therefore have a greater negative effect on the evenness of reproductive output among individuals of this species.

Finally, parasite aggregation can also influence the spread of rare alleles through a parasite population. The aggregation of adult parasites in hosts will lead to the aggregation of infective stages, seen, for instance, in the clumped distribution of helminth eggs released in host fecal deposits. Not only will the number of infective larvae per fecal patch vary in the same way as the number of adult worms per host, but the relatedness of these eggs is also spatially heterogeneous: eggs from the same fecal patch are, on average, much more genetically similar than eggs from different patches. These will then often be transmitted in packets to the same host. Cornell et al. (2003) have modeled the spread of rare recessive but beneficial alleles through a hypothetical parasite population with the characteristics (simple one-host life cycle, etc.) of trichostrongylid nematodes parasitic in ruminants. Their results show that aggregation can promote the spread of recessive traits because it guarantees some degree of inbreeding, and thus the production of rare homozygotes. This phenomenon could explain the rapid spread of recessive drug-resistant phenotypes in nematode populations (Cornell et al. 2003). More generally, it may also mean that aggregation could, under certain circumstances, allow more rapid parasite adaptation to changing conditions.

6.4.2 Sex Ratio

The highly fragmented nature of parasite populations resulting from aggregation, and, in some parasites, the associated low levels of genetic variation within infrapopulations can favor sex ratios departing from unity. Equal investments in offspring of the two sexes should be selected under a wide range of conditions, but biased sex ratios can also evolve (Charnov 1982; Godfray and Werren 1996; West et al. 2000a). Factors that can push the population sex ratio away from unity over evolutionary time include the probability of mating and the likelihood of inbreeding. Because of the aggregation of parasite individuals among hosts, biased sex ratios are to be expected in parasite populations.

May and Woolhouse (1993) and Morand et al. (1993) have modeled the consequences of aggregation and the type of mating system on the probability of mating in parasites. In monogamous mating systems, such as those of schistosomes, the pairing probability of female parasites will be higher when sex ratios are male biased, especially when the population is highly aggregated. In polygamous mating systems, female-biased sex ratios can increase the probability of mating, particularly when aggregation is not pronounced. The results of these mathematical models suggest that the distribution of parasites can exert a selective pressure on the sex ratio of the parasite population.

In accordance with the models, sex ratios are almost invariably male biased in the monogamous schistosomes (Mitchell et al. 1990; Morand et al. 1993). Schistosome sex ratios are also under some environmental influences (see Moné 1997), but the male bias persists under most conditions. This appears to be a true adaptive phenomenon and not merely a consequence of differential mortality between the sexes, since the sex ratio of cercariae emerging from snails is also male biased (Mitchell et al. 1990). Schistosome sex ratios in the definitive host show density dependence, and tend toward unity as infrapopulation size increases (Morand et al. 1993). Thus the aggregation and population structure of the parasite can also have a proximate effect on sex ratios and interact with genetically programmed patterns of male and female offspring production.

In the polygamous nematodes (Roche and Patrzek 1966; Guyatt and Bundy 1993) and acanthocephalans (Crompton 1985), female-biased sex ratios have been documented repeatedly over the years and are much more frequently observed than male-biased ratios (fig. 6.12). In these worms, the sex ratio of juveniles is closer to unity, and the biases observed in adult populations are apparently caused by mechanisms acting between hatching and adulthood. The key mechanism appears to be differential mortality between the sexes, with females surviving longer than males following infection (Poulin 1997d). Seasonal fluctuations in sex ratio in natural populations and temporal changes in sex ratio in experimental infections both provide evidence for sex-dependent mortality rates. However, sex ratios of nematodes are also density dependent (Tingley and Anderson 1986; Stien et al. 1996, 2005), and aggregation of the population could influence sex ratios and lead to adaptive adjustments in the proportions of males and females in parasite populations. A comparative analysis provided little evidence supporting a role for aggregation or population structure in the evolution of parasite sex ratios in nature, although a similar analysis using data from experimental infections indicated that nematode species characterized by high intensities of infection have sex ratios that are less female biased (Poulin 1997d).

Apicomplexans (protozoans), such as the malaria blood parasites of vertebrates, also display female-biased sex ratios of their sexual stages, or gametocytes (Schall 1989; Read et al. 1995; Shutler et al. 1995; West et al. 2001b). The female bias is moderate in some genera like *Plasmodium*, and very pronounced in others such as *Toxoplasma* (West et al. 2000a, b). These parasites differ from helminths in that they multiply directly within their host; in this situation, prevalence of infection becomes a better indicator of the probability of multiple infections (i.e., the frequency of infrapopulations consisting of mixed parasite genotypes) than aggregation. Usually, as either prevalence or mean infrapopulation size

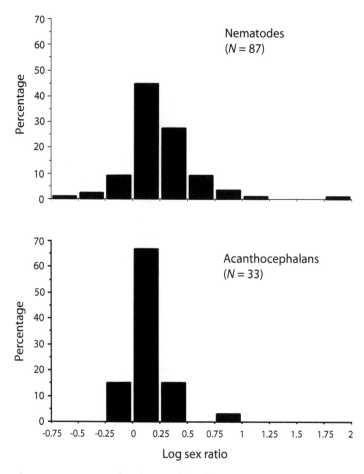

Figure 6.12 Frequency distribution of female-to-male sex ratios among adult populations of parasitic nematodes and acanthocephalans. Values smaller than zero indicate male-biased ratios, and values greater than zero (i.e., the majority of values) indicate female-biased ratios. (From Poulin 1997d)

increases, the frequency of mixed infections increases also. Female-biased sex ratios are favored when the likelihood of inbreeding is high, since in this situation selection favors the production of the lowest proportion of male gametocytes required to fertilize the female gametocytes (West et al. 2000a, b, 2001b; Nee et al. 2002). Because high prevalence indicates that mixed genotype infections are likely and that the probability of inbreeding is reduced, the degree of female bias should vary with prevalence or mean infrapopulation size (Read et al. 1992). Gametocyte sex ratios may be genetically fixed around a locally adaptive value, since in at least one *Plasmodium* species they show no phenotypic plasticity in response to environmental factors, even factors that affect the probability of inbreeding (Osgood et al. 2003; Osgood and Schall 2004). Tests of the relationship between prevalence of infection and sex ratio are therefore best carried out by comparing populations or species. Comparative analyses across populations of two genera of blood

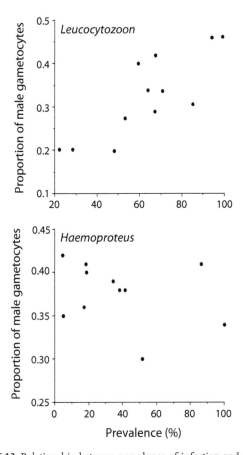

Figure 6.13 Relationship between prevalence of infection and the proportion of male gametocytes in the blood of infected hosts in two genera of protozoans (Apicomplexa) parasitic in birds. The proportion of males tends toward 0.5 as prevalence increases among twelve populations/species of the genus *Leucocytozoon*, but not among eleven populations/species of the genus *Haemoproteus*. (Modified from Read et al. 1995; Shutler et al. 1995)

parasitic protozoans, however, have yielded contrasting results, with only one study supporting the prediction (fig. 6.13). Several phenomena can be invoked to explain this disparity (see Shutler and Read 1998), without weakening the overall link between population structure, inbreeding, and sex ratio in apicomplexans.

Clearly, parasite sex ratios are influenced by several factors, some acting before infection and others acting after. There is mounting evidence that one of the consequences of aggregation of parasites at the adult stage could be a shift in the population sex ratio away from unity; however, other forces acting in the opposite direction can make the role of aggregation difficult to distinguish. Comparative analyses in other taxa of parasites, such as arthropods in which mating systems and sex ratios vary within (Rózsa 1997) and across species (Anstensrud 1990; Heuch and Schram 1996; Rózsa et al. 1996), may provide further insight into this problem.

6.4.3 Macroevolutionary Phenomena

The association between spatial distribution and the probability of extinction is well recognized in ecology and biogeography (Hanski 1982; Hengeveld 1992). In free-living invertebrates, the probability of regional extinction decreases as the number of localities occupied within a region increases (Hanski 1982). For parasite populations, the probability of local extinction may well decrease with the relative number of hosts occupied, or the prevalence of infection.

This can best be seen by considering the hypothetical extreme situations. When aggregation of parasites is at its maximum and prevalence at its minimum, that is, when all parasites are in a single host, the parasite population is in jeopardy. Because the parasite-induced host mortality is usually dependent on the number of parasites per host (Anderson and Gordon 1982), the infected host is likely to die and take the entire parasite population with it. At the other extreme, when prevalence is maximum and there is no aggregation, when the parasites are evenly distributed among all available hosts, the only way the parasite population can go extinct is if the entire host population goes extinct. Actual parasite populations show intermediate levels of aggregation (Shaw and Dobson 1995), resulting in a few large infrapopulations having a high risk of extinction and several small infrapopulations having a much lower risk of extinction (Price 1980). The entire parasite population only becomes at risk when aggregation levels are high for long periods of time.

A high probability of extinction for highly aggregated parasites in founder host populations could explain some of the incongruence between the phylogeny of parasites and that of their hosts caused by missing parasite branches (Paterson and Gray 1997; Paterson et al. 1999; see chapter 3). Founder host populations often have fewer parasite species than their population of origin. For instance, the possum *Trichosurus vulpecula*, introduced to New Zealand from Australia, has a poorer parasite fauna in New Zealand than in its country of origin (Stankiewicz et al. 1996, 1997a). A total of only two hundred to three hundred individuals, from several sites in Australia, were released in many localities throughout New Zealand. Since some of the parasite species missing in New Zealand have direct life cycles and since suitable intermediate hosts are available for others, founder effects resulting from parasite aggregation may explain the discrepancy between the New Zealand and Australian parasite faunas. As a further illustration of the founder effect on parasite species richness, possums introduced from the New Zealand mainland to small offshore islands harbor even fewer parasite species (Stankiewicz et al. 1997b).

There is no empirical evidence supporting a role for aggregation in local parasite extinction, and it may prove difficult to obtain any. It may be possible to contrast the regional occurrence of parasite taxa with different aggregation levels. For instance, we might predict that highly aggregated parasite species will be present in fewer host populations, such as populations of a fish species in different lakes, than weakly aggregated parasites with similar life cycles. In any event, with or without these data, the effect of parasite distribution among hosts on the risk of extinction is probably very strong.

6.5 Conclusion

The host population represents a collection of resource patches among which parasites are unevenly distributed. The aggregation of parasites among hosts produces many small infrapopulations and a few large ones, so that the number of conspecifics sharing the same host is extremely variable. The selective pressures exerted on parasites by the very nature of their distribution may have contributed, among other features, to the plastic growth and developmental schedules and biased sex ratios exhibited by many parasites.

It is often difficult to distinguish between the causes and consequences of aggregation among the various genetic, individual, and population processes related with parasite distribution. Nevertheless, almost all parasites are aggregated to some degree among their hosts, to the point that Crofton (1971) proposed that this typical population distribution should become part of the definition of what is a parasite. One is left wondering whether the aggregation level of a parasite population is the product of selection or simply a nonadaptive consequence of various population processes. Shaw and Dobson (1995) believe that the intermediate levels of aggregation shown by most parasite populations are the outcome of selection. Reduced mating opportunities when aggregation is low, and increased competition and extinction when aggregation is high, would have favored the aggregation levels observed today. Adaptive or not, the clumped dispersion of parasites has permeated all aspects of parasite ecology and evolution. Nonetheless, its influence on the biology of parasite individuals and populations has not been considered in many recent studies. The variance in infrapopulation sizes appears a much more important characteristic than their average size. The failure of many studies to address this will need to be corrected if parasite evolution is to be placed in the proper context of highly variable infrapopulation sizes.

7 | Parasite Population Dynamics and Genetics

Parasite populations vary in size over both short and long time scales and are affected by both biotic and abiotic factors. Some of these factors produce changes in parasite numbers, whereas others act to reduce the amplitude of fluctuations around an equilibrium population size. The study of parasite population dynamics has been tightly linked with epidemiology, which is concerned with the spread of disease in host populations (Anderson and May 1991; Anderson 1993). The theoretical framework developed in epidemiology is proving useful for the interpretation of field and experimental observations and has played a key role in motivating parasitologists to obtain accurate measurements of population parameters important for the study of parasite transmission (McCallum 2000).

Parasite populations are invariably fragmented into as many subgroups as there are infected individuals in a host population. For practical reasons, it is easier to consider only parasites at one stage in their life, such as adult parasites only, when defining the population. Thus, a parasite population consists of all adult parasites in all individual hosts of a host population (or, in the case of nonspecific parasites, all individual hosts belonging to a few sympatric species of suitable hosts); it is subdivided into numerous infrapopulations of unequal sizes, each inhabiting a different host individual. Infrapopulations are ephemeral groups, lasting no longer than the host's lifespan; new members are recruited via infection processes, and they normally remain in the infrapopulation until they die. Offspring issued from different infrapopulations have the opportunity to mix outside hosts and reassemble in new combinations to form new infrapopulations in new individual hosts. The fragmentation is thus temporary, and changes continually from generation to generation. The processes that regulate the total number of parasites in a population operate at both the whole population level and at the infrapopulation level. On a larger scale, parasite populations of the same species occurring in different host populations are never fully isolated. Exchanges of individuals among these populations, for instance exchanges mediated by host migration, ensure that different parasite populations form an interconnected network, or a metapopulation (*sensu* Hanski and Simberloff 1997).

This chapter will begin by reviewing some mathematical models of parasite population dynamics, and will progress toward a discussion of selected case studies from the field. The emphasis will be on identifying the key parameters influencing parasite numbers over time, and on how these may have been shaped by natural selection. This will be followed by an examination of large-scale patterns in parasite abundance, in an attempt to uncover some general determinants of local parasite abundance. The chapter then turns to the genetic structure of parasite populations and metapopulations, focusing on how life-cycle characteristics and transmission routes determine the connectedness of different populations.

7.1 Models of Parasite Population Dynamics

Almost three decades ago, Anderson and May produced the first models of host and parasite population dynamics (Anderson 1978; Anderson and May 1978, 1979; May and Anderson 1978, 1979). These have served as a foundation for subsequent models, which are all essentially modifications of the original Anderson and May models adapted for more specific circumstances. These models have proven immensely useful for an understanding of the processes driving parasite population dynamics by allowing these processes to be defined, separated from each other, and quantified (McCallum 2000). In the models, the temporal variation in host or parasite numbers is determined by some key instantaneous rates (table 7.1). Thus, the size of the parasite population will change as a function of the overall rate at which individuals are lost and the overall rate at which new individuals are recruited. The way in which the models can be used is best illustrated by applying them to specific life cycles.

As a first example, let's examine the dynamics of an aggregated population of a helminth parasite with a direct life cycle. Adult parasites inside or on a host produce infective stages that leave the host to infect other hosts. Many nematodes and monogeneans are examples of this type of cycle. Some simplifying assumptions can be made at first, such as supposing that the parasite-induced host death rate is linearly proportional to the number of parasites per host, and that parasite mortality and fecundity within hosts are not density dependent. If we further assume that transmission is virtually instantaneous without development in the external environment, we can express changes in the parasite population as a single differential equation (Anderson and May 1978; May and Anderson 1979):

$$\frac{\Delta P}{\Delta t} = \frac{\lambda H P}{\left(\dfrac{\mu}{\beta} + H\right)} - P(\gamma + \alpha + b) - \frac{\alpha(k+1)P^2}{kH}$$

where changes in adult parasite numbers are determined by the numbers of new parasites joining the population minus the numbers of parasites dying of various causes. All symbols are defined in table 7.1. The flow of individuals in the model can be visualized in figure 7.1.

Table 7.1 Key parameters used in models of parasite population dynamics.

Parameter	Description
P	Size of the adult parasite population in the definitive host, or in the only host if life cycle is direct
P'	Size of the juvenile parasite population in the intermediate host
W	Size of the parasite egg or larval population in the external environment
H	Size of the definitive host population
N	Size of the intermediate host population
γ	Instantaneous death rate of adult parasites within the host, due to natural or immunological causes (/adult parasite/t)
b	Instantaneous death rate of the definitive host, where mortalities are due to causes other than parasites (/definitive host/t)
α	Instantaneous death rate of the definitive host, where mortalities are induced by parasites (/definitive host/t)
λ	Instantaneous rate of production of infective stages, e.g., eggs, cysts (/adult parasite/t)
ϕ	Instantaneous rate of predation of definitive hosts on intermediate hosts (/(def. hosts) (int. hosts)/t)
μ	Instantaneous death rate of parasite eggs or larvae (/egg/t)
n	Instantaneous rate of ingestion of parasite eggs by intermediate hosts (/intermediate host/t)
ν	Instantaneous death rate of juvenile parasites within the host, due to natural or immunological causes (/juvenile parasite/t)
d	Instantaneous death rate of the intermediate host, where mortalities are due to causes other than parasites (/intermediate host/t)
β	Transmission rate of eggs or larvae (/egg/host/t)
k	Aggregation parameter of the negative binomial distribution (see chapter 6).

Parasite population size is influenced by host population size, and vice versa. The reciprocal feedback between the two populations means that the host population needs to be tracked simultaneously, with its own differential equation, and linked with the parasite population in a complete model. In addition to those already mentioned, the above equation rests on other assumptions and is a somewhat naive representation of natural parasite dynamics. For instance, infection rates are unlikely to be constant and may depend on host age, the parasite may be a generalist exploiting several sympatric host species, parasites may reduce host fecundity as well as survival, and host death may not lead to parasite death. Still, this single equation captures some essential features of parasite population dynamics and can readily be adjusted to account for slightly different scenarios. For example, adding an equation coupled to that for adult parasites can easily take into account free-living larvae, if they are long-lived or develop outside the host (e.g., Hudson and Dobson 1997).

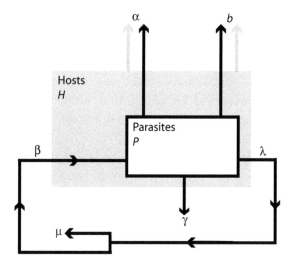

Figure 7.1 Flowchart summarizing the gains and losses of individuals in a parasite population with a direct life cycle. The parameters are defined in table 7.1.

Changes in the size of the parasite population can also be examined mathematically using the expression for the basic transmission rate of the parasite, or R_0 (Anderson and May 1991):

$$R_0 = \frac{\lambda H}{\left(\dfrac{\mu}{\beta} + H\right)(\gamma + \alpha + b)}$$

Basically, R_0 is the average number of offspring produced in the lifetime of a parasite that will themselves live to reproduce in the absence of density-dependent constraints. In chapter 5, a slightly different formula for R_0 was used as a measure of the parasite's lifetime reproductive success and was linked with the evolution of virulence. It is more commonly used to calculate the threshold host population size below which the parasite population cannot survive (i.e., host population size below which $R_0 < 1$) or the present dynamic state of the parasite population (stable if $R_0 = 1$, increasing if $R_0 > 1$). The threshold host population size can also be computed directly from the above equation (McCallum 2000).

The above equations succinctly express the dynamical nature of the parasite population. They are based on simplifying assumptions that can be replaced with more realistic ones. By doing this and by analyzing the influence of each parameter in the model, it is possible to determine the relative contribution of each variable to parasite population dynamics. For instance, in situations where the rate of parasite-induced host death increases, say, exponentially rather than linearly with the number of parasites per host, the stability of both host and parasite populations is likely to be enhanced. Similarly, density-dependent effects on parasite survival or fecundity within infrapopulations will also contribute to the overall stability of both host and parasite populations (Anderson and May

1978). The next section will review the evidence for density dependence in parasite infrapopulations.

Models like the one above are tools that can be put to several uses (see McCallum 2000). The basic model can be adapted for slightly different life cycles and can be used to examine specific systems. For instance, models can facilitate the interpretation of observed population changes (Scott and Anderson 1984; Keymer and Hiorns 1986; Morand 1993) or help to identify key population parameters around which control strategies can be designed (Smith and Grenfell 1985).

The example of a parasite with a direct life cycle and instantaneous transmission without any time spent outside the host represents the simplest possible scenario, one in which the entire parasite population is always found in a single physical habitat, the host population. When complex life cycles are involved, the parasite population becomes compartmentalized into different hosts, with the possibility of one or more free-living stages as well. In a two-host life cycle where transmission from the intermediate host to the definitive host is by predation and where eggs released by adult parasites must be ingested by the intermediate host, the parasite population consists of three types of individuals (eggs, juveniles, and adults) in three distinct habitats (the external environment, the intermediate host, and the definitive host). Changes in these three parts of the population can also be modeled (from Dobson and Keymer 1985; Dobson 1988):

$$\frac{\Delta P}{\Delta t} = \phi HN\left[\frac{P'}{N} + \left(\frac{P'}{N}\right)^2\right] - P(\gamma + b) - \alpha H\left[\frac{P}{H} + \frac{\left(\frac{P}{H}\right)^2 (k+1)}{k}\right]$$

$$\frac{\Delta W}{\Delta t} = \lambda P - \mu W - nWN$$

$$\frac{\Delta P'}{\Delta t} = nWN - P'(v + d) - \phi HN\left[\frac{P'}{N} + \left(\frac{P'}{N}\right)^2\right]$$

As before, symbols are defined in table 7.1 and the flow of parasites among hosts is illustrated in figure 7.2. Changes in the population of adult parasites are determined by how many new parasites are acquired through predation by the definitive host on infected intermediate hosts minus the numbers of adult parasites dying either of natural causes or because of their effects on hosts. Changes in the egg population are defined by the rate at which eggs are produced by adult parasites and the rates at which they are either lost or ingested by intermediate hosts. Finally, changes in the population of juvenile parasites are determined by the number of new ones recruited through ingestion of eggs by intermediate hosts minus the numbers lost either because of natural mortality or transmission to the definitive host. This life cycle is typical of most acanthocephalans and many cestodes and nematodes.

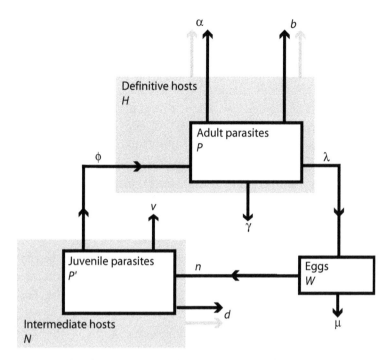

Figure 7.2 Flowchart summarizing the gains and losses of individual adults, juveniles, and eggs in a parasite population with a two-host life cycle. The parameters are defined in table 7.1.

Again, analysis of these equations suggests that density-dependent constraints on adult survival and reproduction can regulate the whole parasite population and lead to long-term stability. The same modeling approach, or a slightly modified one, has also been used to investigate the consequences of either parasite manipulation of the intermediate host's susceptibility to predation by the definitive host (Dobson and Keymer 1985; Dobson 1988) or the addition of a new intermediate host to the cycle (Morand et al. 1995).

In the models, the key processes include the rate of production of eggs or infective stages (λ), their transmission rates to new hosts (β and n), parasite mortality rates (γ, μ and v), and parasite-induced host mortality, or virulence (α). For the models to be validated or to be used in a predictive context, these parameters need to be quantified in natural systems. Using field data, there are many ways to estimate these parameters (McCallum 2000). For instance, transmission rates can be estimated from the relationship between infection levels and host age, provided that some basic assumptions are satisfied (McCallum 2000). Laboratory experiments are also useful to evaluate the various components of parasite mortality. Not surprisingly, the few host-parasite associations for which all key model parameters have been quantified include mainly systems that have been intensively studied over many years, such as the nematode *Trichostrongylus tenuis* in red grouse (Hudson and Dobson 1997) or the digenean *Schistosoma mansoni* in its snail and rat hosts (Morand et al. 1999a). The careful empirical quantification of population parameters can also lead to improvements in model construction. For example, a reassessment of laboratory data indicates that the per capita rate of transmission of schistosome miracidia to their snail intermediate host is not constant: it decreases with increasing densities of both miracidia and snail hosts (Carter et al. 1982; Fenton et al. 2002). This finding can then serve to construct more realistic models.

The above models apparently assign little importance to abiotic factors, which can cause abrupt changes in population size. There is no doubt that abiotic phenomena cause fluctuations in parasite populations (see Esch and Fernández 1993). Physicochemical variables will determine the survivorship of eggs or other infective stages in the external environment. The growth and fecundity of adult parasites in ectothermic hosts can also be influenced by the external temperature. These density-independent processes usually do not, however, account for the long-term stability of parasite populations. Some would argue that parasite populations are generally unstable and not subject to any intrinsic regulation (Price 1980). The basic models of parasite population dynamics presented above do not include density-dependent regulation either; however, it can easily be added to the basic models. When this is done, the models predict that parasite populations can be regulated to a large extent by density dependence. There is enough evidence in favor of population regulation and density dependence (see Keymer 1982; and the following section) to conclude that the models capture real biological phenomena.

7.2 Density-Dependent Regulation

Mathematical models focus on the dynamics of the whole parasite population (all parasite individuals in all hosts). In practice, though, any density-dependent regulating process leading to changes in population abundance will result from mechanisms acting within infrapopulations (all individuals within one host), which form interacting groups within the otherwise fragmented population. Thus overall population regulation is achieved by the sum of events taking place within the various infrapopulations.

The importance of aggregation becomes clear in this context. In a highly aggregated parasite population, there will be a few large infrapopulations (containing a large proportion if not most parasites in the population) in which density-dependent processes will act strongly (see Churcher et al. 2005). From the perspective of the host, regulation of parasite numbers is only achieved through mechanisms acting in a fraction of the infected hosts. In another parasite population of the same size but distributed more evenly, there may be no infrapopulations of a size greater than the threshold size above which regulating processes become important. This threshold size is not meant as some fixed infrapopulation size; obviously, what is considered crowded in one host may be considered sparse in another. For instance, the threshold beyond which mean worm fecundity begins to decrease with increasing infrapopulation size is three thousand worms per host for the nematode *Trichostrongylus colubriformis* in sheep, but only about one hundred worms per host for the nematode *Heterakis gallinarum* in pheasants (Dobson et al. 1990; Tompkins and Hudson 1999). Various characteristics of the host, such as its size, as well as the presence of other parasite species, will influence the infrapopulation size at which density-dependent processes become important. Still, parasite aggregation remains a key force determining whether there will be enough dense infrapopulations for population regulation to occur. Since aggregation of parasites is the rule (see chapter 6), we may expect density-dependent regulation to be common in parasite populations.

Infrapopulations are short-lived, their maximum longevity being equal to the life span of the host. During their existence, infrapopulations undergo a constant turnover of individuals. New individuals are gained by recruitment of infective stages and old ones are lost through mortality. In some parasites, new individuals can be gained through the direct reproduction of parasites inside the host, and in unusual species of parasites, individuals can even be lost or gained through emigration and immigration; these possibilities are not considered any further here. Thus only density-dependent effects on survival, growth, and fecundity of parasites (e.g., parameters γ, λ, and ν in the models) in infrapopulations will regulate the overall parasite population abundance. In infrapopulations of helminths in their only or definitive host, there are two common ways in which density dependence is expressed. First, survival and establishment success of parasites can decrease with increasing density such that there is an upper limit on the number of individuals per infrapopulation. Second, the growth and body size of parasites, and thus their fecundity, can be reduced at higher densities. Increases in mortality at high parasite density will regulate the size of given infrapopulations, whereas decreases in growth and fecundity at high parasite density will contribute to regulation by influencing the availability of infective stages for all infrapopulations. Other density-dependent effects are also possible. For example, in the nematode *Anguillicola crassus* parasitic in the swimbladder of eels, it is worm development that shows density dependence. There appears to be an upper limit of about two or three adult female worms per eel; new worms arriving in the host do not die, however, but instead enter a stage of arrested development and accumulate in the swimbladder wall, where they wait for a chance to replace a dead adult (Ashworth and Kennedy 1999). Members of one sex may also be more susceptible to density dependence than those of the other sex: in the nematode *Protospirura muricola* parasitic in mice, the mass of individual male worms declines with increasing numbers of worms per host, whereas that of females does not change (Lowrie et al. 2004). Density dependence can thus affect several aspects of parasite biology. The relationship between density, or the number of parasites per infrapopulation, and the magnitude of changes in survival, growth, development, or fecundity may be roughly linear. But it may also assume other shapes. If for instance a threshold density must be reached before mortality increases or fecundity decreases, then the relationship may approximate an exponential function.

The mechanisms responsible for density-dependent regulation of parasite numbers are numerous, but rarely identified in specific studies. Here, I consider four possibilities: (1) exploitation competition, (2) interference competition, in which individuals secrete chemicals that harm their competitors, (3) host-mediated restriction, and (4) parasite-induced host mortality. These may operate simultaneously, which explains why it is proving difficult to elucidate precisely which mechanism is operating in a given host-parasite system.

The presence of one parasite in a host and its exploitation of the host mean that there will be fewer resources available for other parasites of the same species. However, as long

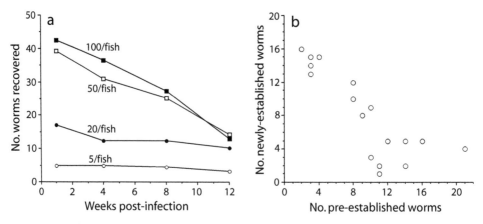

Figure 7.3 Evidence for density-dependent establishment and survival of the acanthocephalan *Pomphorhynchus laevis* in experimentally infected rainbow trout, *Oncorhynchus mykiss*. (a) Recovery of worms over time in four treatments of fish hosts given different initial numbers of infective cystacanths. Each point is the mean of four to fourteen fish. (b) Establishment of new worms in fish harboring various numbers of pre-established, twelve-week-old worms. All fish were given twenty cystacanths in the second infection; each point represents a different fish. (Modified from Brown 1986)

as individual parasites obtain sufficient resources, the effect of exploitation competition will be negligible. Only above a certain density, or a certain infrapopulation size, will the depletion of resources by conspecifics affect any given parasite. For instance, Burn (1980) reported finding consistently two and only two adults of the same digenean species in the gallbladder of their fish host, a situation resulting from tight space constraints preventing the establishment of further worms. Similarly, attachment space in the gut of the definitive host becomes a limiting resource for some acanthocephalans beyond a certain infrapopulation size (Uznanski and Nickol 1982; Brown 1986). In some species, the initial establishment and subsequent survival of worms show a strong dependence on infrapopulation size (fig. 7.3).

In other systems, both space and nutrients may be in short supply in dense infrapopulations. Female oxyurid nematodes living in the gut of insects show reductions in per capita egg production as soon as there is more than one female per infrapopulation (fig. 7.4). The large size of these worms relative to the body size of the host means that there are not enough resources to allow females to achieve their full potential fecundity when two or more females co-occur in the same host. These nematodes also provide good examples of the difficulty involved in distinguishing between exploitation competition and interference competition. The mechanism behind the density-dependent fecundity of females (fig. 7.4) is likely to be competition for resources, but could also result from interference competition (Adamson and Noble 1993). There also appears to be interference competition between the sexes (Zervos 1988a, b; Morand and Rivault 1992). Females may secrete a chemical that regulates the number of males per infrapopulation: the first male joining the infrapopulation survives but further males are apparently not allowed to establish (Zervos 1988a, b).

Similarly, the "crowding" effect commonly reported in dense cestode infrapopulations (Read 1951; see chapter 6) may be the product of both competition for nutrients or other

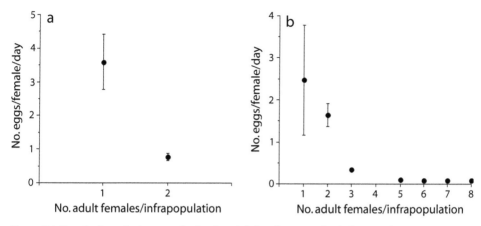

Figure 7.4 Density-dependent egg production by adult female nematodes in insects. Egg output (mean ± standard deviation) of (a) *Blatticola monandros* in the cockroach *Parellipsidion pachycercum* and (b) *Protrellus dixoni* in the cockroach *Drymaplaneta variegata*, as a function of infrapopulation size. There is typically only one male per infrapopulation in both systems. (Data from Zervos 1988a, b)

resources and competition through chemical interference (Roberts 2000). The typical outcome of crowding in cestodes is a reduction in the mean body sizes of individual worms compared to those achieved in smaller infrapopulations. Greater variance in body sizes also characterizes crowded infrapopulations, with many small worms dwarfed by a few very large ones. How can exploitation and interference competition operate at the same time among cestodes? In the cestode *Hymenolepis diminuta*, for example, experimental studies have confirmed that worms in dense infections apparently compete for carbohydrates or other specific nutrients (Keymer et al. 1983; Roberts 2000). In parallel, other studies have identified compounds released by *H. diminuta* that inhibit DNA synthesis and retard growth in other worms of the same species (Zavras and Roberts 1985; Cook and Roberts 1991; Roberts 2000). The same combination of crowding mechanisms may also operate in other cestode species (e.g., Heins et al. 2002). Whatever the nature of the mechanisms, the population consequences are the same.

Host-mediated restriction of parasite numbers in infrapopulations is another potential density-dependent mechanism. For instance, when rodents are first exposed to a dose of schistosome cercariae and then re-exposed to an identical dose a few weeks later, it appears that many more worms are successful at establishing in the first exposure than in the second. Unlike the acanthocephalan example presented earlier, this is not a case of competition for a limiting resource. Instead, the first infection triggers the production of host antibodies that do not affect the recently established worms but that will attack any incoming cercariae (Smithers and Terry 1969; Terry 1994). Similarly, digenean cercariae can face density-dependent host responses when trying to encyst in their second intermediate host. Cercariae of *Cryptocotyle lingua*, for instance, are capable of encysting in cod until the fish host begins to mount an immune response against new cercariae when infrapopulations reach some threshold size (Lysne et al. 1997). Host immune responses play central roles in density-dependent regulation of other kinds of parasites as well. In the nematode *Strongyloides ratti* parasitic in rats, density-dependent effects are mediated entirely by host immunity. Density-dependent effects appear only late in a primary infection, and they are greater in hosts that have been previously exposed to the nematode and absent in immunodeficient hosts (Paterson and Viney 2002). In the monogenean *Protopolystoma xenopodis*, as well as in other polystomatids parasitic in the urinary bladder of frogs or toads, although the proportion of larvae initially establishing in a host may be independent of the number to which the host was exposed, the number surviving to adulthood is strongly regulated and invariably low, typically only one worm per host (Tinsley 1995; Jackson and Tinsley 2003). In this system, too, the lower success of worms in hosts that had been previously exposed to the parasite suggests that host responses play a central role in regulating adult infrapopulation size (Jackson and Tinsley 2001). In the end, the effect of concomitant immunity or other forms of host restriction of parasite infrapopulation sizes may be indistinguishable from actual competitive interaction between parasite individuals.

Parasite-induced host mortality can also serve to regulate parasite populations in a density-dependent manner but it differs from the other mechanisms in at least two ways. First, instead of causing a partial reduction in the number of individuals in an infrapopulation or in their average egg production, it leads to the death of the entire infrapopulation and its loss from the overall population. Because large infrapopulations are the most likely to disappear (if the probability of host death is a function of parasite load), this process can have major impacts on the parasite population. Second, it removes individual hosts from the host population, and thus affects host population dynamics as well. Despite the numerous laboratory examples of large infrapopulations killing their hosts, there are few field demonstrations that hosts harboring more parasites are more likely to disappear, and most of these provide only indirect evidence (see Gordon and Rau 1982; Adjei et al. 1986; Hudson et al. 1992). It is important to note that in many cases host death is necessary for the transmission of the parasite, and that its effect on the parasite population as a whole may be null: the infrapopulation is not lost; it is transferred from one compartment of the population (e.g., juvenile parasites in the intermediate hosts) to another (adults in the definitive host). Parasite-induced host death only serves as a regulating mechanism when it also kills the parasites in the infrapopulation (e.g., Hudson et al. 1992).

As suggested in the preceding paragraph and despite the nature of the examples used so far, there is no reason to believe that density-dependent regulation does not act on parasites in their intermediate hosts. For instance, many studies have shown that cercarial production in snail hosts does not increase in proportion to the number of miracidia penetrating those snails (Touassem and Théron 1989; Gérard et al. 1993). This means that either the establishment success of miracidia and/or their subsequent asexual multiplication are regulated in a density-dependent fashion. The much-studied cestode *Hymenolepis diminuta* provides another example. Regulation of the adult parasite in its definitive rodent host is well documented, and consists of density-dependent effects on worm establishment, survival, growth, and fecundity (Hesselberg and Andreassen 1975; Keymer et al. 1983). In the beetle intermediate host, however, things are different (fig. 7.5). Establishment success is not dependent on density, even at infrapopulation sizes that normally

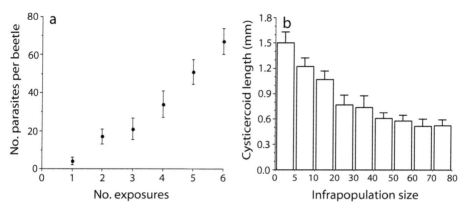

Figure 7.5 Density-dependent effects on establishment and growth of cysticercoids of the cestode *Hymenolepis diminuta* in experimental infections of its intermediate host, the beetle *Tribolium confusum*. (a) Number of established parasites (mean ± 95 percent confidence limits) after beetle hosts had been put through different numbers of exposures to 1,500 parasite eggs; the almost linear relationship suggests that establishment is not dependent on density. (b) Length (mean ± 95 percent confidence limits) of cysticercoids as a function of infrapopulation size, clearly showing that growth is density dependent. (Modified from Keymer 1981)

result in the death of the intermediate host (Keymer 1981). Growth of the larval stages in the intermediate host, on the other hand, is clearly density dependent. Since their body size may determine how likely they are to become established in the definitive host and their future reproductive success, this may be an instance of density-dependent population regulation operating among larval stages in the intermediate hosts. Similar density-dependent effects in the intermediate host have been reported in other cestode species (e.g., Rosen and Dick 1983; Nie and Kennedy 1993), as well as in acanthocephalans (Dezfuli et al. 2001) and in digeneans (Sandland and Goater 2000; Brown et al. 2003; Fredensborg and Poulin 2005). In digenean metacercariae within small crustacean intermediate hosts, density-dependent effects on growth have been shown to lead to reductions in egg output later in life (Brown et al. 2003; Fredensborg and Poulin 2005). Density dependence is thus a force acting on larval helminths in intermediate hosts. In fact, density-dependent regulation could also affect the free-living infective stages in the external environment, for instance if they are subject to predation by nonhosts.

Many of the examples of density dependence currently available come from laboratory studies. It is possible that in nature, infrapopulation sizes are rarely large enough for intraspecific competition to occur, and that threshold densities are only achieved in experimental infrapopulations. This would be consistent with the nonequilibrium view of parasite populations espoused by Price (1980). The largest infrapopulations observed in nature could be either (1) what is left of larger infrapopulations following the action of density-dependent regulation, or (2) the unusual occurrence of many conspecific parasites in the same host, which nonetheless remains below the threshold density for the onset of density dependence. The presence of interference competition mechanisms in many species, however, indicates that intraspecific competition has played a role in the evolution of some parasites, and that it occurs in nature as well as in the laboratory.

Density dependence does not occur in all parasite populations, however (see Marcogliese 1997, and further examples in Combes 1995, 2001). Even related parasite species in the same host can display contrasting patterns. In reindeer, the length and egg output of female worms of the trichostrongylid nematode *Ostertagia gruehneri* are subject to density dependence, whereas these parameters are independent of density in the sympatric trichostrongylid *Marshallagia marshalli*, which is roughly of equal body size and abundance (Irvine et al. 2001). Still, the few examples presented above, and others (see reviews in Keymer 1982; Quinnell et al. 1990; Shostak and Scott 1993), suggest that density dependence is common among host-parasite systems. There have been suggestions, however, that many apparent cases of density dependence may in fact be statistical artifacts (Keymer and Slater 1987; Shostak and Scott 1993). For instance, the variability in the per capita egg production of parasites in small infrapopulations is often higher than that of parasites in larger infrapopulations (e.g., fig. 7.4). A depression in individual egg output in large infrapopulations could therefore result from the chance occurrence of many parasites with low inherent reproductive capacity. A reexamination of published results, using a simulation technique that takes into account the variability among individual parasites (Shostak and Scott 1993), generally supports the existence of density-dependent mechanisms acting in infrapopulations. Therefore, this form of regulation is probably operating in many parasite populations and contributing to their long-term stability.

7.3 Selected Examples of Population Studies

Density-dependent processes acting within infrapopulations may be capable of regulating parasite populations, but demonstrations in natural populations are difficult. The myriad environmental factors also acting on parasite numbers make the presence of regulation difficult to detect, or to discern from the effects of abiotic factors operating independently of density. The following examples illustrate how parasite population abundance results from the complex interactions of both density-dependent and density-independent factors in diverse fish-parasite associations.

7.3.1 *The Cestode* Bothriocephalus acheilognathi

The cestode *Bothriocephalus acheilognathi* has a two-host life cycle involving a copepod intermediate host and a fish definitive host. Infection of fish occurs via predation on infected copepods. The cestode exploits a broad range of fish species, but does not achieve the same reproductive success in all host species. A long-term study of a population of this cestode in a North Carolina lake clearly demonstrates how external factors interact with density-dependent processes to determine population abundance. Typically, prevalence and mean infrapopulation size are lowest in summer months, rise sharply in autumn, peak in winter, and decrease in the spring (Granath and Esch 1983a). These changes are in part controlled by seasonal events in the life cycle of the parasite. Recruitment of new worms by fish takes place only from late spring to the autumn, when copepods are abundant and acquire the infective stages hatched from eggs. Also, during early summer, many adult worms have reached the end of their life and are lost from infrapopulations.

These events are not enough, however, to explain the decline in infrapopulation sizes occurring in spring and early summer. Granath and Esch (1983b) suggested that the rise of temperature in the spring triggers the growth and maturation of cestodes and that this is accompanied by an intensification of exploitation competition for space or nutrients. In winter, almost all parasites are immature, unsegmented worms only a few millimeters long. As they grow rapidly in the spring toward an adult size of a few centimeters, many are eliminated by competition in the densest infrapopulations. Therefore, the population is regulated by density-dependent survival through exploitation competition, which in turn is initiated by a density-independent, seasonal change in ambient temperature.

The study population allows for the role of density-independent, abiotic factors to be examined further. The lake in which the study was performed is used as a cooling reservoir for a power station, and parts of it are thermally altered. In these areas, water temperature fluctuates seasonally but within a higher range, with summer values almost reaching 40°C. Recruitment of new parasites begins sooner in spring and lasts longer into the autumn in thermally altered areas of the lake, but is interrupted for several weeks in summer when water temperature exceeds 35°C because fish predation on copepods declines. As a result, average infrapopulation sizes were smaller in the thermally altered areas than in the unaffected areas of the lake. In addition, effluents from the power plant created a gradient in selenium pollution within the lake. Adult worms in selenium-contaminated hosts produced fewer and less viable eggs than worms in hosts from unpolluted sites (Riggs and Esch 1987; Riggs et al. 1987). These results clearly demonstrate that external factors can override density-dependent processes by preventing infrapopulations from reaching sizes at which competition becomes important.

7.3.2 *The Nematode* Cystidicola cristivomeri

The nematode *Cystidicola cristivomeri* also has a two-host life cycle. It uses a single species of mysid shrimp, *Mysis relicta*, as intermediate host. The life cycle is completed after an infected intermediate host is ingested by a suitable fish host (*Salvelinus alpinus* or *S. namaycush*). There the nematode migrates to the swimbladder of the host where it matures (Black and Lankester 1980). This parasite is long-lived, and may survive more than ten years in the definitive host. For this reason, the dynamics of infrapopulations do not show seasonal fluctuations but must be examined over longer time spans.

Typically, infected shrimps each harbor a single infective larva. Therefore, a single new recruit is added to an infrapopulation in a fish each time the host consumes an infected prey. In the related *C. farionis* in a Norwegian population of *S. alpinus*, the fish host increases its consumption of crustacean prey as it gets older, such that mean infrapopulation sizes increase steadily with fish age (Amundsen et al. 2003). In the Canadian study of *C. cristivomeri*, however, mysid shrimps are only important food items in the diet of small host fish, and tend to become secondary as fish grow and become increasingly piscivorous. As a consequence, recruitment rates are high in young infrapopulations harbored by young fish hosts, but decrease with age. The pattern observed in natural populations shows an increase in mean infrapopulation size with increasing fish age up to a certain age, which varies among populations in different lakes, after which infrapopulation size stabilizes (Black and Lankester 1981). This probably results from a parasite recruitment rate higher than the parasite mortality rate in the first few years of an infrapopulation's existence, followed by a period in which both rates are roughly identical until the death of the host.

Whereas infrapopulations may not often reach sizes at which parasite survivorship becomes negatively affected, there is evidence of density-dependent reproductive output in this nematode. Both the proportion of female nematodes reaching sexual maturity and the average length of females are inversely related to infrapopulation size. Since fecundity is strongly correlated with body length in *Cystidicola* nematodes (Black and Lankester 1981; Black 1985), per capita egg production is therefore reduced in large infrapopulations. Thus both truly density-dependent effects and temporal changes in recruitment rates, which coincide with increases in infrapopulation size but are dependent only on host feeding behavior, contribute to the long-term stability of the nematode population.

7.3.3 *The Nematode* Cystidicoloides tenuissima

The nematode *Cystidicoloides tenuissima* is a common parasite of salmonid fish in the Holarctic. Adults are found in the stomachs of their fish definitive hosts, and juveniles live in the body cavities of aquatic insects, particularly mayflies, whence they reach the definitive host when the latter ingests infected intermediate hosts.

In the River Swincombe, in England, where adult parasites infect brown trout, *Salmo trutta*, and juvenile Atlantic salmon, *S. salar*, maturation of the parasite is strongly linked with water temperature. Gravid female worms are only present, and eggs are only released in the water, from summer to early winter (Aho and Kennedy 1984). The timing of egg release coincides with the appearance in the river of a new cohort of the mayfly *Leptophlebia marginata*, the only species in the River Swincombe in which the nematode can develop to become infective to fish. Insect hosts become infected by eating nematode eggs throughout the summer and autumn; the majority of eggs ingested by insects are not eaten by *L. marginata* (fig. 7.6). Nematode juveniles acquired by mayflies in the summer soon become infective to fish; those acquired later may only reach infectivity the following spring as there is little or no development during the winter months.

Trout consume more mayflies than salmon, and as a consequence they harbor most of the adult parasite population (fig. 7.6). Fish hosts do not feed much on *L. marginata* during the summer; the mayfly becomes more common in fish diet in autumn and throughout winter and spring. Consequently, mean infrapopulation size of adult parasites in fish increases beginning in winter and peaks in spring before declining rapidly in summer. At that time, not only does recruitment stop until the autumn, but also adult parasites start dying. There is no evidence for density-dependent competition acting in this

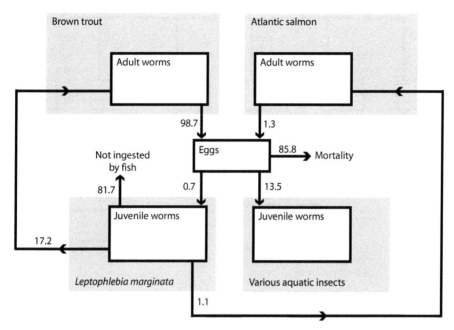

Figure 7.6 Flowchart summarizing the circulation and the transmission dynamics of the nematode *Cystidicoloides tenuissima* in the River Swincombe, England. All numbers are percentages. Most adult worms live in trout, and thus trout parasites release much more eggs than salmon parasites. The majority of eggs die or are ingested by insects in which they cannot develop further. Of the very few that are eaten by the mayfly *Leptophlebia marginata*, the only suitable intermediate host, only 18.3 percent (17.2 percent in trout, 1.1 percent in salmon) ever establish in definitive hosts, with the majority being recruited by adult infrapopulations in trout. Numbers in the figure are from one site in the river but are representative of other sites. (Data from Aho and Kennedy 1987)

system. The onset of high mortality of adults coincides with a peak in infrapopulation sizes, but it appears to be caused by a temperature-dependent host response (Aho and Kennedy 1984). Thus, in this system, infrapopulation sizes are controlled by a density-independent rejection induced by high summer temperatures, and are kept below the threshold, if any, at which density-dependent competition would start operating.

7.3.4 *The Acanthocephalan* Acanthocephalus tumescens

As suggested by the previous example, parasites capable of utilizing many host species at the same stage in their life cycle can have their population partitioned unequally among different hosts. The acanthocephalan *Acanthocephalus tumescens* is found in the intestine of several fish species, both native and introduced, in the freshwater systems of Patagonia, Argentina. For example, *A. tumescens* uses six of the eight fish species found in Lake Moreno as definitive hosts. How is the flow of individuals partitioned among these six host species?

The acanthocephalan is transmitted via predation from its intermediate host, the amphipod *Hyalella patagonica*, to its fish definitive hosts. Postcyclic transmission also

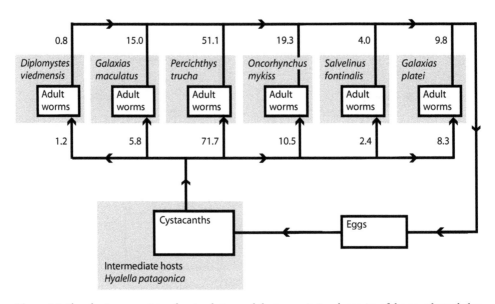

Figure 7.7 Flowchart summarizing the circulation and the transmission dynamics of the acanthocephalan *Acanthocephalus tumescens* in Lake Moreno, Argentina. All numbers are percentages of the total number of individual parasites at a particular life stage. One of the six definitive host species, the fish *Percichthys trucha*, harbors more than 70 percent of the adult worm population, from which come half of the total number of parasite eggs released in the lake. An unknown proportion of those eggs is ingested by the intermediate host, the amphipod *Hyalella patagonica*. (Data from Rauque et al. 2003)

occurs: the larger piscivorous fish species can acquire adult *A. tumescens* by feeding on small fish already infected by the parasite (Rauque et al. 2002). Based on the proportion of fish infected, the mean number of parasites per fish, and the proportion of female parasites that are mature, the fish *Percichthys trucha* ranks as one of the least favorable for the parasite (Rauque et al. 2003). However, it is much more abundant than the other host species in the lake system. When considering the relatively high abundance of *P. trucha*, this host species turns out to be the most important for the parasite population (fig. 7.7). Overall, it harbors more than two-thirds of the adult parasite population, accounting for about half of the total number of parasite eggs produced. In contrast, the mean number of parasites per fish is several times higher in the fish *Diplomystes viedmensis* (Rauque et al. 2003), but the rarity of this host in the fish community of the lake means that it plays only a minor role in the parasite's population dynamics (fig. 7.7). Consequently, the large fraction of adult parasites occurring in *P. trucha* are found in many relatively small infrapopulations rather than a few large ones, which constrains the role of density-dependent regulation. Another interesting feature of this system is the role played by the two introduced salmonid fish, *Oncorhynchus mykiss* and *Salvelinus fontinalis*: together, they account for almost one quarter of *A. tumescens* eggs released in the lake (fig. 7.7). This suggests that the parasite population benefits from the presence of these introduced fish, apparently achieving a greater total egg output than it would otherwise. Clearly, introduced host species can have substantial impacts on the dynamics and total size of local populations of nonspecific parasites.

7.4 Patterns of Parasite Abundance

The size of natural parasite populations is difficult to measure. For starters, one needs to know both the mean number of parasites per host and the number of hosts in a host population; while the former parameter is relatively easy to estimate from a sample of hosts, the latter parameter is often more problematic. A useful way to approach this problem is simply to consider only measures expressed in a standard "per host" manner. Thus, for comparative purposes, it is preferable to use parasite abundance, defined as the mean number of parasites per host, rather than total parasite population size. Abundance includes uninfected hosts, and is thus not a measure of mean infrapopulation size. However, abundance is often decomposed into its constituent parts, prevalence (the percentage of infected hosts in a population) and intensity of infection (the mean number of parasites per infected host); intensity is thus another name for mean infrapopulation size. These standardized measures can be used to answer two basic questions: (1) Are parasite abundance, prevalence, or intensity more or less constant among different populations of the same parasite species, that is, are they parasite species traits? And (2) what host or parasite features are responsible for the huge variability in abundance, prevalence, and intensity observed among parasite species?

Comparative approaches can yield answers to the second question above, for instance by determining the characteristics of parasite species occurring at small infrapopulation sizes compared to those of species occurring at high infrapopulation sizes. However, to perform such comparisons, one must assume that mean infrapopulation size, that is, intensity, or other parasite population descriptors (abundance or prevalence) are true species traits. In other words, the values used for a given parasite species must be typical of that species. There is no doubt that prevalence, abundance, and intensity of infection by most parasite species vary, often considerably, in both time and space. For example, Janovy et al. (1997) monitored the summer prevalence and abundance values of seven parasite species in a fish population over a fourteen-year period. Each species displayed substantial fluctuations in both prevalence and abundance in apparent response to temporal variation in water flow in the river. Nevertheless, Janovy et al. (1997) detected long-term stability superimposed on the wild short-term oscillations in both population descriptors. The prevalence and abundance of all seven parasite species were highly variable over time, but the variation was constrained within species-specific boundaries. Average values do not capture this dynamic pattern, but still provide an indication of the typical values associated with each parasite species. Spatial variation among populations of the same parasite species also appears to be bound by species-specific limits. Arneberg et al. (1997) examined variation in intensity, abundance, and prevalence within and among species of nematodes parasitic in mammals. They found that estimates of intensity from different populations of the same parasite species display significant repeatability, that is, they are generally more similar to each other than to estimates from other nematode species. The same applies to estimates of abundance, but not to prevalence, which varies considerably among different populations of the same species (Arneberg

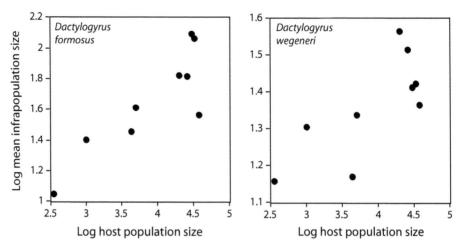

Figure 7.8 Relationship between mean infrapopulation size, i.e., intensity of infection, and host population size, for two monogenean species parasitic on carp, *Carassius carassius*. The data are from nine isolated ponds in Finland, each holding a different number of fish. Prevalence of both species is 100 percent in almost all ponds, so that the relationships remain the same when abundance is used as a descriptor instead of intensity. (Data from Bagge et al. 2004)

et al. 1997). These studies suggest that the mean infrapopulation size is a species character: values for a given parasite species may vary over time as well as in space, but this variation is constrained within species-specific limits.

The next step is to explain interspecific variation in mean infrapopulation size (i.e., intensity), abundance, or prevalence. Why is one parasite species consistently achieving higher infrapopulation sizes than other, similar species? Two kinds of explanations can be used. The first invokes host features that influence either the resources available for parasites, and thus how many parasites can be supported by one host, or the probability of infection. Host body size is an obvious indicator of the resource potential of an individual host, such that we might expect parasite species exploiting larger-bodied hosts to achieve larger infrapopulation sizes than related species using small-bodied hosts. Indeed, this rule seems to apply to many taxa of nematodes. Oxyurid nematodes infect hosts ranging from insects to large mammals, depending on the species, with infrapopulation sizes typically increasing with host body size. The same applies to trichostrongylid nematodes, which exploit mammals ranging from small rodents to large ruminants. The density or size of the host population is another potentially important host feature, affecting the likelihood of host-to-host transmission for parasites with direct life cycles (see section 7.1). For instance, among natural ponds holding different numbers of fish hosts, the mean infrapopulation size of two monogenean species increases with host population size (fig. 7.8). Host population size or density can account for both intra- and interspecific variation in parasite population parameters. Comparative studies of parasitic nematodes of mammals have shown that abundance or prevalence of directly transmitted species in the orders Strongylida and Oxyurida correlates positively with host density, whereas there is no relationship in nematode orders where transmission involves intermediate hosts (Arneberg et al. 1998a; Arneberg 2001). Thus, to a large extent, differences in population parameters between parasite species are a reflection of the different host species they use.

The second way to explain interspecific variation in parasite population descriptors is to look for parasite features that covary with their population parameters. In free-living organisms, species with small body sizes generally attain higher abundance than larger species within the same taxon (Cotgreave 1993; Blackburn and Lawton 1994; Blackburn and Gaston 1997). Although there are exceptions, this is a well-established macroecological pattern (Brown 1995; Gaston and Blackburn 2000). Does it apply to parasites? Arneberg et al. (1998b) found consistent negative relationships between intensity, or mean infrapopulation size, and parasite body size among nematode species parasitizing mammals, independently of several other variables considered in their analyses. The same pattern applies to helminth endoparasites of fish (Poulin 1999b). All else being equal, the smaller the parasites, the more of them can fit inside a given host and be supported by the resources available from the host. These results suggest that parasite populations are subject to the same patterns and processes as populations of free-living animals. And as for free-living animals, there are exceptions to the rule. Among copepods ectoparasitic on fish, there is a positive relationship between parasite body size and mean infrapopulation size (Poulin 1999b). The weaker link between body size and density-dependent population regulation in ectoparasites than in endoparasites may provide an explanation, but this remains speculative. This atypical finding illustrates how little we know about the factors driving species-specific infrapopulation sizes and other population parameters, an area where further comparative studies are needed.

7.5 Genetic Structure of Parasite Populations

Within a parasite population, individuals are grouped into infrapopulations. Although exchanges of individuals among infrapopulations are usually not possible, the offspring of an individual parasite will be recruited by other infrapopulations, thus allowing genetic mixing within the population. On a higher spatial scale, different parasite populations occurring in different host populations are more or less interconnected, forming a metapopulation maintained by exchanges of individuals among populations. Although the existence of parasite metapopulations is beyond doubt, there have been few theoretical or empirical investigations of the effect of metapopulation structure on local population abundance and dynamics. The few studies available suggest that migration rate among populations is a key factor determining the regional spread and persistence of a parasite (see Hess 1996; Lopez et al. 2005). Migration among parasite populations also contributes to gene flow, and thus determines to a large extent the genetic structure of parasite populations and metapopulations. The recent proliferation of genetic studies of parasites forms the basis of this section. These studies evaluate the genetic structure of parasites on two spatial scales: (1) among infrapopulations within given parasite populations, and (2) among the parasite populations forming a metapopulation. Using mainly mitochondrial or microsatellite DNA markers, these studies indirectly shed some light on many ecological aspects of parasite population biology.

Genetic structure refers to the statistical pattern of associations between alleles (Hartl and Clark 1997). It can be measured as the distribution of genetic variation among individual parasites over different spatial scales, which is to say, across infrapopulations or populations. A range of structures can be observed, reflecting the extent and freedom of genetic exchanges. At one extreme lies panmixia, a situation in which an individual is equally likely to mate with any other individual in the infrapopulation, population, or metapopulation, depending on the level of study. At the other extreme, we have scenarios in which infrapopulations or populations are completely isolated, resulting in genetically distinct groups of individuals. Most studies have used Wright's F-statistics (e.g., F_{ST}) to assess the nonrandom distribution of parasite genotypes among infrapopulations or populations. F_{ST} measures the relative proportion of total genetic variation that is found, say, among populations rather than within populations; essentially, it is a measure of average genetic differentiation, though it gives no information on the spatial configuration of that differentiation. Although one must be careful when interpreting F-statistics as evidence of genetic structure (see Criscione et al. 2005; Prugnolle et al. 2005a), they provide a powerful tool for parasite population studies.

Various evolutionary forces interact to determine genetic structure. At the molecular level, populations can differ in parts of the genome that are selectively neutral and in parts that are under strong selection. On the one hand, mutation creates novel alleles, selection increases the frequency of favorable alleles, and drift results in stochastic fluctuations in allele frequencies; these forces act to structure groups of individuals into distinct genetic assemblages. On the other hand, gene flow tends to blur any boundaries between these groups. Other factors can also have important effects on genetic structure, especially in parasites (see Nadler 1995). For instance, asexual reproduction or self-fertilization in hermaphrodites can enhance genetic structure. These modes of reproduction lead to the production of spatial packets of clonal or genetically similar infective larvae, each packet then likely to be recruited by the same infrapopulation. The outcome can be a greater genetic similarity among parasite individuals within an infrapopulation than among infrapopulations.

Let's first look at the empirical evidence for genetic structure among the different infrapopulations within a parasite population. Even among parasites with very similar transmission modes, such as nematodes with a direct, one-host life cycle parasitizing mammals, vastly different patterns can be observed. Host behavior may be a key reason for these differences. For instance, in the human parasite *Ascaris lumbricoides*, the genetic similarity of worms from hosts of the same household in Guatemalan villages is greater than expected by chance alone (Anderson et al. 1995). Thus, transmission of this nematode appears clustered within human households, rather than occurring randomly throughout the village; the attachment of human hosts to their home constrains the pool of eggs from which infrapopulations are recruited. In contrast, in the rat parasite *Strongyloides ratti*, there is very little genetic differentiation among infrapopulations from hosts captured in the same rural area, indicating that the mobility of the host allows it to sample from all locally available parasite genotypes (Paterson et al. 2000).

Digeneans provide good illustrations of how host mobility can influence the genetic structure of infrapopulations. Digeneans multiply asexually in their molluscan first inter-mediate host, usually producing free-swimming cercariae that subsequently encyst in an-other host. This phenomenon may generate a spatial mosaic consisting of local clusters of infective larvae, with cercariae in each cluster being genetically identical to each other but different from those in other clusters. Without host movements, these conditions should promote strong genetic structuring. In a lake in northern Germany, Rauch et al. (2005) found that there are on average four different genotypes of the digenean *Diplosto-mum pseudospathaceum* per infected snails. In ponds on the island of Guadeloupe, exten-sive field studies have identified a more extreme situation, where infected snails almost always harbor only a single genotype of the digenean *Schistosoma mansoni* (Barral et al. 1996; Sire et al. 1999; Théron et al. 2004). Yet, in both cases, the next host in the life cycle, the one that is infected by the cercariae shed by snails, harbors a diverse mixture of geno-types (fig. 7.9). One rat was found to harbor 105 *S. mansoni* genotypes, representing just over half of all genotypes identified from the local population (Théron et al. 2004). In fact, the number of distinct genotypes in rats appears to increase linearly with infrapopulation size, suggesting that new genotypes just accumulate steadily as the infrapopulation grows (fig. 7.10). In both *S. mansoni* and *D. pseudospathaceum*, the mobility of the host ex-poses it to infection by cercariae representing a wide range of the genotypes available in its immediate vicinity, thus overcoming the homogenizing effect of asexual multiplica-tion of the parasites in snails (Barral et al. 1996; Théron et al. 2004; Rauch et al. 2005).

Nevertheless, on a slightly larger spatial scale, genetic differentiation still develops among infrapopulations. Given the extremely limited dispersal of cercariae in standing water, and the fact that host individuals do not regularly move throughout the entire area occupied by their population, a significant genetic structure emerges at the population level, even on a scale of tens of meters. This is the case for adult *S. mansoni* populations in rats on Guadeloupe (Sire et al. 2001a). Genetic structure among infrapopulations is also observed in a *S. mansoni* population infecting people in a Brazilian village. Unlike the Guadeloupe example, in Brazil over half of the snail intermediate hosts of *S. mansoni* that were examined harbored multiple parasite genotypes, that is, they released cercariae descended from different eggs, despite the parasite using the same snail species as in Guadeloupe (Minchella et al. 1995). More importantly, the genetic composition of adult infrapopulations indicates that human hosts are more likely to be infected at transmis-sion sites very close to their home, with the probability of acquiring genotypes from other sites diminishing rapidly with increasing distance (Curtis et al. 2002). Similarly, in the di-genean *Fascioloides magna*, Mulvey et al. (1991) found that deer, the parasite's definitive host, tended to harbor infrapopulations of adult worms representing only a very small number of genotypes. In this species, cercariae encyst on semi-aquatic plants, forming patches of genetically identical clones. Prevalence of infection is low in deer, and thus the probability of infection while browsing on vegetation is low; however, when infection does occur, multiple individuals of the same clone are simultaneously recruited to the in-frapopulation. The outcome is that different infrapopulations tend to consist of different clones, resulting in a clear genetic structure among infrapopulations (Mulvey et al. 1991).

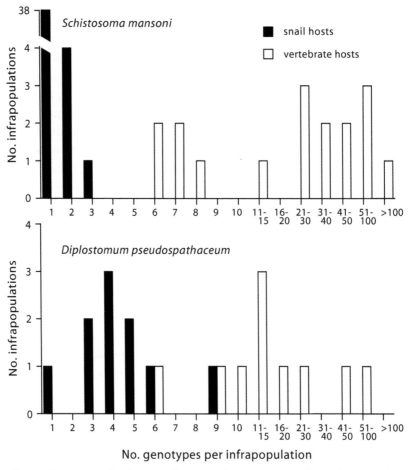

Figure 7.9 Frequency distributions of genotype diversity among infrapopulations of digeneans in their snail intermediate host and in their next host. For *Schistosoma mansoni*, the
next host is the rat serving as definitive host; for *Diplostomum pseudospathaceum*, the next
host is a fish serving as second intermediate host. In these examples, different genotypes in
the same host are issued from different miracidia, i.e., different parasite eggs. (Data from
Théron et al. 2004; Rauch et al. 2005)

In *S. mansoni* and *F. magna*, there is no second intermediate host: although using different routes following their exit from snails, cercariae of both species next infect the mammalian definitive host. The presence of a second intermediate host, or of paratenic hosts,
can serve to mix genotypes prior to the formation of adult infrapopulations in the definitive
host. Vilas et al. (2003) have found very little genetic differentiation between infrapopulations of the digenean *Lecithochirium fusiforme* in its fish definitive host. They attributed this
to the accumulation of a mixture of genotypes in small fish species serving as paratenic
hosts before these mixtures are recruited into adult infrapopulations. Rauch et al. (2005)
suggest that the same phenomenon occurs in *Diplostomum pseudospathaceum*, with fish
second intermediate hosts assembling diverse groups of genotypes that are then transferred to adult infrapopulations in bird definitive hosts. Longer life cycles, with one or more

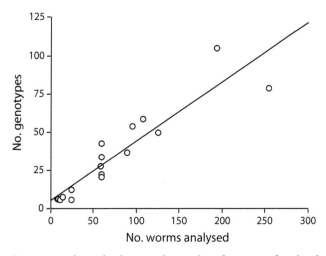

Figure 7.10 Relationship between the number of genotypes found and the number of worms analyzed per adult infrapopulation of the digenean *Schistosoma mansoni* in seventeen naturally infected rats. Infrapopulation sizes ranged between 9 and 1,109 worms, but since only a random sample of worms from large infrapopulations were genotyped, the number of worms analyzed ranges from 8 to 254. The linear relationship suggests that the number of genotypes per infrapopulation does not saturate, but instead grows linearly with increasing infrapopulation size (Data from Théron et al. 2004)

hosts inserted between the snail first intermediate host and the definitive host, can therefore reduce the genetic structure of adult infrapopulations.

Let's now move to a higher spatial scale, and examine the genetic structure among the different populations forming a parasite metapopulation. This involves quantifying the genetic similarity among parasite populations on scales of hundreds to several thousand kilometers. At the upper end of this range, there is sometimes evidence of substantial and stable genetic subdivisions among groups of parasite populations. These discontinuities reveal the existence of different lineages of populations that are more or less independent of each other. For example, populations of the cestode *Paranoplocephala arctica* and of the nematode *Heligmosomoides polygyrus* form distinct clades across their European range, showing some congruence with similar subdivisions existing among the populations of their rodent hosts (Wickström et al. 2003; Nieberding et al. 2004). Genetic divergence on that scale falls within the domain of the emerging field of phylogeography (Avise 2000). Here, the focus will be on the genetic structure of parasite populations that form true metapopulations and among which genetic exchanges occur. In situations where local selection is very strong, it is possible that phenotypic differentiation could exist among populations in spite of gene flow. Since all available studies have centered on genetic differentiation, however, only genetic structure is discussed below.

Anderson et al. (1998) reviewed studies of geographical population structure in parasitic nematodes and concluded that, overall, there is only very little genetic structure, a pattern consistent with high levels of gene flow among populations. This is true even in situations where genetic structure is expected a priori, such as in nematode species that reproduce mainly by parthenogenesis (Fisher and Viney 1998) or in cases of apparent total geographic isolation between populations (Braisher et al. 2004). The maintenance of sufficient gene flow among parasite populations is dependent, to a large extent, on host movements on geographical scales. For instance, Blouin et al. (1995) compared the genetic structure displayed by five species of trichostrongylid nematodes, four from livestock (cattle and sheep) and one from a wild host (deer), across the United States. They found marked genetic subdivisions among populations of parasites in deer, but none among the parasites of cattle and sheep. The frequent long-distance transport of livestock and other husbandry practices facilitate gene flow among populations of livestock parasites, whereas habitat fragmentation limits large-scale movements by deer and thus parasite gene flow. Indeed, the greater the geographical distance between two deer populations, the greater the genetic differentiation between their nematode populations (fig. 7.11a). Gene flow can also be maintained by intermediate hosts in nematode species with complex life cycles. In species of the genera *Contracaecum* and *Pseudoterranova*, the absence of genetic structure among adult worm populations is the consequence of the great mobility of the seal definitive hosts and the fish paratenic hosts (Paggi et al. 1991; Nascetti et al. 1993).

In ticks, genetic structuring among populations is also rare at geographical scales (e.g., Hilburn and Sattler 1986; Lampo et al. 1998; Kain et al. 1999). As for nematodes, though, host vagility is crucial in maintaining gene flow among tick populations. This is beautifully illustrated by a comparison of the genetic structure of populations in two distinct races of the same tick species, *Ixodes uriae*, parasitic on different seabirds (McCoy et al. 2003). One race exploits kittiwakes and the other lives on puffins; the latter host species is known to make more frequent and more distant visits to other colonies than the former. As one might expect, there is stronger genetic structure among ticks of kittiwakes than among those of puffins (McCoy et al. 2003). In kittiwakes, but not in puffins, the genetic differentiation between any two tick populations increases with the geographical distance separating them (fig. 7.11b). The findings suggest that tick dispersal in kittiwakes occurs in a stepping-stone manner, that is, there is a higher probability of ticks being transferred to a nearby bird colony than to a distant one (McCoy et al. 2003, 2005b).

The free-living infective stages of most parasites have a very limited dispersal capability, and host movements are the main routes for gene flow among parasite populations. Host vagility and the nature of the life cycle can interact to overcome geographical distance or other physical barriers to genetic mixing. Two examples using digeneans demonstrate this nicely. First, in *Schistosoma mansoni*, although some genetic differentiation exists among populations inhabiting fragmented marshy habitats on the island of Guadeloupe, there is also evidence of gene flow (Sire et al. 2001b). This parasite uses two hosts, snail first intermediate hosts and rat definitive hosts. Rats are clearly more vagile than snails, and there is good evidence showing that exchanges of parasites among populations are indeed mediated almost entirely by rat movements (Prugnolle et al. 2005b).

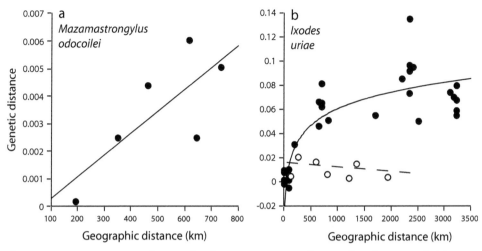

Figure 7.11 Pairwise genetic and geographic distances among populations of (a) the nematode *Mazamastrongylus odocoilei* in white-tailed deer from the eastern United States, and (b) the tick *Ixodes uriae* on either kittiwakes (black circles) or puffins (open circles) along the Atlantic coastlines. Genetic distance measures the relative differences in DNA sequences between populations, based on mitochondrial sequences for the nematode and microsatellite markers for the tick. Isolation by distance is apparent among populations of the nematode and of ticks on kittiwakes, but not among tick populations on the more vagile puffin hosts. (Data from Blouin et al. 1995; McCoy et al. 2003)

Once a *S. mansoni* population is cut off from others by an interruption in host movements, as in populations maintained in the laboratory (Stohler et al. 2004), it can show significant genetic differentiation after only a few generations. In parasites with complex life cycles, gene flow between populations is always likely to be controlled by the more mobile host (see Jarne and Théron 2001).

The second example involves four digenean species all found in salmon, *Oncorhynchus* spp., in the same rivers of the northwestern United States. Three of these species complete their entire life cycle in freshwater hosts, reaching sexual maturity in the digestive tract of salmon. In contrast, the fourth species, *Nanophyetus salmincola*, uses salmon as its second intermediate host, and matures in the intestine of fish-eating terrestrial mammals and birds. Criscione and Blouin (2004) compared mitochondrial DNA variation among populations of each of these species, sampled in the same localities, and found patterns consistent with the fact that only *N. salmincola* uses a host capable of rapidly dispersing it across localities. The three species cycling only in freshwater hosts displayed strong genetic structure among their populations, whereas high gene flow resulted in a lack of structure in *N. salmincola* (Criscione and Blouin 2004).

Finally, the cestode *Echinococcus granulosus* provides a further example of how host vagility can compensate for parasite features that might otherwise lead to genetic structuring of parasite populations. The cestode has a two-host cycle, using a wide range of domestic or wild, herbivorous or omnivorous mammals as intermediate hosts, and carnivores as definitive hosts. The parasite multiplies asexually in its intermediate host. However, individual intermediate hosts contain *E. granulosus* cysts derived from genetically different embryos, such that there is some genetic heterogeneity in intermediate hosts despite the asexual multiplication of cestode larvae (Lymbery and Thompson 1989). Still, this phase of asexual reproduction can cause a reduction in the genetic diversity of adult infrapopulations, and could facilitate genetic structure of cestode populations on a geographic scale. That is not the case, though. Across mainland Australia, there is no evidence of genetic structuring among *E. granulosus* populations, and only weak differentiation between populations from the mainland and those from Tasmania (Lymbery et al. 1997). This and previous examples highlight the powerful role of host dispersal as a vehicle for parasite migration and exchanges among populations, and its central importance in maintaining the fabric of the parasite metapopulation.

7.6 Conclusion

To date, the population biology of parasites has been investigated on three fronts. First, the dynamics of parasite populations can be modeled mathematically, with a few simplifying assumptions. This is the epidemiologist's approach, and it has proven to have predictive value. Second, empirical studies of field populations have highlighted the many density-dependent and density-independent mechanisms acting to regulate parasite abundance over time in specific systems. This is the approach of the ecologist and it has served to validate some aspects of the models. Third, genetic structure among infrapopulations and among populations allows us to determine transmission processes and estimate the frequency of exchanges of individuals among populations. This is the geneticist's approach, one that has served to delineate the boundaries of parasite metapopulations.

But, from the evolutionary biologist's perspective, some of the main questions remain completely unanswered, even unasked. For instance, what is the evolutionary significance of density dependence? Is it an unavoidable consequence of crowding, or have the mechanisms behind it been actively shaped by selection? Clearly, in cases of interference competition, selection has favored individuals capable of eliminating their competitors. But even in situations where exploitation competition is the operating mechanism, why is density dependence detected in some species but not in others? When it is present, why does it manifest itself at different densities among closely related species? Is the occurrence of density dependence related to parasite phylogeny? It may be that density-dependent mechanisms have evolved differently in various parasite lineages. There are very few comparative analyses of population-level phenomena among parasite species. Ecologists studying free-living organisms have performed several such analyses but have not often considered possible phylogenetic effects (Harvey 1996); there exist several methods making this feasible for parasites (Morand and Poulin 2003). As more examples of density-dependent regulation in parasite populations become available, we may be able to tackle these important questions.

Another particularly exciting possibility will be the development of metapopulation models for parasites. Metapopulation biology has become a core discipline within ecology

(Hanski and Gilpin 1997), but it has yet to be applied widely to metazoan parasites. Population-genetics studies are yielding estimates of migration rates among parasite populations, and these provide the empirical basis on which metapopulation models could be constructed. When they become available, these models will generate valuable insights into the impact of habitat change on extinction probabilities or dynamics of local populations, as well as into the spread of resistance genes or the effectiveness of control strategies.

8 | Interactions between Species and the Parasite Niche

Previous chapters have addressed several aspects of parasite ecology as though parasites of one species were alone in their host. This is rarely the case as most free-living animals, especially vertebrates, are used as host simultaneously by several species of parasites. Various pairs of these different parasite species, though sharing a host, will not always co-occur on a finer scale: an intestinal worm and an ectoparasitic arthropod on the same vertebrate host will not interact directly and can probably be treated as entirely independent of one another. Often, however, hosts harbor several parasite species belonging to the same guild, which is to say, parasite species that use similar resources such as food or space. These species are likely to interact with one another regardless of differences in other aspects of their biology. This chapter examines interactions among parasite species within the individual host, and serves as a bridge to the next two chapters, which deal with the diversity and structure of parasite communities among host individuals and populations.

Several types of interspecific interactions are possible among parasites within the host. The effect of one species on another can sometimes be positive. For instance, interference with host defense mechanisms by one parasite species can facilitate the exploitation of the host by a second species. More often, however, interactions will be antagonistic such that the presence of one parasite species has negative effects on the numbers, distribution, or reproduction of other species. These antagonistic interactions can include predation; for example, the larval stages of digeneans in snails can prey on smaller larval stages of other digenean species (Sousa 1992, 1993). The most commonly reported interactions and the most studied, however, consist of various forms of interspecific competition.

The sort of evidence used to infer that competition for resources occurs between two species is varied (Thomson 1980). First, a reduction in the number of individuals of one species when the other is present suggests that competition is taking place. Second, individuals of one species can change the way in which they use a resource when the second species is present. This is possible because of some plasticity in morphology or behavior;

among parasites, this often would take the form of a slight shift in the site of infection. Both of these phenomena, the numerical and functional responses to the presence of a competitor, are indications that two species have effects on one another. Ecologists often feel that numerical responses are more convincing demonstrations of competitive interactions than functional responses (Thomson 1980). Here, I will give equal weight to both types of evidence and discuss them in turn. I will also consider cases in which similar parasite species do not interact because of small but consistent differences in resource use. These may represent evolutionary niche shifts or character displacements that resulted from intense competition in the past. They may however be the result of other evolutionary forces, and the fact that they prevent competition may be a fortuitous consequence.

8.1 Numerical Responses to Competition

As emphasized in the previous chapter, the regulation of parasite populations occurs through mechanisms acting in infrapopulations. Similarly, interactions between parasite species take place within individual hosts where two or more species co-occur. To demonstrate that two parasite species compete, one has to show a negative relationship between the numbers of individuals of species A and species B among several host individuals. This can be done using a sample of naturally infected hosts, or ideally using experimental infections. In samples of naturally infected hosts, the negative relationship may be best measured among young hosts than across all age groups, since long-term exposure to both parasite species can eventually mask the true interactions between them (Bottomley et al. 2005). The experimental approach has the advantage of allowing the infrapopulation sizes of one species alone in the host to be compared with its infrapopulation sizes when it shares the host with a presumed competitor, while doses of infective larvae and other variables are kept constant. As long as the experimental infection process accurately reproduces the natural one, that is, exposure to few infective larvae per unit time over a long period as opposed to a single massive dose, then experimental studies are a very powerful means of quantifying the effect of competition among parasite species.

Dobson (1985) reviewed published studies of concurrent infections of parasites and found that reductions of infrapopulation sizes by as much as 50 percent were common in mixed infections of gastrointestinal helminths in mammals. Examples of numerical effects of competition among closely related and unrelated helminths of mammals, under experimental conditions, are shown in figure 8.1. Typically, competition between species is asymmetrical, with one species suffering severe losses and the other being almost unaffected. In other words, the infrapopulation of the winning species is mostly unchanged but it reduces the size of the infrapopulation of the losing species, sometimes excluding it entirely from the host individual. This may indicate one-sided interference rather than exploitative competition (as defined in chapter 7), but can also result from host-mediated effects involving immune responses. Asymmetrical interactions are

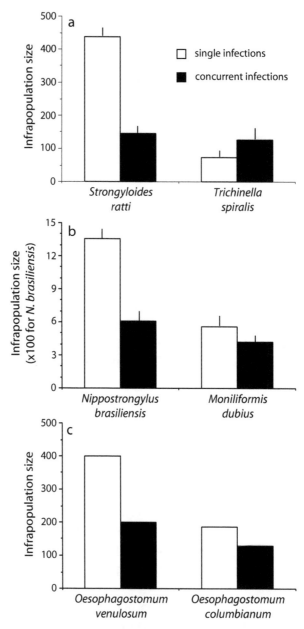

Figure 8.1 Infrapopulation sizes (means + standard errors, if available) of helminths of mammals in single and concurrent infections in experimental studies: a, two nematode species in rats; b, a nematode and an acanthocephalan in rats; c, two congeneric nematode species in sheep. The outcome of competition is typically asymmetrical, with only one species (on the left) suffering a significant numerical decrease. (Data from Moqbel and Wakelin 1979; Dash 1981; Holland 1984)

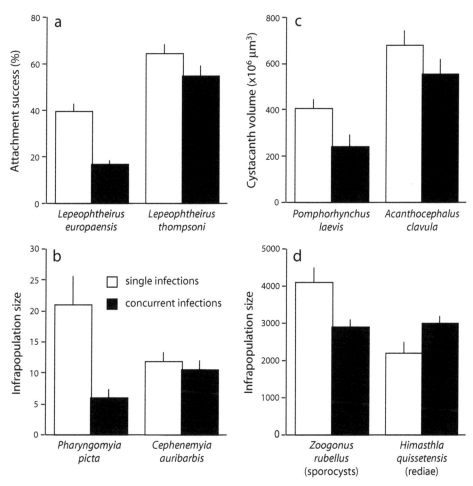

Figure 8.2 Examples of asymmetrical competition between pairs of parasite species: a, attachment success of two copepod species on fish in experimental conditions; b, infrapopulation sizes (numbers of parasites per host) of larvae of two species of bot flies on deer in natural infections; c, cystacanth sizes of two acanthocephalan species in their amphipod intermediate host in natural infections; d, infrapopulation sizes (numbers of rediae or sporocysts per host) of two digenean species in their snail intermediate host. All data are means (+ standard errors). In all four examples, the species on the left incurs a statistically significant cost in concurrent infections relative to single infections, whereas the one on the right does not. (Data from Dawson et al. 2000; Vicente et al. 2004; Dezfuli et al. 2001a; Hendrickson and Curtis 2002)

not limited to reductions in the size of adult helminth infrapopulations; they are also observed in other ways and in a range of parasite taxa. They may be manifested as reductions in the establishment success of ectoparasitic arthropods on their vertebrate hosts, impaired growth of acanthocephalan cystacanths in their intermediate hosts, or lower asexual multiplication rates by digeneans in their snail intermediate hosts (fig. 8.2).

Given the smaller size of intermediate hosts, extreme forms of asymmetrical competition are probably more likely in mixed-species infections of intermediate hosts. For instance, the establishment success of cysticercoids of the cestode *Hymenolepis diminuta*

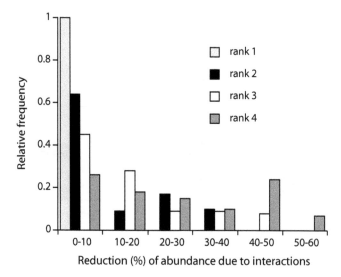

Figure 8.3 Relative effect of antagonistic interactions on the abundance of individual parasites of different digenean species exploiting the same snail population. The data come from twenty-five studies on littorine snails. The relative (percent) reduction in abundance compared to infrapopulations not sharing their snail host with other species was calculated for each species in each study. Species were ranked in order of competitive dominance (rank 1 being most dominant) based on evidence available for each host-parasite system. Dominant species incur very small reductions in numbers, if any, due to interspecific interactions; subordinate species regularly suffer severe losses. (Modified from Kuris and Lafferty 1994)

in its insect intermediate host is severely reduced by the presence of another cestode, *Raillietina cesticillus*, but the reciprocal effect is not observed (Gordon and Whitfield 1985). In the same vein, the development of the acanthocephalan *Leptorhynchoides thecatus* in its amphipod intermediate host is markedly retarded by the presence of another acanthocephalan, *Pomphorhynchus bulbocolli*, in the same host (Barger and Nickol 1999). There is also evidence that competition between different species of malaria parasites takes place within the mosquito vector, and not in the vertebrate host (Paul et al. 2002). Finally, in a comprehensive review of published studies on larval digeneans in snail intermediate hosts, Kuris and Lafferty (1994) showed that subordinate digenean species suffered severe losses from interspecific antagonism compared to dominant species (fig. 8.3). The main determinant of a species' rank within these hierarchical interactions appears to be size, with the larvae (rediae) of large sizes preying on and excluding larval stages of other digenean species from their snail host (Sousa 1992, 1993). Intense one-sided competitive interactions between parasite species are also possible inside the much larger vertebrate hosts, but are probably limited to those rare situations in which there is a pronounced shortage of resources (e.g., Jackson et al. 1998).

Asymmetrical competition between pairs of parasite species appears to be widespread. The generality of this phenomenon suggests that two co-occurring parasite species are rarely equal: small interspecific differences in abilities to exploit host resources typically translate into marked differences in competitive success. Dawson et al. (2000) suggested that, in pairwise interactions, generalist species—those with lower host specificity—should be more sensitive to competition than specialist species. Although differences between specialists and generalists may explain some cases of asymmetrical competition, they do not apply to other instances, and thus there must exist other explanations for the universality of this phenomenon.

As indicated above, numerical responses other than reductions in infrapopulation size are commonly observed. Often the species suffering the most from interspecific competition will incur a reduction in body mass (e.g., Poulin et al. 2003c) or fecundity (e.g., Moqbel and Wakelin 1979; Silver et al. 1980; Holland 1984) when co-occurring in a host with its competitors. These can lead to a decrease in recruitment rates and affect the parasite population as a whole. Like effects on survival, effects of competition on fecundity are often asymmetrical, with one species affected much more than the other. It must be pointed out, however, that the outcome of competition may be influenced by the relative densities of the different parasite species as well as the order in which they become established in the host. The species that suffers most from competition when it enters the host simultaneously with a competitor can perform much better if given a head start, if allowed to establish in the host well before the competing species arrives. Priority effects have been observed in some of the systems where asymmetrical competition occurs in concurrent experimental infections.

Despite the often large numerical effect of interactions among parasite species in experimental infections, the magnitude of interspecific competition between parasites in nature may often not be important. In the laboratory, large infrapopulations of different species are placed in the same host individuals, a situation that does not arise often in natural systems. The reason for this is that the typical distribution of parasites among hosts creates few opportunities for two or more species of parasites to co-occur in the same host individual in sufficient numbers for competition to become measurable. Assuming that different parasite species display aggregated distributions (see chapter 6), and that these distributions are independent of one another, potentially severe antagonistic effects may never be realized and parasite populations may coexist in the same host population. The restraining influence of aggregation on interspecific competition is also predicted by mathematical models (Dobson 1985; Dobson and Roberts 1994; Roberts et al. 2002).

When moderate numbers of parasites of different species end up in the same host, they can also alter their resource use and avoid the numerical effect of competition; this will be addressed in the next sections. Parasitologists in recent years have sometimes devoted more effort to the study of functional responses to competition than to the measurement of numerical effects. Inferring the existence of competition from functional responses without demonstrating numerical effects can be misleading (Thomson 1980), but the many examples of niche adjustments shown by co-occurring parasite species are often quite striking (see section 8.3) and suggest that these are indeed responses to antagonistic interactions.

8.2 The Parasite Niche

If competition results in a change in the way parasites use host resources more often than in a change in the number of parasites, we need a method of quantifying resource use by parasites to detect competition. One approach is to determine the ecological niche of individual parasite species, which is the multidimensional habitat volume occupied by parasites and defined by several physical and biotic variables (Hutchinson 1957). Most of these variables, such as the exact type of food particles ingested by parasites, are extremely difficult to quantify, just as they are for the majority of free-living animals. Parasitologists have therefore focused on the spatial dimension of the niche. Just as they tend to be highly host specific (see chapter 3), parasites are also restricted to particular sites on or in the host. Detailed measurements of parasite attachment sites can be made and used to compare the niches of parasites in single and mixed infections.

These measurements can be simple to obtain. If the parasite habitat can be defined along a linear axis, such as the vertebrate intestine, or in some vertebrates the entire, undifferentiated gastrointestinal tract extending from the esophagus to the anus, then a measure of the niche can be taken as the mean or median position of individual parasites. The niche itself would be the region encompassed by the range of their positions. Of course, the intestine is not really a linear or one-dimensional habitat; parasitic worms can make different use of intestinal villi and extend into the intestinal lumen to various extent. Still, despite many simplifying assumptions (Bush and Holmes 1986b), this one-dimensional view of the intestine as a habitat provides useful information on parasite niches. Indices of niche width and other n-dimensional measures of site specificity have also been used (e.g., Anderson et al. 1993; Rohde 1994b). These or other more complicated measures may be necessary when parasites occupy structurally more complex microhabitats in the host, such as the gills.

Whether or not the habitat of a parasite can be reduced to a single dimension, its niche is not necessarily continuous in space. For example, adult infrapopulations of the acanthocephalan *Corynosoma cetaceum* occupy three distinct parts of the digestive tract of their definitive host, the dolphin *Pontoporia blainvillei*. The parasites are located in the dolphin's stomach, which consists of three chambers separated by sphincters. Not only are worms in different stomach compartments exposed to different physicochemical conditions, but they are also more or less physically isolated from each other (Aznar et al. 2001). Similarly, monogeneans are often found in distinct and physically separated locations on the body of their fish hosts. The monogenean *Anoplodiscus cirrusspiralis*, for instance, inhabits the fins and the nasal cavities of its host (Roubal and Quartararo 1992). In these and many other cases, the niche is separated into discrete spaces, and the parasite infrapopulation is thus spatially segregated into smaller units.

Ultimately, a parasite exploits the niche that has been favored by selection over countless generations because it maximizes its fitness. An individual settling outside its species' niche would achieve lower fitness; for instance, it would experience reduced resource intake, greater risk of mortality resulting from host immune responses, or a lower probability of mating. While the ultimate reasons for site selection within the host are obvious, the proximate cues used by parasites are not always clear. Richardson and Nickol (2000) have investigated whether the physiological or biochemical properties of the pyloric ceca of fish hosts might explain why the acanthocephalan *Leptorhynchoides thecatus* only settles in that part of the host's digestive tract. Earlier studies indicated that the ceca do not contain more food, or experience less peristaltic motion, than the rest of the fish's intestine. Richardson and Nickol (2000) experimentally changed the conditions (e.g., pH, protein concentration) in the fish's gut to eliminate the differences between the ceca and the intestine. In the end, they could not identify a specific factor responsible for microhabitat selection in *L. thecatus*. Thus, the proximate mechanisms acting to maintain the integrity of the parasite niche may often rely on something more complex than simple responses to basic stimuli.

Defining the spatial boundaries of a parasite's distribution in its host leads one to believe that there might be a maximum number of parasite species, or parasite niches, that can be fitted in a host. How can we estimate the number of niches in, say, an intestine? Dividing the length of the intestine by the average niche width of parasites does not provide an answer since there are gradients in habitat quality along the intestine and some parts may be unsuitable for all parasites. Another way would be to take the maximum number of species co-occurring in one host as equivalent to the number of niches available in hosts of that species (e.g., Kennedy and Guégan 1996). This can only be a lower estimate, however, as it assumes that hosts are saturated with parasite species and that there are no niches left unoccupied. We can only guess at how many parasite species could be accommodated in a given host individual.

A distinction must be made between the fundamental niche and the realized niche. Strictly in terms of spatial location in the host, the fundamental niche is defined as the potential distribution of the parasite in the host, that is, the range of sites in which the parasites of one species can develop. Realized niches are subsets of the fundamental niche, and consist of the sites actually occupied by parasites in the host. Realized niches represent either the optimal sites within the fundamental niche if interactions with other species are unimportant, or the portions of the fundamental niche actually available to them because of antagonistic interactions with other parasite species. The best way to measure the fundamental niche of parasites would be to quantify their distribution in controlled, single-species infections (e.g., Holmes 1961; Patrick 1991). Since this is often impossible, another way to estimate the boundaries of the fundamental niche consists in summing up the distributions of parasite individuals across all hosts in which the species occurs, in a sample of naturally infected hosts containing mixed infections (Bush and Holmes 1986b). The functional response of parasites to interspecific competition can then be examined by comparing the realized spatial niche of infrapopulations alone in the host with that of infrapopulations sharing the host with presumed competitors.

8.3 Functional Responses to Competition

Holmes (1973) refers to adjustments in infection site in response to the presence of a competitor as interactive site segregation. This happens when the fundamental niches of

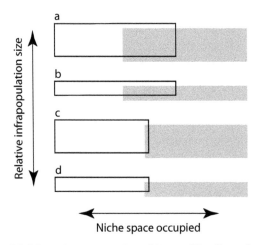

Figure 8.4 Schematic representation of the possible effects of concurrent infections on the infrapopulations of two parasite species. Hypothetical infection doses, i.e., the number of parasite larvae of each species administered to hosts, are kept constant in all cases. The vertical dimension of the rectangles illustrates the relative size of each infrapopulation, while the horizontal dimension indicates the use of niche space by each infrapopulation: a, the infrapopulation sizes and fundamental niches of the two species when in single infections; b, a numerical response to concurrent infection, in which infrapopulation sizes are reduced; c, a functional response to concurrent infection, in which the realized niches of both parasites are adjusted to reduce niche overlap; and d, a joint numerical and functional response. In these examples, all effects are symmetrical, i.e., both species display comparable responses (Modified from Poulin 2001b)

two parasite species overlap; when they co-occur in the same host individual, one or both of them adopt sites other than their preferred sites, but still within their fundamental niche, in order to minimize the overlap and thus competition. Such a functional response may occur with or without numerical effects of competition, that is, with or without reductions in infrapopulation sizes (fig. 8.4). It requires that parasites are able to detect heterospecifics or at least a change in the environment caused by heterospecifics, and that their behavior is plastic enough to adjust by changing the site of attachment. A slight shift away from the preferred site can be associated with a decrease in fitness if conditions in the alternative site are suboptimal. The cost of competition must therefore outweigh the cost of this shift in attachment site for interactive site segregation to be favored by selection. Indeed, substantial spatial niche overlap does not mean that competition occurs, even between related species (e.g., acanthocephalans in eels; Kennedy and Moriarty 1987; Kennedy 1990), and site segregation will not be favored if competition is not costly. If competition is very severe, however, if it is maintained over several generations, and if competing species are very likely to co-occur in the same host within each generation, we may expect the niche shift to become genetically fixed. This would mean a permanent reduction of the overlap between fundamental niches instead of between realized niches, and has been labeled by Holmes (1973) as selective site segregation. It is really just a form of ecological character displacement, a common evolutionary consequence of competition for limiting resources by sympatric species (Schluter 2000). This evolutionary phenomenon will be considered further in the following section; here, the discussion centers on proximate adjustments to the realized niche in response to the immediate presence of a competitor.

Intestinal helminth communities in many birds often consist of large numbers of individuals of several species, and provide good evidence for functional responses to the presence of competitors. In four species of grebes, for instance, Stock and Holmes (1988) found that each helminth species was usually restricted to a predictable portion of the intestine and that the various species were arranged along the entire length of the intestine. There was much overlap between the fundamental spatial niches of the different parasite species, however, when the fundamental niche of each species is determined from the pooled distributions of conspecific parasites in all hosts (fig. 8.5). A reduced overlap between the realized spatial niches of pairs of helminths co-occurring in individual birds relative to that between their fundamental niches would be evidence for interactive site segregation as a functional response to competitors. Stock and Holmes (1988) indeed found that among all pairs of species with fundamental niches that overlap by at least 5 percent, significant reductions in the overlap of realized niches were the general rule.

Bush and Holmes (1986a, b) also found evidence of functional responses to interspecific interactions in rich communities of helminths in birds. Among the more common species of intestinal parasites of the lesser scaup duck, *Aythya affinis*, overlaps of realized spatial niches between pairs of species were typically less than half those between fundamental niches (fig. 8.6). Interestingly, whereas the breadth of the realized niche of many species within individual hosts increased with infrapopulation size, the overlap between the realized niches of adjacent species did not increase with their combined infrapopulation sizes. This was due to asymmetrical expansions of the realized distribution in the intestine, that is, expansions in one direction only, away from the main competitor's niche.

Interactive site segregation also occurs in gastrointestinal helminth communities characterized by lower species richness, such as those found in fish hosts. Friggens and Brown (2005) investigated the cestode communities living in the spiral valve intestine of two elasmobranch host species. They observed that, across all pairs of cestode species, the spatial distribution of the two species in a pair along the spiral intestine tended to overlap significantly less than expected from a null distribution model. Along a second niche dimension—the size class of conspecific hosts used by the parasites—niche overlap was sometimes considerable; however, niche overlap was never high along both niche dimensions considered (fig. 8.7). Sometimes, interactive niche segregation is not pervasive throughout the entire community, but restricted to interactions involving one or a few dominant species. For example, the presence of the acanthocephalan *Neoechinorhynchus golvani* in its cichlid fish host causes all other common intestinal helminth species to shift their distribution to avoid interacting with the acanthocephalan. However, other pairwise associations between species other than *N. golvani* show no evidence of site segregation (Vidal-Martinez and Kennedy 2000).

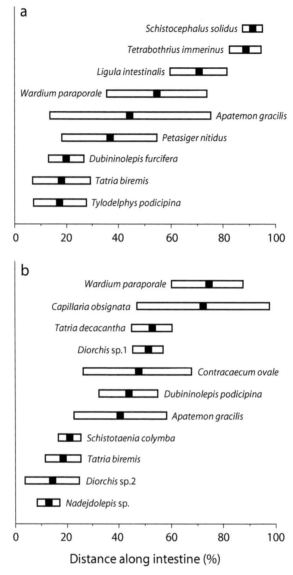

Figure 8.5 Fundamental niches of intestinal helminths in two species of grebes: a, *Aechmophorus occidentalis*, and b, *Podiceps nigricollis*. The mean position (± standard deviation) along the length of the intestine is the average of the positions of the medianth individual in each infected host individual. Helminth species with prevalences lower than 25 percent are not shown. (Data from Stock and Holmes 1988)

In spite of the striking examples discussed above, interactive site segregation is not a universal phenomenon among communities of gastrointestinal helminths. Within any taxonomic group of host species, it is possible to find examples of clear-cut site segregation and examples where site segregation is completely absent, where the realized niche of any species is independent of that of other species (e.g., for mammal hosts, compare Hauk-isalmi and Henttonen 1994 with Ellis et al. 1999). Thus, the influence of interspecific

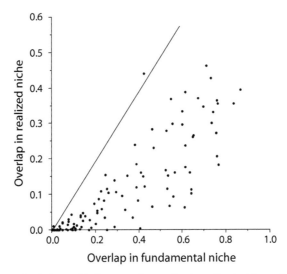

Figure 8.6 Relationship between the overlap in realized niche and the overlap in fundamental niche in all pairwise associations (120 in total) among sixteen intestinal helminth species in the lesser scaup duck, *Aythya affinis*. All parasite species were common ones, found in at least one-third of birds examined. The niche is defined as the linear distribution of the parasites in the intestine, and overlap between the niches of two species is measured as the proportion of the individuals of the two species found in the area of overlap. Points falling below the line represent overlaps between realized niches smaller than between fundamental niches. (Data from Bush and Holmes 1986b)

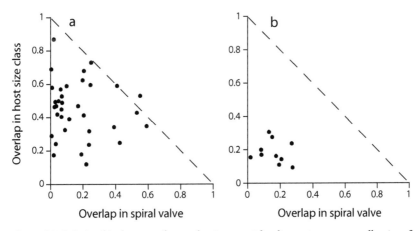

Figure 8.7 Relationship between the overlap in two niche dimensions among all pairs of cestode species living in the spiral valve intestine of two elasmobranchs: a, nine cestode species in *Urobatis halleri*, and b, five cestode species in *Leucoraja naevus*. Niche overlap is calculated using Pianka's index (0 = no overlap, 1 = complete overlap) considering either which host size class is used by the parasites, or in which spiral valve they occur (there are thirteen valves in the spiral intestine of *U. halleri* and nine in *L. naevus*). The fact that almost all points fall below the broken line indicates that niche overlap is never high along both niche dimensions simultaneously. (Modified from Friggens and Brown 2005)

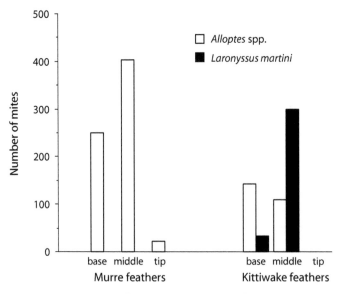

Figure 8.8 Distributions of mites along the length of feathers on murres (*Uria* spp.) and kittiwakes (*Rissa* spp.). A single species of mite of the genus *Alloptes* occurs on murres, where it is found mainly on the middle sections of feathers followed by the base of feathers. On kittiwakes, *Alloptes* occurs with another species, *Laronyssus martini*, which shows a strong preference for the middle sections of feathers. The realized niche of *Alloptes* on kittiwakes differs from that on murres, possibly because of a shift of its sites of attachment toward the base of feathers in response to interspecific competition. (Modified from Choe and Kim 1989)

competition or other interactions on the spatial structuring of helminth communities is variable; it cannot simply be assumed to exist without empirical demonstration.

Spatial niche adjustments in response to competition occur in parasites other than intestinal helminths of vertebrates, although the latter have received more attention in previous studies. Arthropods ectoparasitic on birds, for instance, often form communities rich in individuals and species (Choe and Kim 1988, 1989). Most species are site specific, and the presence of one species can influence the realized distribution of another species (fig. 8.8). Similar observations have been made on oxyurid nematode species parasitizing cockroaches (Adamson and Noble 1992), and interactive niche segregation no doubt occurs in other types of parasite communities. In communities of monogeneans ectoparasitic on fish, however, the situation is very different. The typical pattern observed is that the niche breadth of any given species is independent of the presence or abundance of other species (e.g., Geets et al. 1997). Competition, and thus interactive site segregation, do not generally play important roles in these communities; some of the determinants of the monogenean niche will be discussed in the following section.

Many more examples could be given of an apparent narrowing of realized niches and of reductions in spatial niche overlap when competitors are present (see Combes 1995,

2001). These all suggest that competition is frequently avoided through active site segregation. There exists at least one other explanation for these results when they are obtained from naturally infected hosts. Parasites may select attachment sites along the whole breadth of their fundamental niche but survive only in those portions of the fundamental niche not occupied by competing species. This would create an illusion of site segregation but would in fact be the simple result of differential mortality along the spatial dimension of the niche. Although they can only focus on few pairs of parasite species, experimental studies in which known numbers of parasites of different species are used for single and multiple infections of hosts provide unequivocal arguments in favor of interactive site segregation (fig. 8.9).

Spatial segregation of the realized niches of competing species may be achieved at a larger scale: a parasite species could reduce niche overlap with competitors by excluding them from the individual hosts in which it has established. This phenomenon is common among larval trematodes exploiting snail hosts (Kuris and Lafferty 1994). But is this interactive exclusion a functional response of parasites to competition, or an extreme numerical effect? The latter is probably the process responsible for the exclusion of one species by another. In the case of larval trematodes, parasites are recruited by hosts harboring competing species and by hosts not harboring competitors, and they might simply fail to establish and survive in hosts harboring competitors. Thus partial or total exclusion is a case of interspecific competition but, even if it results in the spatial segregation of parasite species, it is not really the outcome of interactive site segregation.

Functional responses along niche dimensions other than spatial location in the host or among hosts have also been reported. For instance, congeneric species of digeneans of similar sizes have been reported to diverge in body size when occurring together, perhaps in order to minimize the overlap in resource use (fig. 8.10). Such examples reinforce the case that interspecific interactions can shape the spatial or size structure of helminth communities, even if actual competition is avoided through niche shifts.

The importance of interspecific interactions in the above examples ranges from undeniable to insignificant. Interactions appear to be a consequence of the broad and overlapping fundamental niches of the co-occurring species. This is not a characteristic of all parasite communities, however. In some communities, referred to as isolationist parasite communities, parasite species have narrow fundamental niches that are often identical to realized niches. In these communities the breadth and position of the niche of one species is independent of the presence of other species (Rohde 1979; Price 1980). Interactions play no detectable roles, and the characteristics of the niche have other determinants. Holmes and Price (1986) emphasized that interactive and isolationist communities are at two ends of a continuum of parasite communities, such that interspecific interactions range from important to insignificant in determining the sizes and distributions of parasite infrapopulations. The existence of the continuum is open to debate, however, as the detection of species interactions depends to a large extent on how one chooses to measure interactive effects, that is, numerical or biomass effects, functional responses, and so forth. Different tests applied to the same parasite community can yield different results (e.g., Moore and Simberloff 1990). Based on the general likelihood that

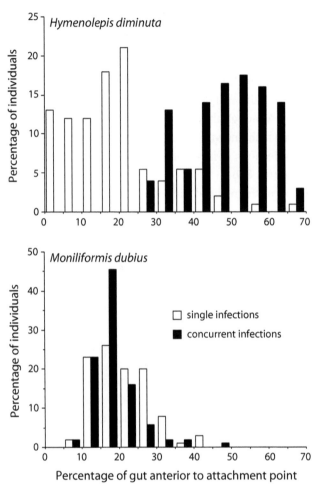

Figure 8.9 Distributions of the cestode *Hymenolepis diminuta* and the acanthocephalan *Moniliformis dubius* along the intestine of rats in experimental infections. The fundamental niches of the two worms, i.e., their distributions in single infections, overlap completely. Their realized niches in concurrent infections, however, show only minimal overlap due to the distribution of *H. diminuta* shifting down the intestine. This is a true example of interactive site segregation and not an artefact of differential mortality along the length of the intestine since similar numbers of worms of both species were recovered in single and concurrent infections. (Modified from Holmes 1961)

different species will co-occur in the same individual hosts, different helminth communities appear to cover the entire spectrum of possibilities, from cases where interactions are extremely unlikely to take place, to situations where they appear unavoidable (Poulin and Luque 2003). In any event, species interactions, whether or not they currently play meaningful roles in parasite communities, may have been important earlier in the evolution of these communities and helped shape the modern niche of their component species, as explored in the next section.

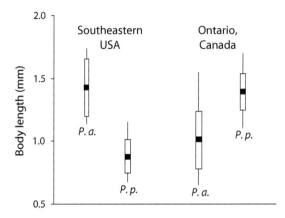

Figure 8.10 Body length (mean ± standard deviation, and range) of two congeneric digenean species parasitic in raccoons, *Pharyngostomoides adenocephala* (*P.a.*) and *P. procyonis* (*P.p.*), in two areas where they occur in sympatry. In areas of North America where they do not co-occur, both species attain only small sizes. When they co-occur, one of them (*P. adenocephala* in the southeastern USA, *P. procyonis* in Ontario) reaches a significantly larger size than the other. (Modified from Butterworth and Holmes 1984)

8.4 Evolutionary Niche Restriction

In some communities, parasite species have restricted fundamental niches that overlap very little or not at all with those of neighboring species, and interspecific interactions appear unlikely and are probably unimportant. These narrow and highly specific niches may have had at least two evolutionary origins: (1) they may be the product of intense competition between parasite species in the past, or (2) they may have other origins and have coincidentally prevented competition from occurring. In other words, either specialized niches are the consequence of intense interspecific competition in the past, or interspecific competition in the past never occurred because of the lack of overlap between specialized niches. Both historical and current ecological forces may be responsible for the niches of extant parasite species (Brooks 1980; Holmes and Price 1980). Distinguishing between the two scenarios is practically impossible (although comparative studies would probably offer useful insights). Whatever the evolutionary reason for the lack of niche overlap in isolationist communities, however, the influence of present competition in these communities is apparently negligible.

Evolutionary niche restriction driven by interspecific competition can occur in some circumstances (Holmes 1973). Selective site segregation over evolutionary time will be favored when abundant and prevalent parasite species share the same host population for numerous generations, and when competition among these species leads to reductions in fitness. If these conditions are met, parasite species will exert selective pressures on one another and coevolve such that their niches diverge. After enough time, the community would consist of species with nonoverlapping fundamental niches, and only the ghost of past competition would remain (Connell 1980). There is evidence supporting this scenario in some communities. In the intestinal helminth community of lesser scaup ducks, *Aythya affinis*, the prevalent and abundant species (or core species; Hanski 1982) have nonoverlapping niches more predictable and more evenly dispersed along the intestine than expected by chance (Bush and Holmes 1986b). Core species are the ones most likely to encounter one another generation after generation, and to interact if their niches overlap. In contrast, less prevalent and abundant species (satellite species; Hanski 1982) have niches randomly distributed along the intestine of their duck host. In other helminth communities in some avian hosts, species most likely to compete because of similar resource requirements, that is, congeneric helminth species, typically have disjunct niches or niches with minimal overlap (Stock and Holmes 1988). Since this applies to both fundamental and realized niches, an evolutionary niche divergence is the likely explanation.

There are however many parasite communities in which interactions are not detectable now and may never have been important because the various species are present at low abundance and do not co-occur regularly (e.g., Lotz and Font 1985; Goater et al. 1987; Kennedy and Bakke 1989). These are characterized by low infrapopulation sizes and low recruitment rates. Realized niches are not influenced by the presence of other species, whether or not fundamental niches overlap. In these isolationist communities, the various species are not abundant enough to exert mutual pressures on one another, and any observed niche restriction has evolved independently of interspecific interactions (Price 1980, 1987).

Thus we must consider other selective factors in any attempt to reconstruct the evolution of parasite niches. Euzet and Combes (1998) provide a good case study with their analysis of habitat selection by monogeneans that have become more or less endoparasitic. They argue convincingly that escape from predation, in the form of cleaner fish, for instance, has driven some monogenean taxa to invade protected sites on their fish hosts. In addition, they propose that polystomatid monogeneans parasitic in amphibians have shifted to an internal niche to allow them to survive in the terrestrial environments that their hosts have colonized. These examples illustrate quite well that forces other than avoidance of interspecific competition can select for restricted parasite niches.

Based mainly on observations of monogeneans and other ectoparasites of marine fish, Rohde (1979, 1991, 1994b) proposed that niche restriction in parasites often evolves to facilitate intraspecific encounters and mating. Several lines of evidence suggest that interspecific competition has no great ecological or evolutionary importance in these communities. First, many apparently suitable niches are vacant and not utilized. Second, the common occurrence of congeneric parasite species on the same fish species (fig. 8.11) suggests that interactions are not strong enough to restrict the number of morphologically and ecologically similar species using the same host (Rohde 1991). Third, the presence of one species, even at high numbers, usually has no numerical or functional effects on other species, that is, no detectable effects on their numbers or the extent of their realized niches (e.g., Geets et al. 1997). Other factors, such as the low mean infrapopulation sizes and the highly aggregated distributions of monogeneans, make it unlikely that

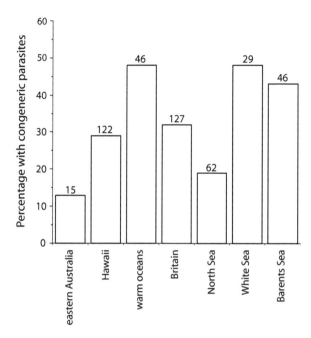

Figure 8.11 Percentage of marine fish species harboring congeneric species of parasites in different surveys made in various localities around the world. One survey (Australia) focused on coccidian parasites, the others on metazoans, especially ectoparasites. The number of fish species examined in each survey is indicated above the columns. Results of one survey (not shown) of deepwater fish found no fish harboring congeneric parasites; however, parasite diversity was very low in that survey. Since not all fish species examined were sampled sufficiently, the above values are minimum estimates; still, they suggest that congeners frequently coexist on the same host species. (Data from Rohde 1991)

large infrapopulations of different species will co-occur on the same host (Rohde 1991; Reversat et al. 1992). If opportunities for intense competition are few or nonexistent, there will be no selection for evolutionary site segregation. Encounters between potential competitors are also rare in other types of parasite communities, such as certain assemblages of larval digeneans in snails (Curtis and Hubbard 1993), and co-evolved adaptations against competition are unlikely in these communities as well.

Various observations on monogeneans and copepods ectoparasitic on fish point to another force acting to restrict the niche of parasites. The probability of mating in monogeneans may be low on average because of the typically small infrapopulation sizes shown by these parasites, especially on marine fishes. Rohde (1991) presented evidence that the realized niches of adult parasites are often more restricted than those of juveniles, and that they become even narrower at the time of mating. In the monogenean *Benedenia lutjani*, for instance, infective stages attach anywhere on the body of the fish host, but then migrate to the fish's pelvic fins at a time coinciding with their male reproductive organs becoming functional (Whittington and Ernst 2002). A few days later, when the worms reach sexual maturity—when both male and female organs are fully developed—they move again to aggregate on the fish's branchiostegal membranes, small skin folds behind the opercula (Whittington and Ernst 2002). Thus, as they mature, the parasites concentrate on an increasingly small surface area.

Other observations support the notion that a narrow niche favors mate finding in monogeneans. For instance, the niches of sessile species are narrower than those of related but more mobile species (Rohde 1991, 1994b). In a study of the monogeneans *Pseudohaliotrema* sp. and *Tetrancistrum sigani* on their shared fish host, Geets et al. (1997) found that both species had very narrow niches within the gills of the host, but that their niche breadth was completely unaffected by the presence of other ectoparasite species on the host. However, both species showed highly clumped distributions within their niche space, and different individuals of the same species co-occurred more frequently on the same gill filaments than expected by chance alone. In the monogenean *Haliometra* sp., Lo (1999) observed that many more worms were found close enough to one another to have physical contacts without having to detach and move than expected based on a uniform usage of the suitable niche space available to them. These findings are consistent with the idea that highly restricted niches in monogeneans facilitate mate finding. Of course, there may be exceptions. In monogeneans such as *Anoplodiscus cirrusspiralis*, which occupy the fins as well as the nasal cavities of their fish host (Roubal and Quartararo 1992), the disjunct nature of the niche and the resulting splitting of the infrapopulation are hardly compatible with the idea that niche restriction favors mate finding. In ectoparasitic copepods, which generally occur in low-density infrapopulations just like many monogeneans, mate finding is also believed to have played a significant role in the evolution of restricted niches (Rohde 1991, 1994b). As for monogeneans, mate finding is not the universal driving force behind niche evolution, however. In the copepod *Lernanthropus cynoscicola*, co-occurrences of males and females on the same portion of the same gill arch were not greater than expected by chance alone, a result suggesting that mating is not the main determinant of site selection (Timi, 2003). Nevertheless, the above and other observations suggest that, in general, niche restriction often serves to increase the chances of encountering a potential mate in low-density infrapopulations.

Reproduction may therefore have played a role as great as or greater than competition in the evolution of restricted niches in some parasites. There are at least two other ways in which the reproductive biology of parasites may have driven the restriction of the niche. First, monogenean and other ectoparasite species of the same genus often coexist on the same fish hosts (fig. 8.11), but the overlap of their realized niches is typically very limited. The overlap between the realized niches of congeneric species of fish ectoparasites is generally lower than that between noncongeners (Rohde 1991). Since all species, whether congeners or not, use similar resources, this observation cannot easily be reconciled with site segregation as a means to avoid competition. Niche restriction may instead serve to maintain or reinforce reproductive barriers between similar parasite species. This conclusion is supported by the interesting observation that congeneric monogenean species with overlapping niches have copulatory organs that differ much more in shape or size than those of congeneric species with segregated distributions on the host (Rohde

1991, 1994b). If niche restriction evolved to reduce interspecific competition, surely it would have acted similarly on all congeneric species regardless of the differences in their copulatory organs.

The second way in which reproduction may have shaped parasite niches has to do with gradients along the niche dimensions and their effects on the reproductive success of parasites. Many types of parasites achieve greater reproductive success in some portions of their fundamental niche than in others (e.g., Sukhdeo 1990a; Chilton et al. 1992). It is possible that these differences in fitness have favored a narrowing of the niche around sites where reproductive output is maximized.

The general role of interspecific competition in the evolution of restricted parasite niches is probably not negligible but is difficult to demonstrate. No experiments can separate the effects of competition, reproduction, or other factors; the above arguments are all based on observations, mostly of natural populations. Still, competition will exert selective pressures on the niche dimensions of two species sharing the same host species when (1) their abundances and distributions among hosts make them likely to co-occur frequently in large numbers in the same host individual, (2) their fundamental niches overlap significantly, and (3) one or more host resources required by both species is limiting. These conditions are probably satisfied by many communities of intestinal helminths of vertebrates, even species-poor ones if the right circumstances prevail. Studies on niche restriction are still few, and generalizations cannot be made at this point, but it is likely that interspecific competition has played a role in some types of parasite assemblages, though not in all of them.

8.5 Conclusion

Demonstrating the existence of past or present interspecific interactions among parasites is an important step toward understanding the structure of parasite communities. It is not an easy step, however. Numerical or functional responses observed in experimental infections may be artifacts of infrapopulation sizes larger than natural ones, and their interpretation requires caution. Similarly, the absence of overlap between the fundamental niches of two parasites may be the result of forces other than past or present competition. These caveats still allow some conclusions to be drawn. When competition does occur, numerical effects are often asymmetrical, with one species bearing the brunt of the competition. Often but not always, realized niches are narrower than fundamental niches, presumably to reduce overlap with the niches of other species. When niche segregation has evolved to fixation, past competition may only be the cause if opportunities for competition existed among ancestral species.

Interspecific competition has long been assumed to play an important role in structuring communities of free-living animals. Competition can also structure parasite communities, but it is not the only process capable of this; for instance, simple patterns in the recruitment of new parasites can also confer some nonrandom structure to parasite communities. However, assemblages of parasites sharing a host may have a more predictable structure if the component species interact with one another, and they may be likely to look like random collections of species if interactions are negligible. To some extent, it is the interactions between individual parasites of different species in the same host that may determine the species composition of communities and the relative abundance of different species. The next chapter picks up from this one and examines patterns of species associations in various parasite communities.

9 | Parasite Infracommunity Structure

The assemblage consisting of all parasites of different species in the same host individual, whether they actually interact or not, forms an infracommunity (Holmes and Price 1986; Bush et al. 1997). Infracommunities are subsets of the component community, which consists of all parasite species exploiting the host population. The composition of infracommunities, in terms of the number and identity of species and the relative numbers of individuals of each species, will depend on many factors. In theory, infracommunities can range from highly structured and predictable sets of species, to purely stochastic assemblages of species coming together entirely at random. Interactions among parasite species are one of the main forces that can shape an infracommunity and give it a structure departing from randomness. We may thus expect more predictable groupings of species if competitive interactions are strong. In isolationist parasite communities, where interactions are negligible either because of very narrow niches or small infrapopulation sizes, the co-occurrence of species in hosts is not expected to deviate from that expected by chance if interspecific interactions are the main structuring processes in parasite infracommunities. This chapter therefore picks up where the previous one left off and explores how species interactions and other processes determine the composition, diversity, and structure of parasite assemblages.

Interactions are far from the only factor that can influence infracommunity structure. Infracommunities are typically short-lived, their maximum life span being equal to that of the host. They are also in constant turnover of parasite individuals, with new ones being recruited and old ones dying out all the time. The probability of each parasite species being recruited into an infracommunity, and the way in which they join infracommunities, will also affect the composition and size of infracommunities and determine whether or not they have a predictable structure.

The vast majority of available studies on parasite community ecology are based on the examination of patterns observed in one or a few samples of host individuals, patterns existing among different infracommunities sampled at one point in time. These provide a snapshot of what the infracommunities looked like at the time of sampling, but no information on their development through time, from the moment the first parasite arrived in a host. Very few investigations have attempted a longitudinal survey of parasite infracommunities, beginning with uninfected hosts, either young individuals or animals reared in captivity, that were allowed to recruit parasites under natural conditions (e.g., Vidal-Martinez et al. 1998; Bagge and Valtonen 1999). Beyond confirming basic predictions, such that hosts are first colonized by the locally most prevalent parasite species, these investigations are too few to provide robust information on the general processes shaping the dynamics of parasite communities. Studies based on snapshots of infracommunities thus form the basis of the present chapter.

The aim of this chapter is to review the evidence that the structure of parasite infracommunities differs from that of random assemblages, and to discuss and evaluate the role of interspecific interactions and parasite recruitment patterns in structuring infracommunities. The relative ease with which several host individuals from the same population can be sampled facilitates statistical testing for significant departures from random species assemblages. The lack of adequate null models for comparisons with observed patterns has plagued some earlier studies on parasite infracommunities (see reviews in Simberloff 1990; Simberloff and Moore 1997). Throughout the chapter, I will emphasize the need to contrast observed patterns with those predicted by proper null models (Gotelli and Graves 1996). The first few sections of this chapter make use of evidence based either on species richness, or on the presence or absence of parasite species from infracommunities. Data on abundance (number of parasites per host) and biomass of various parasite species within an infracommunity can provide greater insights into the structuring of parasite assemblages, and they are discussed at the end of the chapter.

9.1 Species Richness of Infracommunities

Since each infracommunity is a subset of the parasite species present in the component community, the maximum number of species in an infracommunity is equal to the number of species in the component community. Typically this upper limit on species richness is not realized, at least in rich component communities, and usually no single infracommunity comprises all species that are locally available. In the first study of its kind, Kennedy and Guégan (1996) tested whether species saturation can constrain species richness in helminth infracommunities of eels. Among sixty-four component communities of intestinal parasites in eels from different localities in Britain, the maximum infracommunity richness did not exceed three parasite species per fish, whereas component communities often contained more than three species (Kennedy and Guégan 1996). The relationship between maximum infracommunity richness in eels and component community richness was curvilinear, such that infracommunity richness became increasingly independent of component community richness as the latter increased, and even leveled off at some maximum value. This suggests saturation of infracommunities with parasite species at levels below the component community richness.

There are at least three reasons to doubt the generality of this phenomenon, however. First, a subsequent study of intestinal helminths of eels across localities in England found different patterns. Norton et al. (2004a) observed a strong linear relationship between maximum infracommunity richness and component community richness, with no hint that infracommunities become saturated with species. Second, a curvilinear relationship between species richness on two scales (infracommunity and component community) is not by itself evidence of species saturation on the smallest scale (see Srivastava 1999; Hillebrand and Blenckner 2002; Hillebrand 2005). Independent evidence that species interactions can limit infracommunity richness is necessary; without it, curvilinear relationships must be interpreted with caution. For instance, Rohde (1998) has argued that differences in life span and transmission rate among parasite species in a component community are sufficient to create a curvilinear relationship and a false impression of species saturation in infracommunities. Third, the few other reports of curvilinear relationships between maximum infracommunity richness and component community richness are not particularly convincing. This is either because the curvilinear relationship depends entirely on a single data point, or because it does not explain much more of the variance in infracommunity richness than a linear function (e.g., Morand et al. 1999b; Calvete et al. 2004).

What may be saturated in infracommunities is total parasite biomass, and not species richness. In comparisons of helminth communities across 131 species of vertebrate hosts, total infracommunity biomass increases with host body mass (George-Nascimento et al. 2004). Larger-bodied host species have higher total energetic requirements and can also support a higher parasite biomass. The parasite biomass, however, can be allocated among few or many species, and limits on total infracommunity biomass do not in turn set limits on species richness.

Thus, species saturation of infracommunities is probably the exception rather than the norm. Among published studies on gastrointestinal helminths of birds and mammals, the relationship between maximum infracommunity richness and component community richness is clearly linear (fig. 9.1). The slope of the relationships suggests a process of proportional sampling (Cornell and Lawton 1992), whereby the maximum richness of infracommunities is always about one half that of the component community. Studies on species-rich helminth communities in birds suggest that empty niches are commonly observed in infracommunities (e.g., Bush et Holmes 1986b; Stock and Holmes 1988). Overall, the absence of species saturation and the availability of vacant niches suggest that infracommunity richness is strictly a reflection of the availability of species for recruitment.

The richness of infracommunities can provide other clues to the existence of structure in their composition. Not all infracommunities harbor the same number of parasite species; in any given component community, some infracommunities may include only one or two species whereas others harbor over ten. If hosts are random samplers of the parasites available in their habitat, then the richness values of infracommunities should be distributed as though species were assembled independently and randomly. If, on the other hand, assembly rules such as competitive exclusion or nonindependent species

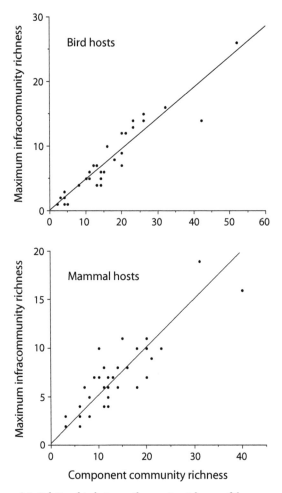

Figure 9.1 Relationship between the species richness of the component community and the maximum observed infracommunity richness, among thirty-one intestinal helminth communities in bird hosts and thirty-seven in mammal hosts. For both types of hosts, a line fitted to the points and forced through the origin had a slope close to 0.5; functions other than linear did not provide a better fit to the points. The relationships remain linear even after log transformation of the richness values. (Data sources given in Poulin 1996g, 1997e)

recruitment are playing important roles, we would expect the distribution of infracommunity richness values to differ from that expected by chance. Comparing the observed frequency distribution of infracommunity richness values with that predicted by a null model can serve as a test for the presence of structuring forces.

The Poisson distribution has been used as a null model for the distribution of infracommunity richness values (Goater et al. 1987); however, it assumes equal probability of infection for all parasite species present in the component community. A better null model should account for the different species occurring with different prevalences (Adamson and Noble 1992; Janovy et al. 1995). The expected frequencies of all possible

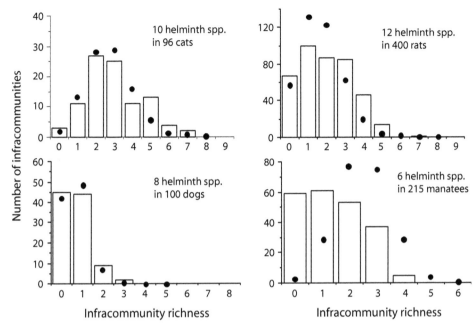

Figure 9.2 Observed and expected frequency distributions of infracommunity richness in four component communities of gastrointestinal helminths parasitic in wild mammals. The expected frequencies (black circles) were computed using the algorithm of Janovy et al. (1995) based on the respective prevalences of the different helminth species, and are what we would expect if the occurrence of species in infracommunities is independent of the presence of other species. Often observed distributions follow the expected pattern, sometimes there are more species-rich infracommunities than expected (as in rats), and sometimes there are more species-poor infracommunities than expected (as in manatees). (Data from Calero et al. 1950, 1951; Palmieri et al. 1978; Beck and Forrester 1988)

combinations of species can be computed, based on their respective prevalences, and used to build the null distribution of infracommunity richness values (fig. 9.2). The results of published studies of parasite communities have been compared with the expectations of the null model, and the outcome suggests that random assemblages are common (Poulin 1996g, 1997e; Norton et al. 2004a). In about two-thirds of component communities for which data are available, the observed distribution of infracommunity richness values did not depart significantly from that predicted by the null model. This applies to intestinal parasites of all kinds of vertebrates and to ectoparasites of fish.

The cases in which the distribution of infracommunity richness values differed from that predicted by the null model may be examples of interactive communities in which species interactions cause assemblages to depart from randomness. Situations where species-poor infracommunities are more frequent than expected by chance (fig. 9.2) could be the product of strong competitive interactions such as exclusion, which would prevent the unchecked accumulation of species within infracommunities. Positive interactions among species, on the other hand, could generate more species-rich infracommunities than expected by chance, with the presence of one species facilitating the recruitment of others. Deviations from the expectations of the null model are not necessarily due

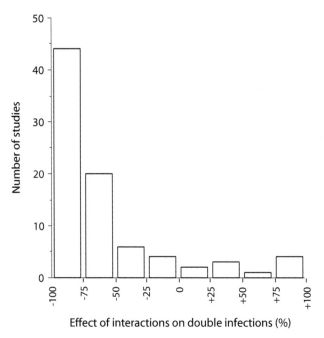

Figure 9.3 Relative effect of antagonistic interactions on the frequency of double infections of snails (i.e., infections by two parasite species) by larval digeneans across published studies in which a total of approximately 300,000 snails were examined. The effect of interactions was quantified as the relative difference between the observed and expected frequencies of double infections; a value of zero indicates no effect, and a value of −100 percent indicates a complete loss of double infections due to interspecific interactions. (Modified from Kuris and Lafferty 1994)

to interspecific interactions, however; other explanations are possible, such as extreme heterogeneity among host individuals in susceptibility to infection.

More complex null models can also be used, models in which variables other than the prevalence of the various parasite species can be included. Janovy et al. (1990) used computer simulations to generate random parasite assemblages and to assess the influence of aggregation and competitive interactions. They found that the basic null model, in which the presence of one species in an infracommunity is determined by its prevalence and independent of that of other species, provided the best fit to field data on metazoan parasites of fish.

In sharp contrast with studies of adult helminths in vertebrates, the evidence from surveys of larval digenean communities in snail intermediate hosts indicates that there are typically much fewer infracommunities harboring more than one species of digeneans than expected by chance (fig. 9.3). Interspecific competition among digeneans is seen by some as the most important structuring force in these communities, with dominant species excluding subordinate ones from individual snails (Kuris and Lafferty 1994). Other processes, such as spatial or temporal heterogeneity in infection rates, may also affect the likelihood of more than one digenean species co-occurring in the same snail (see

Esch et al. 1997, 2001). The use of null models can be a powerful tool to detect the action of structuring processes in infracommunities, but it does not always allow one to discriminate among the potential processes.

Species richness is only one descriptor of the composition of parasite infracommunities. Parasitologists have often used indices of diversity or evenness to quantify patterns in infracommunity structure. These measures combine species richness and the relative infrapopulation sizes of the different species. Diversity increases as both species richness and the evenness in their infrapopulation sizes increase. The merit of diversity and evenness indices in studies of parasite infracommunities has been questioned on statistical grounds (Simberloff and Moore 1997). There is at least one obvious biological reason why their use can be misleading. Parasites of different species in the same infracommunity can be of very different body sizes even if belonging to the same feeding guild. In the intestine of vertebrate hosts, for instance, large cestodes can dwarf other helminth parasites. In a situation like this, estimates of numerical dominance are of no importance. Measures of diversity or evenness would be more meaningful if biomass were used in the computations instead of numbers of individuals. Discrepancies in body size among helminth species in infracommunities also complicate the classification of parasites into core and satellite species, a common practice in parasite community studies (e.g., Bush and Holmes 1986a; Esch et al. 1990). Clearly, a species that is a satellite in terms of numerical abundance can be a core species in terms of biomass, making the use of relative abundance for classifying parasite species totally inadequate. The issue of numerical abundance versus biomass will be revisited in section 9.5.

9.2 Nestedness in Infracommunities

A nonrandom distribution of species richness values among infracommunities is one indication of the possible action of species interactions or other structuring forces. Another clue can be found in the species composition of infracommunities. If the species forming each infracommunity are not simply random subsets of the species available in the component community, then some species assembly rules may exist. For instance, nested subset patterns are common departures from random assemblages in free-living communities occupying insular or subdivided habitats (Patterson and Atmar 1986; Cook 1995; Worthen 1996; Wright et al. 1998). In a parasite component community, this pattern would mean that the species forming a species-poor infracommunity are distinct subsets of progressively richer infracommunities (fig. 9.4). In other words, common parasite species, those with high prevalence, would be found in all kinds of infracommunities, but rare species would only occur in species-rich infracommunities.

What processes can generate nested subset patterns? This phenomenon was first seen as the inevitable consequence of differences among free-living species in their abilities to colonize and persist in different habitats (Patterson and Atmar 1986). For parasite infracommunities, differences in transmission rates among parasite species could thus play a role. Another likely explanation is that individual hosts are not identical habitats,

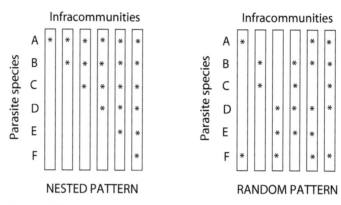

Figure 9.4 Two hypothetical distributions of parasite species among infra-communities (i.e., among individual hosts) illustrating the concept of nestedness. Each rectangle represents a different infracommunity, and infracommunities are arranged from least (at left) to most (at right) species rich. The average infracommunity richness and the average prevalence of the six parasite species are the same for the two examples. In a perfect nested design, a parasite species occurring in a host individual with n species will be found in all host individuals with $n + 1$ species.

and that the heterogeneity among them causes differential recruitment or extinction rates among parasite species. In communities of free-living organisms, habitat characteristics such as area and isolation can contribute to nested patterns (Patterson and Atmar 1986; Worthen 1996), and individual host characteristics could do the same for parasite infracommunities. For example, larger-bodied host individuals may be easier to colonize because of the greater amounts of food they ingest, their larger surface area, or their greater mobility. Larger host individuals are also older than small ones, and have thus had more time to accumulate parasites. It may be that all parasite species colonize larger hosts, whereas only the most vagile species will encounter and colonize smaller hosts. In this scenario, larger hosts would harbor richer infracommunities than small hosts, and a nested subset pattern of parasite species could develop along the gradient from small to large hosts. The size structure of the host population could thus provide the basis for the structure of parasite infracommunities.

The first test of potential nestedness among parasite infracommunities was performed in a community of monogeneans ectoparasitic on a tropical freshwater fish. In this system Guégan and Hugueny (1994) found a strong relationship between fish size and infracommunity richness. They also compared the observed distribution of parasite species among infracommunities to that expected under a null model of random allocation based on prevalence, and concluded that monogenean infracommunities displayed a nested species subset pattern. Guégan and Hugueny (1994) concluded that the size heterogeneity among host individuals could provide a structure to parasite infracommunities. Poulin and Valtonen (2001a) found general support for this conclusion. Across twenty-three samples of fish of different species, each representing a different parasite community, nested subset patterns were found in the endoparasite communities of some

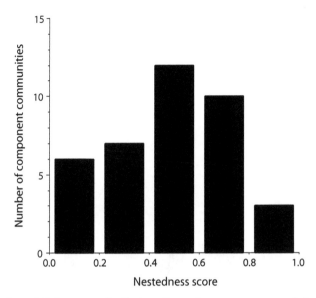

Figure 9.5 Frequency distribution of nestedness scores computed for infracommunities of thirty-eight component communities of metazoan ectoparasites on different marine fish species. The nestedness score is the standardized index C of Wright and Reeves (1992), which has a maximum value of one when nestedness is perfect. These scores (mean ± standard deviation, 0.483 ± 0.243) do not indicate any general tendency toward nestedness in communities of fish ectoparasites. (Data from Worthen and Rohde 1996)

fish species but not others. The correlation between fish size and the species richness of infracommunities was much stronger in parasite communities showing a nested pattern than in those that did not (Poulin and Valtonen 2001a). Nestedness seems to develop in parasite communities where host individuals accumulate parasite species gradually, in a predictable fashion related to their size or age. Nested subsets can be detectable even when size or age differences among hosts are small. Ontogenetic shifts in host diet or habitat can also contribute to the establishment of nested subset patterns in parasite communities (see Zelmer and Arai 2004). Other host features, such as different subsets of hosts specializing on different diets, can prevent the development of clear-cut nested subset patterns, while still producing other types of nonrandom patterns in the species composition of infracommunities. Thus, host biology is a driving force behind the structuring of parasite infracommunities.

Two important questions arise concerning nonrandom combinations of species among infracommunities. First, how common are these nonrandom patterns across different host species or different kinds of parasite communities? Fish and their parasites are by far the best-studied groups in this regard. Worthen and Rohde (1996) examined the component communities of metazoan ectoparasites on thirty-eight different species of marine fish and found that nestedness was extremely rare in these assemblages (fig. 9.5). Their results reinforced previous conclusions that fish ectoparasite communities are unstructured, random assemblages of noninteracting species. Rohde et al. (1998) investigated

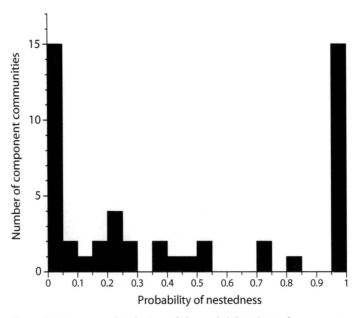

Figure 9.6 Frequency distribution of the probability that infracommunity composition departs from random assembly, across fifty component communities of metazoan ectoparasites on different marine fish species. A probability value less than 0.05 indicates that the observed infracommunities were significantly more nested than those generated by simulations based on a random assembly algorithm. A probability value greater than 0.95 also indicates a departure from random assembly, as the nestedness score of the real community was greater than 95 percent of scores generated by the null model. (Modified from Poulin and Guégan 2000)

an even larger data set on metazoan ectoparasite communities of fish and found that infracommunities form nested subset patterns in less than one-third of these component communities. The proportion of fish species in which nested patterns are observed is even lower for endoparasitic helminths (Rohde et al. 1998; Poulin and Valtonen 2001a).

An interesting trend emerges from these analyses of parasite communities in fish. In these studies, the presence of nestedness is tested by comparing the observed patterns in species composition of infracommunities with those predicted by a null model representing random assembly. The null model generates simulated infracommunities in which the probability of each parasite species being included is equal to its given prevalence in the actual sample. A nestedness score is then computed for each simulated community, and the nested score for the real community is then compared to the distribution of simulated scores. Typically, there are approximately as many communities with a nestedness score significantly lower than the simulated scores, that is, with a significant nested subset pattern, as there are with a nestedness score significantly higher than those obtained from simulations (fig. 9.6). Therefore, for every parasite community showing nestedness, there is one showing a nonrandom assembly of infracommunities that is the exact opposite of nestedness. This nonrandom pattern has been dubbed antinestedness, and it consists of situations where parasite species are always absent from infracommunities richer than the most depauperate one in which they occur (Poulin and Guégan 2000). Component communities with nested subsets of infracommunities consist of species characterized by generally higher prevalences and intensities of infection (i.e., higher infrapopulation sizes) than those in communities showing random or antinested patterns (Rohde et al. 1998; Poulin and Guégan 2000; Poulin and Valtonen 2001a). The biological interpretation of antinestedness is unclear, although several explanations are plausible (Poulin and Valtonen 2001a). The fact remains that for most component communities, that is, for most fish species examined, infracommunities apparently develop by random combinations of parasite species, a conclusion that is unchanged when other types of null model analyses are used (see Gotelli and Rohde 2002).

No other type of parasite communities has received this kind of attention with regard to the existence of nested patterns. Only a few component communities of intestinal helminths of mammals or birds have been studied, and these generally do not show a degree of nestedness greater than expected by chance (Poulin 1996g; Calvete et al. 2004). More empirical evidence is needed from these and other types of parasite communities. Based on the available information, however, there is no good reason to believe that infracommunities in general are something other than mere random assemblages of available species.

The second question we must ask about nonrandom species composition in infracommunities is: Are these nonrandom patterns a true property of certain host and parasite species? In other words, is a nonrandom pattern in infracommunity composition observed in a parasite component community likely to be observed again if the same component community is sampled at some other point in time, or if a community consisting of the same parasite species in the same host species is examined in a different locality? We must determine whether nested subset patterns or other departures from random assembly are local phenomena only, or whether they are predictable and repeatable in time and space. Carney and Dick (2000b) observed significant nested subset patterns in the composition of metazoan infracommunities of a freshwater fish across five lakes, and over two consecutive years in one of these lakes. In contrast, several recent studies on communities of helminth parasites of birds (Calvete et al. 2004) and fish hosts (Poulin and Valtonen 2002; Timi and Poulin 2003; Vidal-Martinez and Poulin 2003; Norton et al. 2004b) have found no predictable and repeatable nested subset patterns. A nested pattern of species composition among infracommunities in one parasite community at one point in time is no guarantee that a pattern will be found in the same community at a later point in time. Similarly, parasite communities comprising the same species in the same host species, but in different localities within the same geographical region, typically do not all display the same patterns of infracommunity composition: nestedness occurs in some but not in others. These findings indicate that local factors, such as the relative availability of intermediate hosts serving as prey to definitive hosts, play fundamental roles in shaping local parasite communities. They also mean that the existence of nested subset patterns, or any other departure from random assembly, does not reveal the action of some universally important ecological processes; if this were the case, then nonrandom patterns would be repeated across all samples of the same host species harboring the same parasite species.

9.3 Species Associations among Infracommunities

Nestedness is not the only nonrandom pattern one may find among a set of infracommunities. Co-occurrences of pairs or larger groups of parasite species that are more or less frequent than expected by chance could make the composition of infracommunities more or less predictable. Positive or negative associations among parasite species are perhaps the strongest suggestion that species interactions can structure infracommunities, that is, determine their species richness and composition. In essence this is no different than looking for a numerical response by one species to the presence of a second species (see chapter 8), except that it involves looking at the matrix of associations among all pairs of species in the community.

Several investigators have examined pairwise associations between parasite species across infracommunities in a component community (e.g., Moore and Simberloff 1990; Lotz and Font 1991; Haukisalmi and Henttonen 1993; Holmstad and Skorping 1998; Dezfuli et al. 2001b; Behnke et al. 2005). Associations between species can be evaluated in at least three ways. First, they can be calculated using presence-absence data; a positive association found with such data only implies that two species co-occur more often than expected by chance, regardless of their numbers. Second, pairwise associations can be quantified as the relationship between the infection intensities of both species, that is, between their infrapopulation sizes. Both approaches can produce spurious associations, and their interpretation must be based on comparisons with the appropriate null expectations. A third approach can be envisaged, based on the fact that numbers of parasites of different species are not always the most appropriate measures to use in tests of associations. Because different parasite species are of widely different sizes, especially among intestinal helminths, pairwise correlations between the infrapopulation biomass of two species across infracommunities can be instructive. Few investigators have used this approach (e.g., Moore and Simberloff 1990), though it may be more sensitive and better able to detect the action of interspecific interactions.

Variance tests on binary presence-absence data such as those on parasite species in infracommunities often make the assumption that the number of positive covariances should equal the number of negative ones if parasite infracommunities are random assemblages (e.g., Schluter 1984). An excess of positive or negative associations can be viewed as evidence for facilitation or antagonism, respectively. Several factors other than interspecific interactions can also bias the sign of pairwise associations one way or the other. For instance, heterogeneity among host individuals with respect to ecological or immunological factors can lead to certain hosts having higher probabilities than others of acquiring all kinds of parasites. In addition, the methods used to detect associations between the occurrences of pairs of species are more sensitive to positive associations than to negative ones (see Haukisalmi and Henttonen 1998; Vickery and Poulin 2002). In other words, these methods are more likely to detect a positive covariance between species than a negative one of comparable absolute magnitude. This may in part explain why positive associations between species regularly outnumber negative ones in parasite community studies (e.g., Lotz and Font 1991; Holmstad and Skorping 1998; Howard et al. 2001).

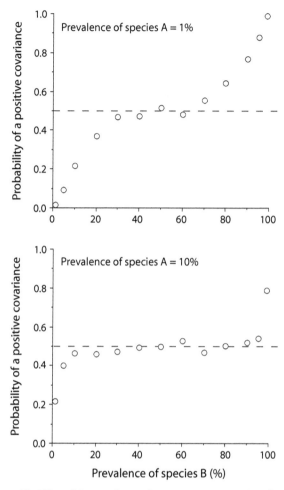

Figure 9.7 Effect of the prevalence of two parasite species, A and B, on the probability of observing a positive association between them. Each point is the average of five hundred Monte Carlo simulations in which sample size (the number of infracommunities sampled) was kept constant. The occurrence of each parasite species in an infracommunity was independent of the presence of the other, so that deviations from the expected value of 0.5 (the broken line) represent spurious associations. (Modified from Lotz and Font 1994)

These potential biases must be taken into account using a precise null model designed specifically for the parasite community under study. Lotz and Font (1994) did just that in their study of intestinal helminths of bats. They used computer simulations to demonstrate that the prevalence of two parasite species can influence the probability that a spurious association will be detected between them even if their respective occurrence in infracommunities is truly independent of one another (fig. 9.7). A rare species with low prevalence is likely to show a spurious negative association with other rare species and spurious positive associations with common species. The rarer the species, the more pronounced the bias. The null expectation for pairwise co-occurrences involving rare species is thus not one in which positive and negative associations are equally likely.

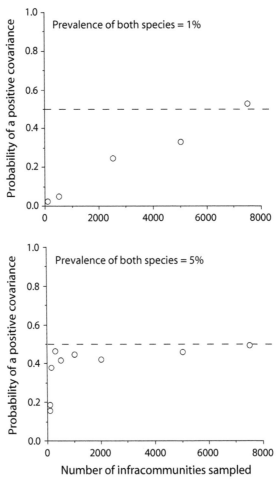

Figure 9.8 Effect of the number of infracommunities sampled on the probability of observing a positive association between two parasite species. Each point is the average of five hundred Monte Carlo simulations in which the prevalence of both species was kept constant. The occurrence of each parasite species in an infracommunity was independent of the presence of the other, so that deviations from the expected value of 0.5 (the broken line) represent spurious associations. (Modified from Lotz and Font 1994)

The number of infracommunities sampled from a component community can also affect the probability of obtaining false pairwise associations between parasite species (Lotz and Font 1994). When computing an association between two rare parasite species, a large sample size is needed before the probability of a positive covariance approaches that of a negative covariance (fig. 9.8). This is especially true for species with very low prevalence; but even for two species each with a 10 percent prevalence, about one hundred infracommunities must be sampled before biases in the sign of spurious associations disappear. Rare species therefore introduce a bias to the analysis of co-occurrences that can only be overcome by large sample sizes.

Lotz and Font (1994) used this information to quantify the true species associations in several component communities of intestinal helminths of bats. Using data from all infracommunities, including those in which one or both members of a species pair did not occur, they found that positive associations were more frequent than expected from randomizations based on the prevalence of the species in the component communities. When only infracommunities where both members of a species pair occur were used, however, the number of positive associations did not depart from the predicted one. These results indicate that detecting an association between species can be the simple consequence of joint presences or absences, rarity, or sample size, and have nothing to do with interspecific interactions.

Spurious positive or negative associations affect not only presence-absence data, but also correlations between the infrapopulation sizes of pairs of parasite species. In a matrix of all pairwise correlations among species in a community, the different correlation coefficients are not mutually independent (see Poulin 2005c). If there are strong positive correlations between the infrapopulation sizes of species A and B, and between those of A and C, then there must also be a positive correlation between B and C. Similarly, if there are strong negative correlations between species A and B and between A and C, then it follows that the correlation between species B and C must be positive. In any matrix of pairwise correlations between the species in a parasite community, the range of values that any correlation coefficient can take is constrained by the values of the other correlations (Poulin 2005c). This phenomenon reflects the structure and interdependence of the network of associations among species. Partial correlation analyses cannot fully account for this problem. Because it affects our ability to detect weak associations, it needs to be taken into account in any attempt to infer patterns of community structure from pairwise associations.

Sometimes, clear patterns emerge from an analysis of pairwise species associations. For example, Bush and Holmes (1986a) found only positive associations between all pairwise combinations among the sixteen most common species of intestinal helminths in the lesser scaup duck, *Aythya affinis*. Only a quarter of these associations were statistically significant, but the fact that they all go in the same direction suggests that some force acts to provide a predictable structure to the different infracommunities. In this case, the use of common intermediate hosts by subsets of the helminth species accounted for the positive associations. Typically, however, a mixture of positive and negative associations will be observed, some of which may be statistically significant. Unless a precise null model similar to that proposed by Lotz and Font (1994) is used for comparison, the validity of emerging patterns cannot be judged. Few studies to date have used null models that take into account the biases mentioned above.

At present it is therefore difficult to evaluate the generality of interspecific associations as determinants of infracommunity structure. As with analyses of nested subset patterns, one must ask whether pairwise associations between parasite species are repeatable in space across comparable component communities, and stable over time within a given component community. The few comparisons among component communities consisting of the same parasite species in the same host species but from different localities suggest that patterns of pairwise associations between species are not consistent in space (Kehr et al. 2000; Dezfuli et al. 2001b; Vidal-Martinez and Poulin 2003; Behnke et al. 2005). In addition, pairwise associations between parasite species often show little consistency among different subsets of the same host population, as seen in comparisons between host age classes or between male and female hosts (e.g., Behnke et al. 2005). All this strongly suggests that particular interspecific associations are not properties of those species that will be manifested wherever they coexist. Instead, it appears that local factors dictate whether or not various associations will develop. In contrast, Lello et al. (2004) found associations among gastrointestinal helminth species in a rabbit population that were consistent over a twenty-three-year period. In this system, these predictable associations provided a permanent force determining the composition of infracommunities. Interspecific associations might only become established in certain host populations, for instance where two or more parasite species occur at relatively high abundances; in those situations, interspecific associations will contribute to community structure.

9.4 Species Recruitment and Infracommunity Structure

Most empirical studies of parasite infracommunities have been performed on the intestinal helminths of vertebrates, and much of the theory on parasite communities is derived from these studies. Assemblages of intestinal helminths develop from the moment an individual host acquires its first parasite until the death of the host. An unstated assumption behind the hypothesis that interspecific interactions between parasites can structure infracommunities is that parasites join the infracommunity independently and in a random fashion, and that their initial establishment and subsequent survival are determined by host responses, the presence of other parasites, priority effects, and the strength of interspecific interactions. In fact, many intestinal helminths arrive in an infracommunity when the definitive host consumes their intermediate host. Because intermediate hosts can harbor many larval parasites of one or more species, new recruits join infracommunities in "packets" rather than singly. One parasite species may frequently come associated with another species in the intermediate host, and the recruitment of new parasites into an infracommunity is thus a nonrandom process.

Bush and Holmes (1986a) found evidence that the recruitment of parasite species following different routes can have impacts on the composition and richness of infracommunities. They observed that there were two suites of helminth species, each using a different crustacean species as intermediate host, among the intestinal parasites of lesser scaup. Six cestode species and one acanthocephalan species used the amphipod *Hyalella* as intermediate host, and three cestodes and one acanthocephalan used the amphipod *Gammarus*. Within each suite of species, there were consistent positive associations among species inside the definitive host. Variability among infracommunities was due to a large extent to differences among individual definitive hosts in the relative contribution of each amphipod to the diet.

The recruitment of new parasites into an infracommunity as packets rather than as individuals may not be limited to parasites with complex life cycles using intermediate hosts. It may extend to parasite species with one-host life cycles that release eggs in their host's feces. Often, a fecal patch will contain eggs from different species of parasites whose adults happened to co-occur in an infracommunity. Acquisition of one parasite species by a host grazing on one patch of eggs or infective larvae may often be coupled with the acquisition of other species, so that infracommunities do not recruit new members singly and independently even if these species have simple one-host life cycles.

Intermediate hosts (or patches of infective larvae) can be seen as source communities for definitive hosts (Bush et al. 1993). Infracommunities in definitive hosts may simply be the sum of nonrandom packets of larval parasites arriving from intermediate hosts, minus the individuals that fail to establish and survive. The structure of infracommunities in definitive hosts could represent the combined structures of the source communities acquired by the definitive host. Searching for order in the resulting assemblage may be pointless. One process that could provide a regular structure to infracommunities in definitive hosts following the stacking of source communities would be for strong interspecific interactions to override the acquired structure. Strong competition and facilitation could eliminate any patterns of association transmitted from intermediate hosts and reshape the infracommunity entirely. From the evidence presented in the preceding sections, it looks as though this does not happen often.

The frequently observed positive or negative associations between parasite species in definitive hosts could merely reflect associations existing in intermediate hosts and transferred during transmission. There may be no need to invoke interspecific interactions in the definitive host to explain these patterns. Lotz et al. (1995) used computer simulations to assess the contribution of species associations within intermediate hosts to the overall structure of infracommunities in definitive hosts. They found that pairwise associations between parasite species in intermediate hosts can easily be transferred to the infracommunities in definitive hosts (fig. 9.9). The establishment and survival rates of the two associated species in the definitive host influences the extent of the association's transfer. For instance, the transfer of associations between species is nullified when the survival rate of one or both species is low because poor survival means that parasites are not transmitted as intact packets of recruits, in other words, low survival of one species destroys the integrity of the association.

In Lotz et al.'s (1995) simulations, the strength of the association between parasite species in definitive hosts was never as great as that in intermediate hosts (fig. 9.9). This weakening of the association is expected from the imperfect establishment and subsequent survival of parasites in their definitive host. In tests of interspecific interactions among parasites in the definitive host, associations in the intermediate hosts could serve as a rough null model specifying the maximum strength of associations in the absence of interactions. Departures from the null model, for instance stronger associations in the definitive host than in the intermediate host, could indicate the action of post-transmission processes such as facilitation. Another important result of Lotz et al. (1995) is that positive associations in intermediate hosts are more readily transferred to infracommunities in

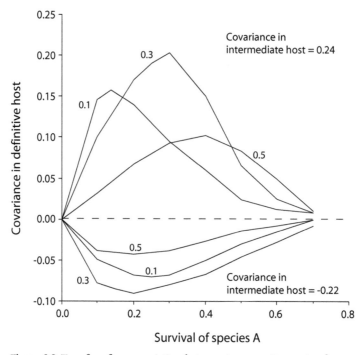

Figure 9.9 Transfer of an association between two parasite species from a source community (in intermediate hosts) to a target community (in definitive hosts). The results are from simulations in which two parasite species, A and B, share the same intermediate and definitive hosts. The two species are aggregated in the intermediate host population ($k = 0.2$, mean number of parasites per host = 5), and either positively or negatively associated, i.e., their infrapopulation sizes in intermediate hosts are either positively or negatively correlated. The covariance between the two species in definitive hosts is presented as a function of the survival rate of species A in the definitive host, which includes establishment success; the values above the curves indicate survival of species B. The transfer of both positive (above broken line) and negative (below broken line) associations is nil when survival rates of both species are low or high, and is greatest at intermediate values when both species have similar survival rates. (Modified from Lotz et al. 1995)

definitive hosts than negative associations (fig. 9.9). This could account for the many instances in which positive associations among intestinal helminths in vertebrate hosts outnumber negative ones (e.g., Lotz and Font 1991, 1994).

Vickery and Poulin (2002) have extended the approach pioneered by Lotz et al. (1995) to include the possibility that one of the two species in an association is capable of manipulating the behavior of the intermediate host in ways that enhance its susceptibility to predation by the definitive host (see chapter 5). They found that not only existing associations involving a manipulative species are readily transferred from the intermediate host to the definitive host, but they can be amplified: the magnitude of the association becomes greater in the definitive host (Vickery and Poulin 2002). Parasite infracommunities in intermediate hosts have received very little attention, but a few recent studies have identified pairwise associations between larval helminths in intermediate hosts, in which both species also share the same definitive host and one species in the pair is known to manipulate the behavior of the intermediate host (Thomas et al. 1997, 1998a; Dezfuli et al. 2000; Poulin and Valtonen 2001b). This way, associations between parasite species can be passed on from one host to the next, up the food chain, without the need to invoke species interactions, the other great structuring force. The only way to distinguish between the two sources of structure would be to examine infracommunities in both intermediate and definitive hosts in the same system. There have been no such studies to date, and they may prove logistically complicated, but they are essential if we are to get the complete picture on infracommunity structure.

9.5 Species Abundance and Biomass in Infracommunities

The failure of previous attempts to find universal and predictable patterns in parasite community structure may be due in large part to their focus on presences or absences of parasite species in different infracommunities. There have been comparatively much fewer investigations of parasite community structure based on abundance data, that is, on the mean infrapopulation sizes of the different parasite species in the community and how they compare to each other. Usually, one or very few species account for the vast majority of individuals in parasite communities (fig. 9.10). The typical infracommunity thus consists of very few numerically dominant species, and a few to several species represented by only a few individuals.

When species in a community are ranked based on how many individuals they contribute to the community, and their relative contributions, that is, their relative infrapopulation sizes, are plotted against their rank, as in figure 9.10, we obtain a curve whose shape can provide insights about the factors structuring the community. In particular, the steepness of the curve directly reflects the dominance of some species over others. Although this approach is common practice in ecology, it has rarely been applied to parasite communities. Norton et al. (2003) found that communities of intestinal helminths in eels fitted the log-normal distribution. This distribution has long been considered one that provides a generally good fit to species abundance distributions in communities of free-living organisms (see Sugihara 1980). The log-normal distribution resembles a normal, bell-shaped frequency distribution of species abundances within a community when the abundance values are log transformed. When relative abundances are plotted against species ranks, the log-normal distribution results in an S-shaped curve, characterized by a steeper drop in relative abundance among species of high rank and those of low rank compared to species of intermediate ranks. In figure 9.10, several intestinal helminth communities follow the pattern expected from a log-normal distribution of abundances, such as those in *Oncorhynchus kisutch*, *Xiphias gladius*, *Procyon lotor*, or *Somateria mollissima*. However, helminth communities in other host species, such as *Pelecanus occidentalis*, *Aythya affinis* or *Eudocimus albus* (see fig. 9.10), do not fit well the expected lognormal distribution because they include an excess of species with low abundance. Hubbell (2001) pointed out that this phenomenon, an excess of rare species, is common

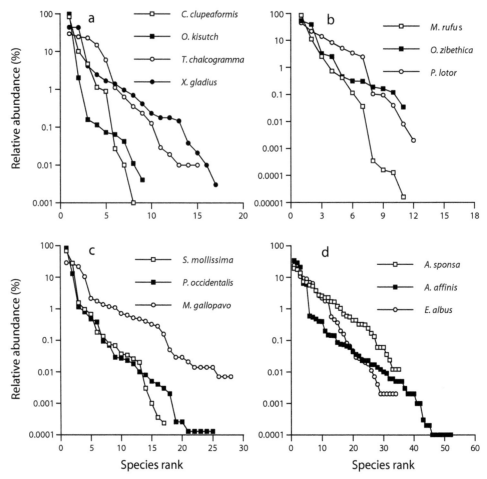

Figure 9.10 Patterns of relative species abundance in the gastrointestinal helminth communities of: a, four fish species (*Coregonus clupeaformis, Oncorhynchus kisutch, Theragra chalcogramma,* and *Xiphias gladius*); b, three mammal species (*Macropus rufus, Ondatra zibethica,* and *Procyon lotor*); c, three bird species with moderately rich helminth communities (*Somateria mollissima, Pelecanus occidentalis,* and *Meleagris gallopavo*); and d, three bird species with very rich helminth communities (*Aix sponsa, Aythya affinis,* and *Eudocimus albus*). Relative abundance is the percentage of the total number of parasite individuals belonging to a given species, averaged across all infracommunities sampled; helminth species are ranked from most to least abundant. A minimum of one hundred infracommunities (i.e., one hundred host individuals) was examined for each host species except *A. affinis,* for which forty-five were sampled; in all host species, estimates of relative abundance are based on the recovery of a total of at least 8,300 individual worms. (In order, data from Leong and Holmes 1981; Arthur 1984; Hogans et al. 1983; Arundel et al. 1979; McKenzie and Welch 1979; Snyder and Fitzgerald 1985; Bishop and Threlfall 1974; Courtney and Forrester 1974; Hon et al. 1975; Thul et al. 1985; Bush and Holmes 1986a; Bush and Forrester 1976)

in communities of free-living organisms. This served as a basis for his neutral theory of biodiversity, which predicts that species abundances will follow the zero-sum multinomial distribution (Hubbell 2001). This distribution is similar to the log-normal except that it has a longer tail of rare species. The neutral theory of biodiversity rests on assumptions that are not compatible with the known biology of parasites (see Poulin 2004), and it will not be discussed further here. There is, however, a family of mechanistic models proposed recently by Tokeshi (1999) that predicts a range of shapes for the

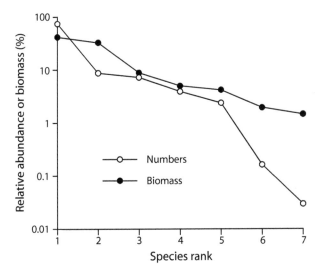

Figure 9.11 Patterns of relative species abundance in the gastrointestinal helminth communities of the fish *Hemigymnus melapterus*. Only the seven most abundant parasite species are included. Abundance is expressed both in terms of number of individuals and biomass, the latter estimated from the volume of each parasite. Relative abundance (or biomass) is the percentage of the total number of parasite individuals (or total biomass) belonging to a given species, averaged across all infracommunities sampled; helminth species are ranked from most to least abundant. (Data from Muñoz and Cribb 2005)

relative-abundance-versus-rank curve. The curves observed in real parasite communities can be compared with those predicted by the different models, to see which one provides the best fit. As each model assigns different weight to various structuring forces, such as interspecific interactions, this approach can identify key factors shaping parasite infracommunities.

First, though, we must consider which measure of species abundance should be used to generate curves of relative abundance versus species rank. Numerical abundance, or the mean infrapopulation size of a species, is the most likely candidate; it is easy to obtain and is equivalent to the abundance measures used in many studies of free-living animals. There is one major problem, however: different species of parasites can differ widely in size, such that two species with equal numerical abundance can differ by one or more orders of magnitude in terms of biomass. Even within parasite species, body sizes can vary substantially (see section 4.1). Intra- and interspecific variation in body size weakens the relationship between abundance and biomass. For example, among the intestinal helminth parasites of the fish *Hemigymnus melapterus*, the difference in individual body volume between the largest and smallest species is of over three orders of magnitude (Muñoz and Cribb 2005). Using biomass to compare species instead of numerical abundance produces a different pattern in this community (fig. 9.11). Based on biomass, the dominant species is less dominant and the rare species are less rare than what is suggested by patterns in numbers of parasites (fig. 9.11). The curve is flatter when biomass is used, suggesting a more even representation of the different species in the community.

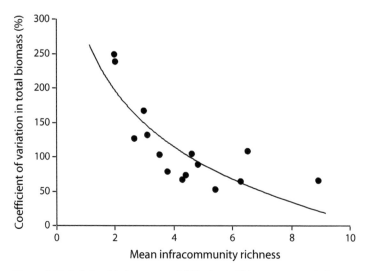

Figure 9.12 Relationship between variability in total biomass among infracommunities and mean infracommunity species richness, across the component communities of metazoan parasites in fifteen species of marine fish hosts. Variability in biomass is measured as the coefficient of variation in biomass (standard deviation expressed as a percentage of the mean); biomass was estimated from the volume of each parasite. (Modified from Mouillot et al. 2005)

True patterns of community structure may be completely missed when an analysis focuses on numbers rather than biomass, especially if one is interested in resource partitioning and how it affects the community. The functional importance of a parasite species in a community is more likely dependent on its biomass than on the number of individuals present.

Biomass should therefore be used to measure relative abundance in parasite communities in order to test models such as those of Tokeshi (1999). He introduced a series of niche-oriented stochastic models to fit species abundance patterns. Tokeshi's niche-apportionment models are mechanistic models that allow simulations of the processes thought to be assembling communities. In these models, the abundance of each species is proportional to the fraction of the total niche that they apportion; each model is based on different apportionment rules, representing a spectrum of possibilities ranging from complete independence among species to strong species interactions (Tokeshi, 1999). When the communities simulated according to a particular apportionment model have patterns of species abundance that differ significantly from that of the real community under investigation, then this model and the ecological mechanisms it represents can be excluded as possible structuring forces. Mouillot et al. (2003) applied these niche-apportionment models to parasite community structure, using relative biomass as an estimate of species abundance. They found that one model, the random-assortment model, provided a good fit to the observed patterns of species abundance in the communities of metazoan parasites on three of the six host fish species investigated; no other model fit the observed patterns (Mouillot et al. 2003; see also Muñoz et al. 2006). The random-assortment model simulates a situation where the abundances of the different species in a community vary independently of each other (Tokeshi, 1999). These findings support many previous studies based on presence-absence data, reviewed earlier in this chapter, which concluded that interspecific interactions play very minor roles in structuring parasite communities.

A focus on parasite biomass reveals other facets of infracommunity structure that otherwise may have gone unnoticed. Mouillot et al. (2005) have shown that there exists a clear negative relationship between variability in total parasite biomass among infracommunities in a component community, and the mean species richness of these infracommunities (fig. 9.12). Total parasite biomass per infracommunity is more stable, and less variable, in species-rich parasite communities. Although this greater stability is not necessarily associated with greater structure and predictability in terms of species composition, it nonetheless indicates that high species richness can produce more homogeneous infracommunities. This kind of insight is only made possible by the measurement of parasite biomass in addition to the more traditional counts of individuals, and it strongly argues in favor of further studies of biomass patterns in parasite communities.

9.6 Conclusion

Structure in a species assemblage means a predictable, or repeatable departure from random assembly. It can only be demonstrated by the presence of consistent statistical associations between species or other nonrandom patterns. Much of this chapter has emphasized the need to contrast observed patterns with those predicted by accurate null models. This is an approach only recently adopted by parasite ecologists, and more empirical studies are needed. Based on the evidence currently available, however, it appears that the richness and composition of parasite infracommunities are often the product of stochastic events and that infracommunities are often simply unstructured assemblages.

There have been attempts to identify the factors responsible for the development of either isolationist, unstructured communities, or interactive, structured communities. Host characteristics could be important determinants of community structure. For instance, isolationist communities of intestinal helminths may be more likely in hosts with simple alimentary tracts, specialized diets, and low vagility (Kennedy et al. 1986; Goater et al. 1987). It has been proposed that the distinction between isolationist and interactive communities amounts to a distinction between ectothermic and endothermic hosts (Goater et al. 1987). There are exceptions to this paradigm, however, that raise questions about its generality (Moore and Simberloff 1990). Although there may be broad differences between the intestinal infracommunities of fish or amphibians on the one hand, and birds or mammals on the other, parasite communities within these host groups differ widely. Among intestinal helminth communities in fish, for instance, there is evidence that communities span the full range from isolationist to interactive (Poulin and Luque 2003). The likelihood that a community is influenced and structured by species interactions may depend on its productivity, that is, on the average biomass of infracommunities in fish hosts (Mouillot et al. 2005). Thus, host features alone do not constrain the sorts of parasite communities they harbor. The only parasite communities for which a general conclusion is possible are larval digeneans in their mollusc intermediate host. In these communities, interspecific interactions are generally strong enough to override the stochastic assembly of species, creating assemblages whose structure regularly departs dramatically from randomness, typically with fewer species co-occurrences in the same host individual than expected.

In the end, infracommunities can only be understood by examining the way in which new parasites are recruited, that is, the way in which parasites are acquired by hosts. Recent studies have emphasized that infracommunities in definitive hosts are nothing but the combination of infracommunities from intermediate hosts, on which other processes may act. This view adds a new level to the hierarchical arrangement of parasite communities and promotes the study of parasite communities at several levels simultaneously. There is no doubt that future studies of infracommunity structure will need to address species recruitment as well as species interactions in order to elucidate the structuring forces acting on parasite assemblages.

10 | Component Communities and Parasite Faunas

The infracommunities of parasites discussed in the previous chapters are subsets of a larger assemblage of species known as the component community, or the ensemble of populations of all parasite species exploiting the host population at one point in time. Component communities are longer-lived assemblages than any of their infracommunities; they last at least a few host generations and usually much longer, as long as the host population persists in time. This creates an important distinction between the structure and dynamics of infracommunities and component communities. Whereas infracommunities are assembled over ecological time scales by infection and demographic processes, component communities are formed over evolutionary time scales by processes such as invasions, speciation, extinctions, and colonizations or host switches.

The various parasite component communities of the different populations of one host species are all subsets of the entire set of parasite species exploiting that host species across its entire geographical range. I will refer to this larger collection of parasite species as the parasite fauna. The term *community* cannot apply to the parasite fauna because usually no single host population will harbor all species in the fauna, that is, some species in the fauna never actually co-occur in nature. This makes parasite faunas artificial rather than biological assemblages, but the processes generating them are biological ones and worthy of discussion. Parasite faunas are formed over evolutionary time as new parasite species join component communities, and as others become extinct from all component communities. Just as gene flow maintains the integrity of a parasite species across its different interconnected populations, exchanges of individuals and species among component communities maintains the community network that is the parasite fauna. The parasite faunas of different host species will evolve over their phylogenetic history, with related species inheriting original faunas from a common ancestor that can then be modified independently in different lineages.

In this chapter, I examine parasite assemblages at higher hierarchical levels and greater spatial and temporal scales than in the previous chapter. The evolution of faunas and component communities of parasites is discussed jointly because of the clear links between these assemblages. Processes mentioned with regard to one level can easily be applicable to the other. An advantage of looking at these two levels together is that we get a more complete picture of the evolutionary history of assemblages of parasite species.

10.1 Richness and Composition of Component Communities

Theoretically, the maximum number of parasite species that can be found exploiting a population of a host species equals the number of species in the parasite fauna. In practice, however, the richness of the component community rarely approaches that of the parasite fauna. There is evidence that component communities often reach a saturation level of species below that of the parasite fauna (Aho 1990; Kennedy and Guégan 1994, 1996). This emphasizes the artificial nature of the fauna; it is not a real assemblage from the parasites' perspective because some species may never physically meet in the same component community. Typically, in comparisons across related host species, the maximum or average richness of component communities increases with the richness of the parasite fauna until leveling off beyond a certain point (fig. 10.1). A linear relationship would indicate a process of proportional sampling, with component communities consisting of a fixed proportion of the species in the fauna regardless of the latter's richness. The curvilinear relationship observed instead suggests that component communities become saturated with species where the curve reaches an asymptote, and that populations of host species with very rich parasite faunas do not harbor proportionately richer parasite communities than those of host species with less diverse parasite faunas.

There are at least two potential explanations for this apparent saturation. First, the number of niches available in an individual host may set a limit on the number of parasite species that can coexist in a host population. If this were the case, once the component community becomes saturated with species, further additions of species would be impossible unless new species outcompete and replace established ones. This would result in the parasite species richness of the fauna being a poor predictor of the richness of component communities (Kennedy and Guégan 1994). Second, the apparent saturation of species in component communities might simply reflect the fact that they cannot sample the entire pool of parasite species available in the parasite fauna. Many parasite species are not distributed throughout their host's entire geographic range, and are thus excluded from some component communities. Thus, it might not be lack of niche space that limits species richness in component communities, but lack of new colonizing species within the region. This second explanation appears more plausible because there is no evidence that additional species cannot be accommodated in component communities. In many plant and animal communities, when species are ranked by body size, the size ratio of adjacent species declines steadily with increasing size (Ritchie and Olff 1999). This pattern is expected from tightly packed communities where resource partitioning results in limiting similarity in body size among coexisting species. Communities of ectoparasites and internal helminths of fish are clear exceptions to this pattern, suggesting that they are far from packed with species (Rohde 2001). If new species could

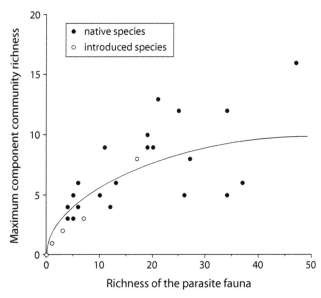

Figure 10.1 Relationship between the number of helminth parasite species in the richest observed component community and the richness of the parasite fauna, obtained from published surveys, across thirty-two species of British freshwater fish. Native and introduced fish species are indicated by different symbols. The curvilinear function illustrated provides the best fit to all the data points. Note, however, that a linear relationship provides the best fit to the data for introduced fish species. (Modified from Kennedy and Guégan 1994)

be introduced to a component community, and if they were capable of exploiting the host species, they could probably become established. This is illustrated by the component communities of Hawaiian freshwater fishes, which have increased in species richness by 30 to 50 percent following the introduction of exotic fish species, and their parasites, to the islands (Font 1998).

Thus, the richness of a component community is limited by how many species from the parasite fauna of the host species can reach it. Though regional availability of species places an upper limit on component community richness, local processes are important influences on whether or not a parasite species becomes established. In general, the physical or biological characteristics of the habitat, combined with various historical events leading to gains or losses of parasite species, will determine local parasite species richness and shape the evolution of component communities.

Frequent contacts and exchanges of parasites between adjacent host populations of the same species should lead to highly homogeneous component communities all consisting of the same species. In contrast, if different host populations are physically isolated from one another, we may expect them to develop very different component communities. These would also tend to be poor in species, since they cannot fully sample the pool of species available regionally. Distance among component communities indeed proves a good predictor of similarity in species composition and species richness among

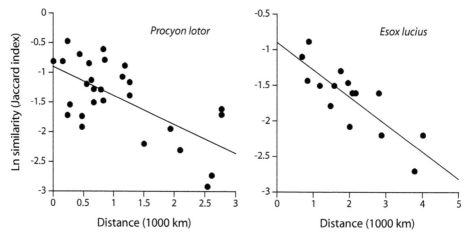

Figure 10.2 Relationship between similarity in species composition and distance between component communities, for all pairwise comparisons among eight helminth component communities of a mammalian host, the raccoon *Procyon lotor*, and among six helminth component communities of a fish host, the pike *Esox lucius*. The Jaccard index corresponds to the number of parasite species shared by both component communities divided by the total number of species found in the two communities put together; it ranges from zero (no species in common) to one (the two component communities have exactly the same parasite species). (Modified from Poulin 2003b)

parasite component communities of fish hosts (Poulin and Morand 1999). In multivariate analyses, it is often the best predictor of species overlap between two component communities: the closer they are situated to each other, the more likely they are to consist of the same parasite species. Similarity in species composition among component communities tends to decay exponentially with increasing geographical distance, for a range of different kinds of hosts and parasites (Poulin 2003b; Fellis and Esch 2005; Krasnov et al. 2005; Oliva and González 2005). In some host species, the relationship between similarity and distance between component communities is quite obvious (fig. 10.2); in others, there is no detectable effect of distance on similarity between component communities (e.g., Kennedy 2001). Decay in similarity with increasing distance is either not a universal phenomenon, or, more likely, it becomes masked by other processes that tend to homogenize the composition of component communities regardless of the distance separating them.

Vagility, dispersal, or migratory habits should be important traits of hosts with respect to the richness and composition of parasite component communities. Mobile hosts have the potential to disperse parasites widely, thus promoting a greater homogeneity of parasite component communities within a region. This should become apparent when the compositions of different component communities from the same host species are compared. Parasitologists have used two ways of comparing component communities. First, the Jaccard index can be computed as the proportion of parasite species shared by different component communities. This gives a measure of the qualitative similarity between communities. Second, a more quantitative index can be used to measure the similarity of component communities based on the relative abundance of their various parasite species. Using either similarity measure, it is generally clear that component communities of host species in which populations are isolated from one another, such as freshwater fish or amphibians, are much less homogeneous than those of vagile hosts like marine fish or birds (Poulin 1997e). In addition, in host species where similarity of component communities decays with increasing distances among them, rates of decay tend to be higher for freshwater fish hosts than for marine fish hosts (Poulin 2003b; Oliva and González 2005). This is indeed expected, as host and parasite dispersal should be more constrained among fragmented freshwater habitats than in open and continuous marine systems. From the few available estimates, it looks like rates of decay in similarity are higher among parasite communities of terrestrial mammals than among those of freshwater fish (Poulin 2003b; Krasnov et al. 2005); perhaps the fragmentation of terrestrial habitats, though not as obvious as that of freshwater habitats, is also limiting exchanges of parasite species among component communities. Many host characteristics other than vagility may also influence the richness and composition of component communities (Esch and Fernández 1993). For instance, the habitat preferences and specialized diets of some freshwater fish appear responsible for their component communities having predictable species composition on continental scales, with the same suite of parasite species occurring in most component communities (Carney and Dick 1999; Nelson and Dick 2002). These features of the host species restrict the parasites that can be encountered to a predictable set of species. Some other host characteristics important for the evolution of component communities will be addressed in the discussion of parasite faunas (section 10.3).

It is assumed in the above discussion that exchanges of parasite species between component communities are dependent mainly upon host movements. Exchanges can take other routes, however. Parasites may use other host species as intermediate, definitive, or paratenic hosts, and use them as vehicles for the colonization of other nearby or distant component communities. Among species of freshwater fish, for example, parasite faunas comprising several species of helminths using birds as definitive hosts consist of more homogeneous and predictable component communities than faunas comprising mostly parasite species incapable of dispersing from one water body to another (Esch et al. 1988). In this case, the parasites have a colonizing ability independent of the vagility of the fish host, a fact that increases the similarity of component communities. Fish parasites using bird definitive hosts, or allogenic parasites, probably form a more predictable species subset among component communities than autogenic species, those completing their life cycle in water and incapable of crossing land barriers between freshwater bodies. In the whitefish *Coregonus lavaretus*, for example, component communities of helminth parasites in different interconnected lakes share allogenic species independently of the distance separating them. In contrast, similarity in the composition of their autogenic species decreases significantly with increasing distance (Karvonen and Valtonen 2004). The same phenomenon has even been observed on a much smaller spatial scale, with the similarity of communities of autogenic parasites decreasing with increasing distance along the shore among fish samples from the same lake (Karvonen et al. 2005). In contrast, Fellis and Esch (2005) found that the similarity of communities of allogenic parasites in sunfish, *Lepomis macrochirus*, decreased with increasing distance among isolated ponds, whereas similarity in autogenic parasites was independent of interpond distances. As opposed to the interconnected lakes studies by Karvonen and Valtonen (2004), the physical isolation of the ponds in Fellis and Esch's (2005) study and the inability of autogenic parasites to disperse readily meant that the distribution of these parasites across the landscape was random at the study scale.

Findings such as these do not mean that autogenic species are incapable of dispersing and colonizing new host populations. The six acanthocephalan species using freshwater fish as definitive hosts in the British Isles have all successfully spread to most freshwater systems (Lyndon and Kennedy 2001). They have achieved this feat in large part thanks to their ability to use either migratory eels or trout as suitable definitive hosts, and consequently as dispersal agents. As a rule, though, allogenic species are dispersed more readily, and thus contribute more to the maintenance of homogeneity among component communities, than autogenic species.

Habitat characteristics are also sometimes correlated with the composition or richness of component communities. These associations can sometimes be spurious, but they can also reflect the action of causative mechanisms (Hartvigsen and Halvorsen 1994). For instance, the richness or structure of parasite component communities in fish populations of the same species often correlates with selected physicochemical characteristics of the water bodies they inhabit, such as surface area, depth, altitude, or pH (e.g., Kennedy 1978; Marcogliese and Cone 1991, 1996; Fellis and Esch 2005; Goater et al. 2005). Larger water bodies are more likely to be colonized by new parasite species over time. They contain larger host populations in which parasite extinction may be less likely, and more diverse communities of free-living invertebrates that can serve as intermediate hosts to a wider variety of parasites. The pH and other chemical characteristics can determine whether essential intermediate hosts can exist in given lakes; for instance, snails and the digeneans they transmit to fish are excluded from lakes with low calcium ion concentrations (Curtis and Rau 1980). In some circumstances, local habitat variables can, therefore, be more important determinants of component community richness than broader-scale factors such as the richness of the parasite fauna.

Biological characteristics of the habitats can also be important. For example, the presence of many other host species can result in a large pool of available parasite species that can lead to a richer component community. There is often a relationship between the number of host species of a given taxon in a habitat and the number of parasite species exploiting that taxon. Unionid mussels, for instance, whose larvae are ectoparasitic on fish, are more diverse in water bodies containing many fish species (fig. 10.3). Across different river systems, the relationship between local unionid species richness and fish species richness is clearly linear (Watters 1992). Across different localities in the same river system, however, the relationship weakens considerably (fig. 10.3). The number of potential host species in a locality apparently sets an upper limit on unionid species richness, but other local variables also act to determine how many parasitic mussel species actually occur at that locality (Vaughn and Taylor 2000). Still, as a rule there are more unionid species where there are many fish species; this general pattern probably applies broadly to all kinds of host-parasite associations. This may not always lead to exchanges of parasite

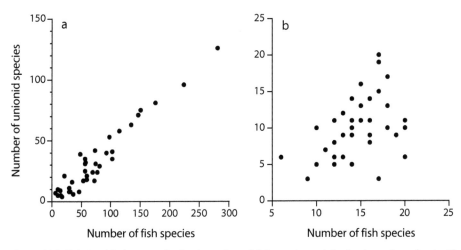

Figure 10.3 Relationship between the local number of fish species and the local number of unionid mussel species: a, across thirty-seven riverine systems of the Ohio River drainage area in the central United States, and b, across thirty-six sampling sites within a single riverine system, the Red River of the south central United States. The linear relationship observed at the larger geographical scale is replaced by a roughly triangular pattern at the smaller scale (Data from Watters 1992; Vaughn and Taylor 2000)

species between sympatric component communities, but often a larger pool of locally available species will facilitate the increase in species richness of given communities.

Even in a habitat favorable to high species richness, component communities will require time to develop to their full potential. Host populations are often splintered and displaced by natural events over time. Most recently, host populations of many species have been introduced to new habitats by humans. The parasite component communities of populations immediately following these displacements are likely to be species poor for many reasons (Torchin et al. 2003). For instance, essential intermediate hosts may be absent from the new habitat, or not enough parasite individuals of the same species accompany the founder host population to allow the species to maintain itself in the new habitat. The component community may reacquire these lost parasite species over time, when new host individuals migrate into the population, and may also gain new parasite species via host switching from sympatric host species. The rate at which parasite species join the component community may be in part determined by the characteristics of the habitat or host species, but absolute time since the displacement may be the key factor.

The increase in component community richness over time is supported by evidence from freshwater fish in Britain. As seen in figure 10.1, the maximum richness of component communities of helminths in introduced species increases linearly with the richness of the parasite fauna. These newly introduced fish hosts do not display the species saturation pattern characteristic of native fish species, probably because their component communities are still in a phase of species acquisition. In fact, time since the arrival of a fish species in Britain appears to be the best predictor of the richness of both its parasite fauna and component communities (Guégan and Kennedy 1993). The same thing can happen on much shorter time scales. Ebert et al. (2001) studied the development of parasite component communities in rock pool populations of the crustacean *Daphnia magna*. These populations last from a few to several years, and originate from the colonization of an empty pool by a few parasite-free host individuals. It was found that the age of the host population explained half of the variance in species richness among component communities. Even more than fifteen years after its establishment in a rock pool, a *Daphnia* population still accumulated new species of endoparasites, whereas the number of epibiont species in a *Daphnia* population stopped increasing and leveled off after just a few years (Ebert et al. 2001). This study did not include metazoan parasites, for which rates of colonization may be lower, but still illustrates how rapidly component communities can develop from a "clean slate" beginning.

The time since the displacement of a host population is not the only variable that may influence the composition or richness of component communities. The distance between the new habitat and the original habitat is also likely to be important. The farther a host population is displaced from its area of endemism, its heartland, the less likely it is that its specialist parasite species will be locally available in the new habitat. For example, salmonid fish populations in their heartland harbor rich component communities comprising a high proportion of salmonid specialist parasite species (Kennedy and Bush 1994). Many salmonid fish have been introduced to new areas, sometimes far away from their heartland. They are generally introduced as eggs, and thus completely free of metazoan parasites. As the distance from the heartland increases, component communities become increasingly species poor and consist of increasingly fewer salmonid specialist parasites (fig. 10.4). Salmonid specialists are gradually replaced by parasites specializing on other taxa, or by generalist parasite species. Component communities in the new areas generally include the same diversity of taxa as in the salmonid heartland, but they are a mixture of species acquired from local fish species (Poulin and Mouillot 2003b). The relative contribution of traditional specialists and new acquisitions to the component community will therefore change with increasing distance from the heartland.

The importance in a parasite assemblage of parasite species that are not host specialists can therefore depend on the scale of investigation. This can be seen in eels, *Anguilla rostrata*, which are more vagile than freshwater salmonids (Barker et al. 1996). Component communities on the Atlantic coast of Canada consist mostly of eel specialist parasites whereas the parasite fauna across the whole continent contains many generalists. As we increase the geographical scale at which the fauna is considered, the species richness increases but the relative number of specialist species decreases (fig. 10.5). The few specialist species of the fauna are well represented in most component communities, whereas each local component community only includes a small proportion of the large pool of nonspecialist species. This illustrates that despite the relative vagility of the host species, nonspecialist parasite species acquired by a host population do not necessarily spread to other component communities of the same host species.

The preceding paragraphs have emphasized that local habitat characteristics and host properties can influence the rate at which parasite species will be gained or lost by component communities over time. Two things must be pointed out, however. First, stochastic events during the evolution of component communities, such as chance colonizations

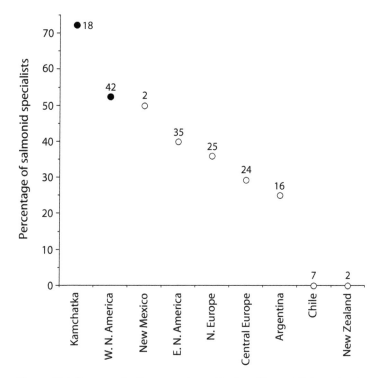

Figure 10.4 Decrease in the percentage of salmonid specialist species in helminth component communities of the rainbow trout, *Oncorhynchus mykiss*, as a function of their distance from the host species' heartland. The trout originates from the Pacific region of North America and Kamchatka (black symbols); other localities are arranged approximately in increasing distance from the heartland. Numbers above the points indicate the richness of the component communities in the various localities. (Data from Kennedy and Bush 1994)

or extinctions, can mask any role played by the local habitat or the host species on the development of parasite assemblages (Esch et al. 1988; Kennedy 1990). Historical contingencies are also recognized as important influences during the evolution of communities of free-living organisms (Ricklefs and Schluter 1993). Second, most of the innovative tests of hypotheses regarding patterns in component community richness have been performed using data on temperate freshwater fish. Patterns observed in these hosts may not be representative of what occurs in tropical fish (e.g., Kennedy 1995) or in groups of hosts other than fish. Similar studies are needed on component communities in other hosts to validate the trends emerging from the research on fish hosts.

10.2 Evolution of Parasite Faunas

The processes described in the previous section and leading to the gradual development of component communities also shape parasite faunas. At both levels, they can be summarized as evolutionary events leading to the acquisition or loss of parasite species

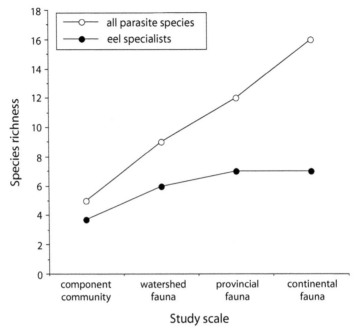

Figure 10.5 Relationship between the species richness either in all metazoan parasites or only in specialist parasite species of eels, *Anguilla rostrata*, and the scale at which the parasite assemblage is examined. Total species richness increases linearly with scale, whereas richness in eel specialists reaches its maximum value at a geographical scale smaller than the largest one studied. (Modified from Barker et al. 1996)

(Paterson and Gray 1997; Vickery and Poulin 1998; Poulin and Morand 2004). These apply as well to component communities in different populations of the same host species, following the breaking up of an ancestral population, as to the faunas of different host species during their phylogenetic history. The evolutionary history of the host species and its populations represent the history of the parasites' living habitat, and we can map the changes in parasite species richness onto host phylogeny (fig. 10.6).

During the divergence of two parts of a host population in a host speciation event, many parasite species of the ancestral host will be inherited by the two daughter host species. These parasites may or may not cospeciate with their hosts (see chapter 3), depending on whether the gene flow between parasites is interrupted. Cospeciation could result in different congeneric parasite species exploiting different but related host species. From a phylogenetic perspective, these congeneric parasites represent the same evolutionary lineage, and cospeciation per se does not influence species richness. These parasites are evolutionary baggage: their presence is not the outcome of ecological processes, and they do not reflect the divergence in ecological features between their hosts. The inheritance of ancestral parasite species during host speciation creates a form of phylogenetic inertia causing related host species to have similar parasite faunas.

The parasite faunas of related hosts are not identical, however. Over evolutionary time, they acquire or lose species at different rates. There are two main ways in which a

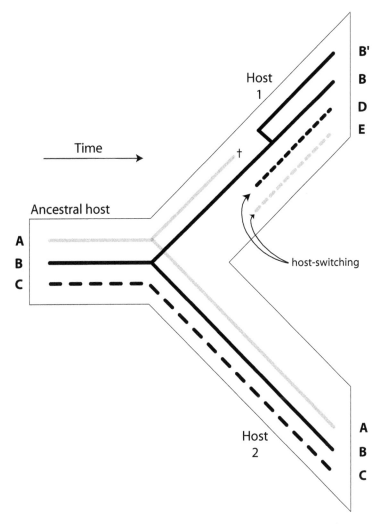

Figure 10.6 Illustration of the evolutionary events leading to the diversification of parasite component communities or faunas. Following a host speciation event, parasite lineages can be inherited by each of the two new host lineages (e.g., parasite B). They can also "miss the boat" by being absent from the part of the host population giving rise to a new host lineage (e.g., parasite C not branching into host lineage 1), or go extinct some time after the host speciation event (e.g., parasite A in host lineage 1). New parasite lineages can also join the host lineages through colonization or host switching (e.g., parasites D and E) or as the result of an intrahost parasite speciation event (e.g., parasite B′ arising from parasite lineage B). (From Vickery and Poulin 1998)

fauna can lose species (fig. 10.6). First, a parasite lineage may go extinct in a host lineage. Extinction of parasites can be caused by several factors. For instance, the host can evolve resistance to a particular parasite, the parasite can be displaced by a colonizing parasite species, other hosts necessary for the completion of the parasite's life cycle may disappear, or environmental changes may lead to inhospitable conditions for the free-living stages of the parasite. Second, because of the aggregated distribution of parasites among host individuals in a population (see chapter 6), the part of the ancestral host population splitting off to give rise to a new host species may harbor no parasites during speciation. A founder host population getting abruptly isolated from the main body of the population may thus be free of certain parasite species, or contain too few individual parasites to allow the parasite species to survive. The two above processes leading to the loss of parasite species are practically indistinguishable because they differ mainly in timing. Studies of host-parasite coevolution suggest that these events are not very frequent but happen regularly (e.g., Hafner and Page 1995; Paterson and Gray 1997; Paterson et al. 1999), and could potentially overcome the inertia mentioned above to create variability among the parasite faunas of related host species.

There are also two main ways in which new parasite species can be acquired by parasite faunas (fig. 10.6). First, hosts can be colonized by new parasite species. This involves host switching—parasites moving in from other sympatric host species provided that the new hosts are immunologically, physiologically, and ecologically compatible with the parasites. Comparisons between reconstructed host and parasite phylogenies suggest that colonization events may have been frequent in some systems (e.g., Barker 1991; Clayton et al. 1996; see chapter 3). In addition, sympatric host species belonging to the same taxon commonly share parasite species, suggesting either inheritance from a common host ancestor or, more likely, the exchange of parasites among host species (Goater et al. 1987; Stock and Holmes 1987; Navarro et al. 2005). Among four sympatric species of grebes, for instance, all but one of the fourteen most common helminth species occur in more than one host species; sharing of parasite species is particularly apparent between the two fish-eating grebe species and between the two invertebrate-eating grebe species (Stock and Holmes 1987). In other systems, however, exchanges of parasite species, or the colonization of a host species by parasites from another host species, appear much less frequent (e.g., Andersen and Valtonen 1990). As a rule, these exchanges are probably common only among related host species, as indicated by the fact that parasites of speciose host taxa often exploit a large proportion of these hosts (Poulin 1992a).

The second way in which new parasites can be acquired by a parasite fauna involves intrahost parasite speciation (fig. 10.6). It happens when parts of the parasite population become genetically isolated without gene flow being interrupted between parts of the host population. This way, an ancestral parasite species gives rise to two or more daughter species without host speciation. This could coincide with new niches becoming available in the host as it undergoes evolutionary changes in morphology or physiology, or with fine-scale resource partitioning among parasites. Intrahost parasite speciation could explain the presence of many congeneric parasite species within the same parasite fauna or even the same component community (Kennedy and Bush 1992; Poulin 1999c; Mouillot and Poulin 2004). Indeed, in some systems, large numbers of congeners in the same host population form true species flocks. Examples include oxyurid nematodes in tortoises, strongyloid nematodes in horses, cloacinid nematodes in kangaroos, and dactylogyrid monogeneans on freshwater fishes (Schad 1963; Kennedy and Bush 1992; Bucknell et al. 1996; Beveridge et al. 2002). Intrahost speciation may not be the sole driving force behind these species flocks, but it certainly contributed to their diversification (Poulin and Morand 2004).

The rate at which parasite faunas acquire or lose parasite species over evolutionary time may be related to the ecological characteristics of the host species. During their phylogenetic history, related host species will diverge with respect to body size, diet, geographical range, or other traits. Several studies have attempted to identify the key variables determining the richness of parasite faunas. For an effect of host ecology on the evolution of parasite faunas to be detected, however, rates of acquisition and loss of parasite species must be relatively high and strongly linked to host ecology. This conclusion is derived from the results of computer simulations in which an ancestral host and its parasite fauna are allowed to change through several rounds of host speciation before a correlation between host ecology and parasite fauna richness is computed (Vickery and Poulin 1998). The strong inertia of parasite faunas over time is likely to be enough to mask the influence of host ecology, if any. The next section discusses attempts to identify influential host traits that may overcome this inertia to shape parasite faunas.

10.3 Species Richness of Parasite Faunas

The comparative approach is the best, if not the only way, to assess the role of host traits in the evolution of parasite faunas. Relating an ecological variable with the richness of the parasite fauna across host species can serve to pinpoint potential reasons why certain host species have evolved richer faunas than others. As apparent from the previous section, there will often be a strong phylogenetic component in parasite faunas, and the effect of ecological variables must be disentangled from that of phylogeny (Brooks 1980; Poulin 1995f; Gregory et al. 1996). Consider two sister host species that have diverged, say, in body size, with species A having a body mass of 30 g and species B having a mass of 60 g. If species A harbors twenty-three parasite species and species B harbors twenty-six species, one might be tempted to conclude that a twofold difference in body mass has almost no influence on parasite species richness. However, if each host species has inherited twenty parasite species from their common ancestor, they would have acquired three and six new parasite species, respectively, since diverging. This is also a twofold difference. Without taking into account the similarities between the two host species that are due to their common phylogenetic origins, one could have erroneously dismissed host body size as a determinant of parasite species richness. The same argument can be made for any other host trait, such as geographical range or diet, and phylogenetic influences on all these traits must be accounted for when comparing different species (Harvey and Pagel 1991). Not everyone agrees that host phylogeny is as important if not more so than ecological forces (Holmes and Price 1980; Price 1987; Bush et al. 2001). However, the development of comparative methods that allow the influence of phylogeny, if any, to be removed (Poulin 1995f) makes the debate irrelevant: why not simply analyze data as though phylogeny mattered? Recent findings leave little doubt that the phylogenetic component of parasite faunas is important, and the use of proper comparative methods is absolutely necessary for future studies.

Phylogenetic influences are not the only confounding factor that requires control in comparative analyses of parasite faunas. The other problem plaguing comparisons of

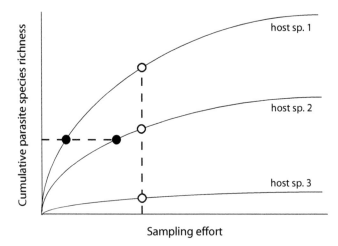

Figure 10.7 Accumulation curves for three hypothetical host species (or host populations). Sampling effort can be measured as the number of individual hosts examined for parasites, the number of parasite surveys performed on each host species, or any other measure of sampling intensity. For each host, the number of parasite species recorded approaches the true parasite species richness (the asymptote) as sampling effort increases. Typically, for a given sampling effort, more parasite species will be recorded from hosts harboring richer parasite faunas (open circles). However, when the same number of parasite species is reported from different host species (filled circles), the apparent similarity in parasite richness may be an artefact of uneven sampling effort. (Modified from Walther et al. 1995)

parasite faunas (or of component communities, for that matter) is unequal sampling effort. As more individuals or populations of a host species are examined, the number of known parasite species in the parasite fauna increases asymptotically toward the true richness value (fig. 10.7). A parasite fauna with few known species may be truly species poor, or it may include a large number of species yet to be recorded. Since different host species have not been sampled evenly, sampling effort often explains most of the variability in richness among parasite faunas (Kuris and Blaustein 1977; Gregory 1990; Walther et al. 1995). In fact, ecological variables often correlate closely with sampling effort. For instance, host species with large body sizes or wide geographical ranges are the subject of more independent parasite surveys than host species with small body sizes or restricted ranges. The larger the geographical range of an animal species, the more university campuses it is likely to include; with many university research projects focusing on the local fauna, species with large geographic ranges will be the subjects of more investigations than those within small ranges. In addition, many biologists show a preference toward large charismatic animals as study organisms. Thus, the close association between sampling effort and some host ecological traits make the separate effects of these variables difficult to distinguish (Guégan and Kennedy 1996). Nevertheless, multivariate approaches or other statistical corrections must be used to discern true ecological influences from sampling artefacts (Walther et al. 1995).

In comparative studies, two related methods are commonly used to control for the confounding effect of uneven sampling effort when trying to evaluate the independent effect of ecological variables (Walther et al. 1995). First, one can include sampling effort as a predictor variable in a multiple regression model (e.g., Gregory 1990; Gregory et al. 1996). Second, the residuals of a regression of parasite species richness against sampling effort can be used as estimates of richness independent of sampling effort (e.g., Poulin 1995f). Both methods, however, assume that the relationship between the number of species recorded and the number of hosts examined has the same shape in all parasite assemblages (see Walther et al. 1995). This is unlikely to be true. The shape and slope of the rising portion of the species accumulation curve will depend on the prevalence of each parasite species in the assemblage and should thus differ between different parasite assemblages, as illustrated in figure 10.7. An alternative approach is possible when precise information is available on which parasite species are found in each individual host in a sample. With this information, a range of extrapolation methods can be used to estimate the number of rare species missing from the sample. These methods are widely used in community studies of free-living organisms (see Palmer 1990; Baltanás 1992; Colwell and Coddington 1994), and readily applied to parasite communities (Poulin 1998c; Walther and Morand 1998). Nonparametric estimators of species richness are easy to compute, and most require only presence-absence data for all observed parasite species in each host examined. From these, the methods extrapolate how many species are likely to have been missed by inadequate sampling and add this number to the observed species richness. The performance of various estimators has been evaluated using both real and simulated data on parasite communities (Walther and Morand 1998; Poulin 1998c). One that performed consistently well was the bootstrap estimator, S_b:

$$S_b = S_o + \sum_{j=1}^{S_o} [1 - (h_j/H)]^H$$

where S_o is the observed species richness, the number of parasite species actually occurring in the sample; H is the number of host individuals in the sample; and h_j is the number of host individuals in the sample in which parasite species j is found. Because even common species contribute to the extrapolation, S_b is always greater than S_o, but only marginally when there are no rare parasite species in the sample. Using simulated data sets, the bootstrap method outperformed the others either at larger host sample sizes (Walther and Morand 1998) or when many rare species, those with low prevalence, are present (Poulin 1998c). Richness estimators should be used to improve the observed species richness value obtained from a host sample, so as to get closer to the true richness value without overshooting it (but see Zelmer and Esch 1999). Since we cannot be certain of the existence of missing species, it may be best to err on the side of caution and settle for an estimate of species richness that is improved while remaining conservative. This is exactly what the bootstrap method achieves (Poulin 1998c) since it reduces the gap between observed and true richness with little chance of overshooting. Therefore, when possible, it should be used to correct for the effects of uneven sampling effort in comparative analyses.

Keeping the potential influence of host phylogeny and sampling effort in mind, we can now review some of the proposed ecological determinants of the richness of parasite faunas. Two theoretical frameworks are considered here, both having provided a foundation for predictive hypotheses regarding the role of ecological factors in determining species richness in different parasite faunas (Poulin and Morand 2004). The first is the epidemiological theory that grew out of mathematical models of host-parasite interactions (see chapter 7), and the second is MacArthur and Wilson's (1967) classical theory of island biogeography.

Epidemiological models were first constructed to explain and predict the spread and maintenance of parasites in host populations (Anderson and May 1978, 1991; May and Anderson 1978; Diekmann and Heesterbeek 2000). Applied to parasite communities, variations of these models can be used to determine how many parasite species can coexist at equilibrium in a host population (Dobson and Roberts 1994; Roberts et al. 2002). In theory, when a parasite is introduced into a host population, its dynamics can follow one of two trajectories: it can go extinct locally, or it can spread through the host population until host and parasite reach equilibrium population densities. Epidemiological models allow us to derive a measure of parasite invasiveness, the basic reproductive number, R_0, on which the fate of the parasite depends (see chapter 7). For metazoan parasites where reproduction occurs via the transmission of free-living infective stages that pass from one host to the next, R_0 represents the average number of female offspring produced throughout the lifetime of a female parasite that would themselves achieve reproductive maturity in the absence of density-dependent constraints. When R_0 is less than unity, the parasite cannot maintain itself and declines to local extinction. When R_0 is greater than unity, however, the parasite can successfully invade the host population, and its numbers grow until equilibrium is reached.

The invasibility of a new parasite species in a host population can be analyzed by mathematical modeling (Roberts et al. 2002). A new parasite species can spread into a host population only if its $R_0 > 1$ within the context of the parasite community already established in the host population. Some of the parameters that influence R_0 correspond to intrinsic properties of the parasite species, such as its fecundity or natural mortality rate within the host (see chapter 7). Most importantly, however, the expression for R_0 includes intrinsic host characteristics that will affect parasite transmission, such as host population density or natural mortality rate. These host features determine to a great extent whether a given parasite species can invade the host population, and whether successive parasite species will also succeed and become established. Therefore, the R_0 value of an invading parasite will depend to a large extent on the host equilibrium density, which itself depends on the regulating effect of the other parasite species already established in the host population.

Thus, all else being equal, we would expect that interspecific variation among hosts in features such as population density or life span should affect the likelihood that various parasite species can achieve threshold R_0 values (i.e., $R_0 \geq 1$) in these hosts. Host species

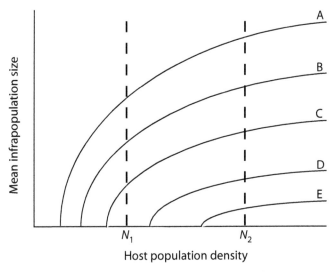

Figure 10.8 The influence of host population density on the mean number of parasite individuals per host (i.e., mean infrapopulation size) for five species of parasites, A–E, in the same host population. Parasite species are ranked in decreasing order of R_0 for a given host population density, with species A having the highest value of R_0 and species E the lowest. The curve for each parasite species intercepts the x-axis at the threshold host density required for that species to be maintained in the host population. Parasite species richness should thus increase with host population density, e.g. from three species at host density N_1 to five species at density N_2. (Modified from Dobson 1990)

with features favouring high R_0 values should be easier to colonize by new parasite species, and should therefore harbor richer parasite faunas. For instance, host species occurring at high population densities should be easier to invade for many parasite species, by promoting higher R_0 values, than host species with typically low population densities (fig. 10.8). The empirical support for this prediction is actually pretty good. Several comparative analyses have found that, after controlling for other variables, host population density proved a strong predictor of parasite species richness across host species. These analyses have focused on protozoans, helminths, and fleas on a range of mammalian hosts (Morand and Poulin 1998; Arneberg 2002; Stanko et al. 2002; Nunn et al. 2003), as well as on ecto- and endoparasites of fish (Morand et al. 2000b; Simkova et al. 2002; Takemoto et al. 2005). Of all the recent analyses controlling for both sampling effort and phylogenetic influences, only Clayton and Walther's (2001) study of chewing lice on birds did not report a positive relationship between host population density and parasite species richness. This is surprising because contact-transmitted lice would seem like the type of parasite that would most benefit from high host population density. Still, the consensus pattern from the other studies, and similar findings regarding the species richness of phytophagous insects across tree species (see Kelly and Southwood 1999; Brandle and Brandl 2001), suggest strongly that high host densities promote the accumulation of parasite species over time by parasite component communities and parasite faunas.

Host population density is only one of two host features in the formula for R_0 that can influence the magnitude of R_0; the other is the natural mortality rate of the host population due to causes other than parasitism. All else being equal, parasite species that invade a population of long-lived hosts are more likely to succeed and become established, because their R_0 value will be higher than if the hosts were short-lived. The prediction issued from the formula for R_0 is therefore that life span and parasite species richness should correlate positively across related host species, if there are no major differences in other parameters. The tests of this prediction are not favorable to date. Among freshwater fish species, there is a positive relationship between endoparasitic helminth species richness and host life span, but it disappears once other variables are accounted for in a multivariate analysis (Morand 2000). Among mammal species, there is actually a negative correlation between host life span and helminth species richness, which may be a byproduct of other covariates (Morand and Harvey 2000). Clearly, the influence of host life span per se is difficult to disentangle from that of related life-history variables, and for the moment appears not to be a factor. Host population density, on the other hand, despite also covarying with a constellation of other host features such as body size or metabolic rate, emerges from multivariate analyses as an important determinant of parasite species richness, as expected from epidemiological theory.

The second theoretical basis for the study of species richness in parasite faunas has been the theory of island biogeography (Poulin and Morand 2004). Because parasite communities are formed by colonization and extinction processes just like other communities, and because of the insular nature of hosts as habitats, MacArthur and Wilson's (1967) theory of island biogeography has been popular and influential in parasite community ecology (Kuris et al. 1980; Simberloff and Moore 1997). The theory and its variants predict that the number of species in an island community is set by the equilibrium between rates of extinction, speciation, and colonization, themselves determined by island characteristics such as size, age, and distance from the mainland. The analogy between an actual island and a host species is far from perfect (see Kuris et al. 1980). Whether the theory proves adequate or not for parasite communities, it has generated much empirical research and has resulted in the identification of host life-history or ecological traits associated with high parasite species richness.

In particular, the body size and geographical range of the host species appear like good potential predictors of faunal richness. Large-bodied host species may provide more space and other resources, and possibly a broader diversity of niches, for parasites. They are more likely to be colonized by parasites for these reasons and also because they consume more prey that may harbor larval parasites and present a greater external surface area for contact with infective stages. They also live longer and are thus less ephemeral habitats than small-bodied, short-lived host species. One could argue that the "size" of host species as islands for colonization would be better estimated by the host biomass available to parasites, the product of host body size and host population density. One elephant per km^2 may be a smaller island for parasites than a few thousand rodents in the same area. Given the demonstrated influence of host density on the richness of parasite faunas (see above), only host body size per se is discussed below. The literature contains numerous examples of positive interspecific relationships between host body size and the richness of the parasite fauna, for different types of hosts and parasites (e.g., Price and Clancy 1983; Bell and Burt 1991; Guégan et al. 1992; Krasnov et al. 1997; Poulin 1997e). However, many other studies found no association between host body size and parasite species richness (Gregory 1990; Poulin 1995f; Watve and Sukumar 1995; Clayton and Walther 2001; Nunn et al. 2003). It is interesting to note that often when correlations are computed before and after controlling for either or both host phylogeny or sampling effort, the strength of the correlations is greatly reduced after corrections are made for the confounding variables, often becoming statistically nonsignificant (Poulin 1995f, 1997e). Thus, when ignoring host phylogeny, the richness of gastrointestinal helminths of birds correlates strongly with host body size; when controlling for phylogeny, the association between the two variables vanishes (Poulin 1995f). Recent reviews of the literature indicate that there are as many positive relationships as there are nonsignificant associations between host body size and parasite species richness published to date in scientific journals (Morand 2000; Poulin and Morand 2004). Even within certain types of parasites, such as trophically transmitted endoparasites or contact-transmitted ectoparasites, there is no consistent pattern across studies. The only general conclusion that can be drawn from the available evidence is that although host body size can play a substantial role in the diversification of some parasite faunas, its importance is far from being universal.

Similarly, a wider geographical range may result in encounter with and colonization by a greater number of parasite species. Host species ranging over vast areas will overlap with the geographical distributions of several other host species, creating numerous opportunities for host switching. Positive correlations between the richness of the parasite fauna and the extent of the host's geographical range have been reported for a range of host and parasite types (e.g., Price and Clancy 1983; Gregory 1990; Guégan and Kennedy 1993). As for host body size, though, correlations between species richness and host geographical range become rather weak when corrections are made for confounding variables (Poulin 1997e), and the significant positive relationships barely outnumber the nonsignificant ones in the literature (Poulin and Morand 2004). This observation suggests that host geographical range is not an overriding force in the evolution of parasite faunas.

Differences in habitat characteristics among different host species can also influence the diversification of parasite faunas. One of the greatest dichotomies of habitat types has to be the aquatic versus terrestrial divide. Comparisons between the parasite faunal richness of aquatic and terrestrial host species appear straightforward at first glance, but they are not. Invasions of land by aquatic animals, or returns to water by terrestrial animals, have occurred infrequently over evolutionary time. Ignoring host phylogeny in such comparisons can result in a kind of pseudoreplication that artificially increases the power of statistical tests. The comparisons must be made between the lineages descended from these rare evolutionary events, and not between the resulting species treated as independent replicates; in other words, they must be made between basal branches in the phylogenetic tree, and not between branch tips (see the beginning of this section). For instance, if

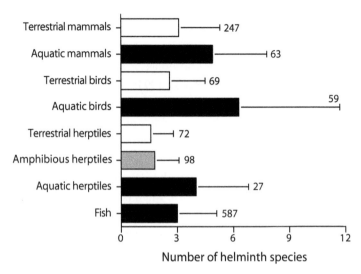

Figure 10.9 Average (+ standard deviation) richness of component communities of intestinal helminth parasites in different types of vertebrate hosts. Numbers next to the bars are the numbers of studies on which the estimates are based. The data include 245 species of fish hosts, 112 species of herptile (reptile and amphibian) hosts, 84 species of bird hosts and 141 species of mammal hosts. These data suggest that aquatic vertebrates (black bars) harbor richer communities of intestinal helminths than their terrestrial (open bars) counterparts. (Data from Bush et al. 1990)

host species are treated as independent observations, the average richness of helminth parasite faunas or component communities of aquatic vertebrates is clearly higher than that of their terrestrial counterparts (fig. 10.9). This can suggest that habitat characteristics are the main determinant of community richness, with host species in aquatic environments encountering a greater diversity of parasites (Bush et al. 1990, 2001). When phylogenetic contrasts between basal branches are used instead to test for a difference in richness between aquatic and terrestrial vertebrates, no significant differences are found (Poulin 1995f; Gregory et al. 1996). Other patterns also fail to materialize once comparative analyses correct for host phylogeny, when few independent evolutionary events are involved. The apparent difference in species richness between the faunas of endothermic and ectothermic vertebrates, presumably due to differences in host vagility and diet (Kennedy et al. 1986), cannot be demonstrated when controlling for host phylogeny: endothermy evolved twice, once leading to mammals and once to birds, and this is insufficient for a proper test of its influence on the diversification of parasite faunas.

Other host characteristics have also frequently been implicated as determinants of the richness of parasite faunas. These include host diet, behavior, metabolic rate, and several life-history traits (e.g., Price and Clancy 1983; Bell and Burt 1991; Gregory et al. 1991; Guégan and Kennedy 1993; Morand and Harvey 2000). Although these variables may indeed affect colonization or extinction rates of parasite species in faunas, more evidence is needed from comparative studies with proper corrections for phylogenetic or sampling effects. Aspects of host genetics, whose links with either parasite colonization or extinction rates are less obvious, have also been shown to correlate with the richness of parasite faunas in specific systems. For example, there is a negative relationship between helminth species richness and host heterozygosity among North American freshwater fish species (Poulin et al. 2000). Also, polyploid species of African cyprinid fish harbor richer monogenean parasite faunas than their diploid relatives (Guégan and Morand 1996). Perhaps polyploidy disrupts host defenses and opens the door to more parasite species colonizing the host. Other mechanisms can also explain this result, but the generality of the relationship remains to be tested in other host and parasite systems.

Some general points must be made about the body of evidence linking observed patterns in parasite species richness with expectations from either epidemiological theory or island biogeography theory. First, predictions derived from both theoretical frameworks are qualitative rather than quantitative. In other words, the theory predicts the direction of a relationship, either positive or negative, but not its strength, which could be measured as the slope of the regression line. Thus, although we would expect that a fivefold difference in host population density between two host species would be associated with a difference in parasite species richness, the magnitude of this difference cannot yet be predicted. Second, even when a significant relationship between some host feature and parasite species richness is found, its predictive power is generally low. There is typically much background noise plaguing these relationships, as seen by much scatter of data points, and host features generally explain little of the variance (i.e., low r^2 values) in parasite species richness. Third, no single pattern is universal: exceptions to all significant associations have been reported. Fourth, all studies present correlational evidence, and one must therefore use caution when inferring causality. For instance, Morand and Harvey (2000) found a positive relationship between the basic metabolic rate of mammalian host species and the richness of their helminth fauna. One interpretation is that host species with higher metabolic rates face greater risks of acquiring parasites because of their higher feeding rates, and thus accumulate a richer fauna over evolutionary time. However, the arrow of causality can point in the opposite direction: perhaps high metabolic rates have evolved in host species already exploited by many parasite species in order to meet the high cost of mounting immune responses against all these attackers. Which is the cause, and which is the consequence? Alternative interpretations are always possible and often too easily dismissed, and unraveling the web of potential interactions is not an easy task.

10.4 Biogeography of Parasite Diversity

The previous section has been concerned with the distribution of parasite diversity among host species. What about its distribution among geographic areas? Are there regions of the world where parasite component communities are consistently more species rich than in other regions? If so, why have these component communities diversified so much? What little is known about the biogeography of parasite species richness is examined below.

One of the best-documented large-scale biogeographical patterns is the increase in species richness with decreasing latitude (Rosenzweig 1995; Gaston and Blackburn 2000). For most plant and animal taxa, species richness peaks in the tropics and decreases toward the poles. This pattern must ultimately be the outcome of higher rates of diversification at low latitudes. Several explanations for this phenomenon have been proposed, not necessarily mutually exclusive, each based on different processes that are most likely all acting in synergy (Gaston and Blackburn 2000). One of the main contenders is the energy hypothesis, which states that higher energy availability in an area provides more resources that can support more species (Wright et al. 1993). More solar radiation is captured in the tropics than at higher latitudes, and this extra energy can be converted into additional biomass in the form of richer floras and faunas. Other explanations include the area hypothesis, which applies to terrestrial habitats and argues that there are more species in the tropics because this is where we find the greatest geographical area, and the time hypothesis, in which tropical areas have a greater total evolutionary age than temperate ones, and have thus had more time for species to evolve to fill all available niches (Rohde 1992; Rosenzweig 1995; Gaston and Blackburn 2000). There is no universal consensus regarding these and other mechanisms, and this may be because all of them are valid, at least to some extent and at certain scales.

Are there also latitudinal gradients in parasite species richness, and if so, what are the likely mechanisms behind these gradients? A priori, we might expect weak relationships between latitude and parasite species richness because, among free-living organisms, the strength of these relationships increases with the average body size of the organisms considered (Hillebrand and Azovsky 2001), and parasites are generally small-bodied. The parasite groups for which biogeographical patterns are best known are without doubt the metazoan parasites of marine fish (see Rohde 1993, 2002). From the small fraction of the parasite faunas of marine fish that have been studied in detail, a clear pattern has emerged: the diversity of monogenean parasites (and other groups of ectoparasites, to a lesser degree) increases with decreasing latitude or increasing water temperature (Rohde 1993, 2002; Rohde and Heap 1998). Thus, the species richness of ectoparasite component communities of marine fish is greater in warm tropical waters than in cold seas (fig. 10.10).

Several processes that could explain the latitudinal gradient in the richness of component communities of monogeneans and other ectoparasites of marine fish have been ruled out. For instance, the host specificity of monogeneans does not vary with latitude (Rohde 2002), so the greater relative species richness on tropical fish is not simply the result of the same number of parasite species exploiting a greater number of host species. Also, the trend is not the product of unequal sampling effort or a phylogenetic artefact that could have resulted from fish lineages harboring diverse parasite faunas colonizing warm waters long ago: the effect of latitude or water temperature on parasite diversity remains after controlling for phylogenetic influences (Poulin and Rohde 1997; Rohde and Heap 1998). One explanation, though, derived from both the energy hypothesis and the evolutionary time hypothesis, may offer a solution. Taxa in warm waters should experience higher rates of diversification because the shorter generation times and higher mutation rates resulting from higher temperatures lead to evolution itself proceeding at a

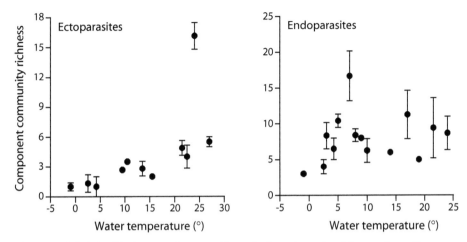

Figure 10.10 Relationship between the mean (± standard error) species richness of parasite component communities and the water temperature at the site of sampling, across different species of marine fish hosts. Results are shown separately for endoparasitic helminths from 55 fish species (62 fish populations) and for ectoparasitic metazoans from 108 fish species (109 fish populations). (Data from Rohde and Heap 1998)

greater pace (Rohde 1992, 1999; see also Allen et al. 2002). Thus, according to Rohde (1992, 1999), over the same time period, and assuming a nonequilibrium state in which fish do not become saturated with parasite species, higher speciation rates in warm seas would lead to a greater parasite species richness than in colder waters. The effective evolutionary time available for speciation in tropical waters would therefore have been greater than in temperate oceans. There is, however, no latitudinal gradient in the species richness of component communities of endoparasites of marine fish, in clear contrast to the pattern observed for ectoparasites (fig. 10.10). Intestinal helminths living in ectothermic fish hosts are also exposed to external water temperatures, but they have not diversified at a higher rate in the tropics. Rohde and Heap (1998) propose that other biological differences between internal and external fish parasites can explain the absence of latitudinal diversity gradients in endoparasites.

Temperature-mediated diversification is an interesting hypothesis that should apply equally well to all organisms, whether parasitic or not (Rohde 1992, 1999). It is not the only potential explanation for the latitudinal gradient in the diversity of ectoparasites of marine fish. Other comparative studies have shown that the body sizes of fish ectoparasites decrease at lower latitudes (Poulin 1995b, 1996c). As discussed in chapter 4, it is possible that diversification is sometimes greater in small-bodied taxa. Thus the greater diversity of fish ectoparasites in the tropics could result from greater diversification rates due to both environmental temperature and parasite body sizes; the two explanations are not mutually exclusive. This example illustrates the sort of challenges impeding investigations into the causes of biogeographical patterns in parasite species diversity.

There have been few other comprehensive investigations of latitudinal gradients in parasite species diversity. Studies on helminth parasites of freshwater fishes have uncovered an unexpected pattern (Choudhury and Dick 2000; Poulin 2001c). Temperate freshwater fish species are hosts to richer component communities of helminth parasites than tropical freshwater fish species. Even after eliminating the potentially confounding effects of uneven sampling effort, host body size, and host phylogeny, one arrives at essentially the same conclusion: given the same sampling effort and for equal body sizes, the component communities of temperate fish taxa comprise more parasite species than those in their tropical relatives (Poulin 2001c). The only exception involves eel species (*Anguilla* spp.), in which tropical species have remarkably richer parasite faunas than their temperate counterparts (Kennedy 1995; Marcogliese and Cone 1998). All other phylogenetic contrasts show higher parasite species richness in temperate fish lineages. Perhaps temperate freshwater fish possess features, such as those discussed in section 10.3, which make them more likely to develop rich parasite faunas. Finally, Cumming (2000) has carried out an investigation of latitudinal gradients in tick species richness in Africa. As the geographical ranges of ticks are better described by climatic variables than by host preferences, Cumming (2000) built regression models in which climate data were used to predict the occurrence of each tick species among cells of a continent-wide grid. These were then used to show that tick species richness increases toward the equator. Unlike the studies of fish parasites described above, though, tick richness in this study was not corrected for host richness, which is to say, it is not expressed on a "per component community" basis. Latitudinal gradients in the richness of mammalian host species follow similar trends, and relative richness of ticks per host species (per component community) may thus be constant across latitudes.

Apart from latitudinal gradients, there are a few other biogeographical patterns in parasite species richness (see Rohde 2002; Poulin and Morand 2004). The only general one worth mentioning applies strictly to marine systems or to large lakes, and it is the bathymetric gradient in parasite species richness. The species richness of most major animal taxa changes with increasing depth. For benthic invertebrates, for example, species richness generally increases with depth down to a certain depth beyond which it decreases again, though this pattern is not without exception (e.g., Gray 1994; Gage 1996). Whatever happens to the diversity of free-living animals in the deep sea, their population density is often low. In particular, fish biomass is thought to decline sharply with depth, although robust estimates are few (Merrett and Haedrich 1997). If potential host species occur at low population density in the deep sea, we might expect that they would harbor a low species richness of parasites since parasite persistence depends on host populations existing above a threshold density (see section 10.3). Indeed, research on deep-sea fish species indicates that they harbor fewer parasite species than shallow-water fish species. This pattern is sometimes not apparent in mesopelagic fish, which live off the continental slope at moderate depth and often migrate vertically to feed, but it is clear in benthic fishes (Noble 1973; Campbell et al. 1980; Gartner and Zwerner 1989; Bray et al. 1999; Marcogliese 2002). A phylogenetic analysis of deep-sea digeneans by Bray et al. (1999)

has shown that there have been few transitions from shallow to deep-sea habitats over the evolutionary history of digeneans, but that these few transitions have sometimes been followed by radiations. Only 17 of the 150 or so recognized digenean families are represented in the deep sea, generally by specialist genera that do not occur in shallower waters (Bray 2004). Indeed, some digeneans, like the aptly named *Profundivermis* sp., only infect their host fish in the deeper parts of its bathymetric range (Bray et al. 1999). Recent studies are even suggesting that there may be hot spots of parasite diversity in host populations living around deep-sea hydrothermal vents (Moreira and López-Garcia 2003; de Buron and Morand 2004), where higher temperatures could promote parasite diversification as they do in tropical waters. Thus, despite its generally low diversity of parasites, the deep sea is still a place where new parasite species can originate. Our knowledge of parasitism in the deep sea is completely inadequate, and we will need much more information before we get a clearer picture of how diverse parasites are in the abyss.

10.5 Host Specificity and the Composition of Parasite Faunas

Despite the many studies on the determinants of species richness of parasite faunas, there has been little attention paid to patterns in the composition of these faunas. There may be no patterns to worry about. It is possible, however, that rich faunas or component communities consist of parasite species with different ecological characteristics than those found in species-poor faunas or communities. For instance, within most families of Canadian freshwater fish, there is a significant tendency for rich parasite faunas to include many specialist, that is, host-specific, parasite species, and for species-poor faunas to consist mainly of generalist parasites with low host specificity (fig. 10.11). This pattern resembles the nested pattern described for infracommunities (see chapter 9), but on a larger scale. This nonrandom distribution of generalist and specialist parasite species among host species could have several possible origins. A likely one could be that accumulation of parasite species, through colonization or intrahost speciation events and low rates of extinction, is facilitated by certain host traits in some faunas whereas it is restricted in other faunas (Poulin 1997a). The observed pattern would fit nicely with the many studies on the determinants of the richness of parasite faunas summarized in section 10.3. Clearly, if colonization is less likely or extinctions more frequent in some faunas, the few species in these faunas should also manage to colonize faunas in which conditions are more favorable. We would thus expect generalist species to occur in all sorts of faunas and specialist ones to be restricted to rich faunas. The nested pattern found in most families was not apparent in salmonid fish, however (fig. 10.11). Fish in this family have been the subject of more extensive introductions to new areas than other fish, which may have resulted in the recent evolution of their parasite faunas differing from that of other fish (see Poulin and Mouillot 2003b). Nested patterns are also not observed among British freshwater fish species (Guégan and Kennedy 1996). These contradicting results

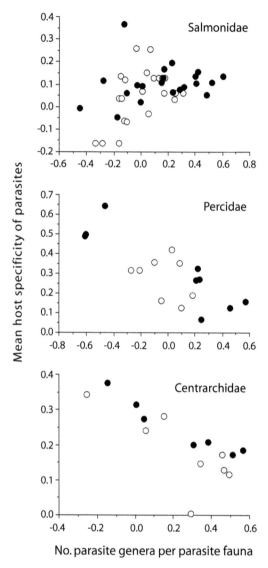

Figure 10.11 Relationship between the average host specificity of parasite species and the richness of the fauna to which they belong, across host species in three families of Canadian freshwater fish. Specificity is defined as the number of host species used by a parasite. Both host specificity and faunal richness have been corrected for unequal sampling effort among fish species, i.e., data are residuals of regressions on sampling effort. Values for ectoparasite (open circles) and endoparasite (filled circles) faunas are presented separately. In Percidae and Centrarchidae (as well as in Cyprinidae and Catostomidae, not shown), the relationship is negative, because species-poor faunas consist mostly of generalist parasites whereas rich faunas include many specialists as well as generalists. The opposite trend is observed in the Salmonidae. (From Poulin 1997a)

indicate that more work is needed to elucidate what determines the composition of parasite faunas.

The pattern observed in most families of Canadian freshwater fish may be the large-scale manifestation of local phenomena. If specialist and generalist parasite species are matched nonrandomly with either species-poor or species-rich component communities, the emergent pattern at the level of the parasite fauna may well look like those in figure 10.11. After all, the parasite fauna is the summation of all component communities across a host species' geographical range. Within a given locality, such as a lake, there will be different component communities coexisting in parallel, each within a different host species. Within a lake, however, there will typically be some overlap between the species composition of different component communities in different fish species, because many parasite species are not strictly host specific. In a way, this means that component communities are artificial divisions of a larger entity. Nevertheless, if we keep them as units of study, and ask whether the proportion of host-specific parasite species in a component community varies among component communities as a function of their species richness, we find very consistent patterns (Vázquez et al. 2005). Within several Canadian lakes, each containing several fish species and thus several parasite component communities, the same pattern emerged: host-specific parasites were more likely to occur in species-rich communities, and species-poor communities were more likely to consist of generalist parasites, than expected from null models (fig. 10.12). The null models used real data on parasite specificity and component community richness to generate networks of associations between hosts and parasites. In the simulated networks, the specificity of a parasite did not affect which type of component community it would be found in (Vázquez et al. 2005). In all cases, observed networks were significantly more asymmetrical than the expectations of the null model, with specialist parasites occurring in rich component communities more often, and generalists occurring in poor communities more often, than expected (fig. 10.12).

The really exciting finding, though, is that the patterns observed in fish-parasite associations are repeated in networks of associations between species of fleas and mammals in the same locality, for each of several localities in Eurasia and North America (Vázquez et al. 2005). Here again, specialist fleas are more frequently found in rich component communities, and species-poor component communities consist more frequently of generalist fleas, than expected from random associations (fig. 10.12). The similarity of these findings from vastly different host-parasite systems hints at common ecological processes shaping the composition of component communities within a locality, with consequences for the composition of parasite faunas. The association between specialist and generalist parasites on the one hand, and species-rich and species-poor communities on the other hand, also resembles the patterns of distribution of specialization reported for mutualistic associations (Bascompte et al. 2003; Vázquez and Aizen 2003). Thus, there may be universal patterns in interaction networks, whether the associations are parasitic or not. Beyond confirming the existence of these general patterns in other kinds of host-parasite associations, the next step will be to elucidate the processes behind them.

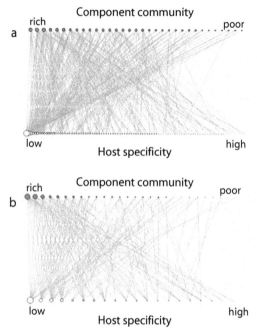

Figure 10.12 Networks of species associations between host and parasite species: a, fish and their metazoan parasites in Lake Huron, Canada; b, small mammals and parasitic fleas in Novisibirsk, Siberia. Each host species, i.e., each component community, is represented by a gray circle, and each parasite species by an open circle; the diameter of the circle is proportional to the richness of the component community or to the number of host species that a parasite can exploit, respectively. Component communities are ranked according to species richness, and parasite species according to host specificity. Connecting lines represent the observed associations between hosts and parasites. Since most associations in the network involve a parasite species with many hosts and/or a species-rich component community, the majority of connecting lines should be vertical links on the left-hand side of the network. However, the excess of diagonal lines illustrates the tendency for specialist parasites to occur in species-rich component communities, and the tendency for species-poor component communities to consist of generalist parasites. (Modified from Vázquez et al. 2005).

10.6 Conclusion

The development of parasite assemblages at large spatial and temporal scales proceeds through a series of evolutionary events, determined in part by the characteristics of the habitat or the host. Stochastic events such as chance colonizations no doubt play a big role, although their influence may sometimes be difficult to detect or quantify. The use of robust comparative methods for analyses of parasite assemblages has been emphasized throughout this chapter, as there is no other way of distinguishing between passive inheritance of parasite species, gain or loss of species related to host or habitat properties, and historical accidents. Many variables have been identified as determinants of high parasite species richness. However, their influence does not extend to all systems, and their predictive power is always limited. The many studies available to date have other shortcomings: most analyses have focused on the richness of helminth parasites in vertebrates, especially fish, and many have used data sets that overlap substantially. Similar studies on parasite assemblages of invertebrates would be instructive, and serve to validate the generality of the conclusions derived from studies of vertebrate parasites.

The ultimate goal of research into the evolution and ecology of parasite species richness is to understand the distribution of this diversity at different spatial scales (Poulin and Morand 2004). This is essentially the same goal as that of the emerging field of macroecology, though its proponents to date have focused mostly on free-living organisms (Brown 1995; Gaston and Blackburn 2000). The progress made by macroecologists toward an understanding of the distribution and abundance of organisms has been extremely rapid, and it provides an inspiration for those interested in the study of parasite biodiversity.

11 | Conclusion

Completed in 1997, the first edition of this book left the reader with many unanswered questions. This was not meant only as a reflection of the state of our knowledge of parasite evolutionary ecology at that point in time, but also as a call to arms, an appeal to parasitologists to tackle unsolved problems. Much was already known about parasites at the time. The parasitology literature of the twentieth century contains a wealth of information on the diverse taxa of parasites and on their relationship with their hosts. There are tens of thousands of pages devoted to descriptive studies, laboratory experiments on development and life cycles, and field surveys of parasite diversity. The next step was to examine this vast amount of information in a broad evolutionary context. It has been very encouraging to see the challenge taken up by so many researchers since the first edition was published. It is becoming difficult to keep up with all the articles on parasite ecology or evolution coming out every month. Nevertheless, after going through this second edition, many readers may feel that most areas would still greatly benefit from further work. The study of the evolutionary ecology of parasites is just coming out of its infancy and still has some growing up to do. The next few years will be exciting ones as we turn to some of the big questions that remain unanswered.

In addition to fundamental questions about the ecology and evolution of parasites, several applied issues are increasingly calling for our attention. Parasite evolutionary ecology permeates several applied sciences, such as epidemiology and medicine, agriculture and veterinary science, conservation biology and wildlife management. Just as Thompson (2005) has argued that society increasingly requires a science of applied coevolutionary biology, I believe that we urgently need to apply evolutionary ecology thinking to parasite issues confronting society. In the political climate prevailing in many countries today, funding for basic scientific research is a low priority, and what little money is available is the subject of intense competition among scientists with different agendas. We should not have to justify a branch of science in terms of its potential tangible benefits for society; knowledge for its own sake should be enough. In the context of limited funding, however, it is crucial to make politicians and society at large aware of what is at stake. Examples of the value for society of research on the evolutionary ecology of parasites are easy to find. Humans have been meddling with parasite evolution for quite some time, either voluntarily or accidentally. We know that parasites can evolve rapidly in response to new selective pressures. We know this from common sense, given the short generation times and high fecundity of many parasites. We also have evidence of this rapid evolution happening, such as the sudden appearance and quick spread of anthelmintic resistance among sheep and cattle parasites throughout the world. We should therefore expect other evolutionary changes in parasites in response to the new pressures we place on them. These are strong arguments to lobby funding bodies and decision-makers to grant a place for parasite evolutionary ecology near the top of the list of priorities.

This final chapter gives two examples of human-induced changes in the selective pressures acting on parasites, and discusses the sort of evolutionary changes they may favor. I will finish by offering my views on the immediate future of parasite evolutionary ecology, and suggest what I think are promising directions for further research.

11.1 Environmental Changes and Parasite Evolutionary Ecology

Human activities have resulted in substantial, large-scale modifications to the natural environment, especially in the past century. This has been paralleled by cultural and societal changes, including increased human encroachment on wildlife habitat, increased international travel, and the globalization of our food supplies (Patz et al. 2000; Petney 2001; Thompson 2001). The impact of these habitat changes on parasite evolution and ecology has not received much attention but it could be very important if the new conditions persist for several parasite generations (Combes 1995, 2001).

Aquatic habitats in particular are being modified extensively in ways that may affect the evolutionary ecology of parasites. For instance, reservoirs created by dams and the slow-moving water of irrigation channels present ideal conditions for the proliferation of snails acting as intermediate hosts not only of human schistosomes but also other digeneans parasitic in domestic animals or wildlife. This can result in foci of high efficiency of transmission to and from the snail intermediate host, and thus lower parasite mortality at the miracidial or cercarial stage. Beyond an immediate increase in the local parasite population, what could be the long-term evolutionary consequences of these new conditions? We do not know. What we do know is that the evolution of life-history traits is driven mainly by an organism's age-specific mortality rates (Roff 1992; Stearns 1992). An increase in larval survival persisting for several parasite generations could select for a different combination of life-history traits, with changes in characteristics of the parasite such as length of the prepatent period, longevity, or fecundity spreading through the population. A greater probability of transmission could also favor an increase in parasite virulence in the definitive host (see chapter 5). At present we know enough to make educated guesses but lack the empirical studies necessary to formulate specific predictions regarding these possible evolutionary changes.

Terrestrial habitats have not been spared. For example, deforestation has created patchy forest habitats separated by vast areas now used for agriculture and other human activities, or simply left vacant. Host populations in these patches may have lower densities than before deforestation, and may be more subject to local extinction. These changes may also affect parasite transmission and population dynamics. In some cases, the effects might be positive, resulting in increased transmission of parasites (e.g., Page et al. 2001). In most cases, however, there is reason to fear that the fragmentation of host populations will also make parasites more prone to extinction (Rózsa 1992; Sprent 1992; Poulin and Morand 2004). Some parasite attributes, evolved for other reasons, can act as hedges against extinction and allow the persistence of parasites under the new conditions (Bush and Kennedy 1994; Poulin and Morand 2004). In any event, the effects of the fragmentation of habitats and host populations can influence the dynamics of parasite populations and the structure of parasite communities. Over time, the new conditions will impact parasite transmission rates and the survival of larval stages, again with potential consequences for the evolution of life-history traits or transmission strategies.

On a global scale, pollution and climate change loom as the most significant human-induced changes to the natural environment. Chemical pollution is perhaps the most pernicious, immediate, and widespread environmental change resulting from human activities. Freshwater and marine habitats are particularly susceptible, receiving industrial effluents, agricultural and domestic waste as well as many accidental spills, all containing toxic chemicals of many types. There is mounting laboratory and field evidence that these pollutants affect both parasites and their hosts (Khan and Thulin 1991; Poulin 1992b). These effects are not trivial: pollution levels can sometimes be accurate predictors of the presence or absence of certain parasite species in given water bodies (e.g., Marcogliese and Cone 1996). Increasingly, parasites are seen as potentially very useful indicators of not only the presence of certain pollutants in a locality, but also of general environmental quality and ecosystem health (Lafferty 1997; Sures 2004; Marcogliese 2005). This came from the realization that parasites are very sensitive to pollution, and that pollutants alter the relative abundances of locally occurring parasite species. The outcomes of pollution are varied and may depend on the host-parasite system, the habitat, or the type of chemical involved. In some cases, the free-swimming infective stages of fish parasites, or the intermediate hosts of fish parasites, are negatively affected by pollutants. In other cases, fish exposed to pollutants incur immunosuppression and are more susceptible to parasite infections. In the first scenario larval parasite mortality is increased, whereas in the second scenario the survival of adult parasites, as well as their growth and fecundity, are enhanced. Again, these are the sorts of environmental pressures likely to lead to evolutionary changes in the biology of parasites, with the new conditions selecting for new optimal combinations of life-history traits. To date, research on the effects of pollutants on parasites has focused on immediate impacts on survival or reproduction, without considering the longer-term evolutionary implications.

Climate change and global-warming scenarios are now accepted as facts by the broader scientific community and by society at large. The predictions for the next few decades include changes in air and sea-surface temperatures, coupled with changes in precipitation, sea level, ocean salinity, and circulation patterns. There is no doubt that these environmental changes will impact natural ecosystems in some way. Weather and climate affect several ecological processes, from the performance of individual organisms, to the dynamics of populations and the distribution of species. Although sometimes treated as a secondary effect or ignored altogether, the extent and intensity of parasitism can also be modulated by climatic conditions (Mouritsen and Poulin 2002a). Parasites can regulate the abundance of their host population, influence the composition and structure of animal communities, and affect the functioning of ecosystems (see Minchella and Scott 1991; Combes 1996; Hudson et al. 1998; Albon et al. 2002; Mouritsen and Poulin 2002b; Thomas et al. 2005). Any influence of climate on parasitism is therefore potentially important for natural communities and ecosystems. Recent surveys have highlighted the causal relationship between climate change and emerging parasitic diseases—diseases suddenly increasing in either local prevalence, geographical distribution, or the number of host species they can exploit (e.g., Harvell et al. 1999, 2002; Patz et al. 2000; Marcogliese 2001; Sutherst 2001; Lafferty et al. 2004). Thus, the many direct and indirect effects of temperature and other climatic variables on parasite transmission are bound to have broad impacts on host communities and natural ecosystems. But what about the evolutionary impacts on parasites? As the relative availability of certain hosts rises or falls, or as the survivorship of parasite infective stages changes, how will natural selection respond? There is little doubt that long-lasting changes to the parasite's perceived environment will be followed by adaptive adjustments to life history and transmission strategies, as well as levels of host specificity and virulence.

The impact of humans on the environment is far more complex than what is described above. Unknowingly, we may be selecting for rapid changes in parasite biology, as we have in so many other organisms (Palumbi 2001). We still do not have enough information to make specific predictions about the direction or magnitude of these changes (Altizer et al. 2003). Nevertheless, the evolutionary ecology approach to the study of parasites provides us with a framework to develop these predictions and design experiments to fill any gap in the knowledge that we need to forecast, and maybe plan for, the changes ahead.

11.2 Parasite Control and Parasite Evolutionary Ecology

From a human perspective, the control strategies we use against human and livestock parasites do not often qualify as habitat changes. From the parasite perspective, however, they truly represent environmental changes causing increases in either larval or adult mortality, or both. For example, the routine use of anthelmintics against parasites of sheep and cattle has imposed new pressures on adult helminths over many parasite generations. Not surprisingly, resistance to some anthelmintics has evolved quickly in many parasite species (Sangster 1999). The onset and spread of resistance to anthelmintics can be retarded by integrating knowledge of parasite epidemiology and population dynamics with grazing management, but in the end the evolution of resistance may be unavoidable (Barger 1999; Stromberg and Averbeck 1999). As a rule, inefficient anthelmintics have quickly been replaced by other drugs so that the increased adult parasite mortality has been maintained over time. We may thus expect other adaptive changes in parasite populations that have been exposed to anthelmintics for several years.

The worldwide use of chemotherapy to combat nematodes of ruminants is equivalent to a large-scale experiment in life-history evolution in parasites (Skorping and Read 1998). This experiment, however, has not been planned, and its potential consequences were not considered carefully by those responsible. If anthelmintics increase the probability of larval nematode mortality, or in other words if the probability of reaching adulthood is decreased, selection may favor a shorter prepatent period, that is, an earlier age at maturity. This would give the parasite a higher probability of producing eggs before the host is treated. It may also coincidentally select for smaller and thus less fecund worms (Read et al. 2000). In contrast, if larval worms are less susceptible to anthelmintic treatment than adults, and if it is adult mortality that is increased by treatment, then selection might favor longer larval stages and delayed maturity (Skorping and Read 1998). For instance, in nematode species in which the larvae undergo extensive tissue migrations before settling as adults in the gastrointestinal tract of the mammalian host, the migrating larvae are relatively safe from anthelmintics whereas adult worms incur higher mortality (Skorping and Read 1998). The pathology associated with these parasites is caused by the larval stages migrating through host tissues, and continuous use of anthelmintics could therefore select for delayed maturity and thus more virulent strains. These and other similar predictions are based on life-history theory and rely on a number of simplifying assumptions (Read et al. 2000). They are starting to receive empirical support, however. Leignel and Cabaret (2001) have recently demonstrated that adult size and thus fecundity in the nematode *Teladorsagia circumcincta* have evolved differently in anthelmintic-resistant and anthelmintic-susceptible strains subjected to different drug treatments. It is therefore very likely that anthelmintics are acting as a new global selection pressure on parasite evolution.

Chemotherapy can affect the evolution of parasites other than nematodes. For instance, treatment of rodent hosts with the antimalarial drug chloroquine accelerates the production of gametocytes in *Plasmodium chabaudi* (Buckling et al. 1997). This is an illustration of the plastic development of the parasite, which begins to divert resources from within-host replication toward the production of transmission stages when conditions within the host deteriorate. Quite possibly, the prolonged use of this drug could select for changes in the life-history schedule of the malaria parasite, with consequences very different from those anticipated from chemotherapy. Other forms of parasite control may have different effects. Vaccines, for instance, can provide the host with immunity against parasites, and if the entire host population gets the vaccine, the establishment of parasites is prevented and no individual parasite has the opportunity to reproduce. Vaccines rarely provide this kind of full protection against parasites, however. A mathematical model of virulence evolution based on an epidemiological framework generates different predictions for the use of different types of vaccines (Gandon et al. 2001). For instance, if a malaria vaccine designed to reduce within-host replication rate is used in a host population over several generations, selection against high virulence will be reduced. Under these circumstances, evolution will lead to higher levels of intrinsic virulence. In contrast, if a vaccine that blocks the infection is used instead, selection will not favor an increase in parasite virulence, and may even favor lower virulence (Gandon et al. 2001). Knowledge of parasite evolutionary biology is the cornerstone on which we build such models, and with universal vaccination against several pathogens now almost a fait accompli, we need detailed predictive models to understand the likely consequences for parasite evolution.

There are many other forms of parasite control, including some that target parasite stages outside their definitive host. Nematophagous fungi, which prey on the soil-dwelling infective stages of nematode parasites of livestock, are a good example. They have been touted as a promising form of biological control against these parasites (Waller 1993). Their effect would be to increase larval nematode mortality, and one obvious evolutionary response to their prolonged action could be an increase in adult growth, life span, and/or fecundity. Other control strategies aimed at larval parasites, such as the eradication of snail populations to control schistosomes, could have similar consequences.

The scenarios described above are only hypothetical, and only time and further research will verify the simple predictions made to date. Their purpose was only to illustrate the kinds of responses we may expect, and to show that we lack the empirical data to predict them accurately. Evolutionary changes in parasites targeted by control strategies are not limited to situations involving drugs or vaccines. There is considerable evidence that the virulence of human parasites and diseases has evolved jointly with transmission routes, and that any form of treatment can influence virulence by favoring more or less virulent strains (Ewald 1994). The current awakening of medical science to the importance of evolutionary thinking in the development of long-term preventive and curative strategies (Williams and Nesse 1991; Stearns 1999) can only help promote further studies of the evolutionary ecology of parasites. Hopefully, new additions to our arsenal against parasites will be designed to account for potential evolutionary responses in these same parasites.

11.3 Future Directions

Over the years, commentators on the future of parasitology have emphasized the need for a greater integration of the various subdisciplines of parasitology into a holistic approach to the study of parasite biology (see Mettrick 1987; Bush et al. 1995). Often, multidisciplinary studies offer new ways of tackling old problems, and I fully endorse the appeal from these earlier commentators. Looking ahead, there are two promising avenues that I think must be followed if we are to elucidate the outstanding questions about the evolutionary ecology of parasites: (1) an experimental approach, especially when applied to model species for which much is already known and complex laboratory manipulations are possible, and (2) a comparative approach applied to large data sets to identify broad patterns and generate new hypotheses. Both approaches are already in use by parasitologists, but not widely.

It is often possible to use natural field experiments, especially to answer the more eco-logical questions relating to parasite populations or communities. In the context of the habitat changes discussed above, for example, before-and-after studies could be performed to follow the changes in parasite biology associated with a modification in their habitat such as the onset of a new control strategy. These studies may not always be easy, but given the numerous opportunities that arise, some may prove fruitful.

Throughout this book, I have often used as examples experimental studies on a range of parasite species that are now maintained in the laboratory. These include parasites with simple life cycles, but also helminths and protozoans with complex life cycles, such as the cestodes *Hymenolepis diminuta* and *Schistocephalus solidus*, the digenean *Schistosoma mansoni*, or the apicomplexan *Plasmodium chabaudi*, to name but a few. There are certain risks associated with laboratory cultures of parasites. For instance, over time, laboratory stocks can lose their genetic diversity and diverge from natural populations. If these risks are avoided, however, laboratory model species offer wonderful opportunities for manipulative studies that can shed light on all aspects of parasite biology, including transmission ecology, host specificity, life histories, virulence or other host exploitation strategies, and population biology. Selection experiments across several generations can also be envisaged for parasites with relatively short generation times or life cycles to evaluate selective responses to a range of conditions. There is no better way to study evolutionary strategies than within an experimental context, where one can be confident about the association between cause and effect. Of course, we will need to establish further model species in all major taxa to achieve higher levels of generality; after all, it would not be prudent to base all our conclusions on a handful of parasite species that do not capture the full spectrum of nature's possibilities.

Another way to achieve this generality, and also a good way to generate hypotheses that can then be tested experimentally, is to look for general patterns or tendencies across all species. This is where the comparative approach comes in. Although well established in evolutionary biology, the comparative method is only slowly being applied to questions regarding parasites (Morand and Poulin 2003). For comparative analyses, we first require the robust and well-resolved phylogenies that are now pouring from molecular biologists and systematists. These are important for their own sake, but can also serve as tools for hypothesis testing and provide the necessary evolutionary framework for studies on the biology of parasites. The second requirement of comparative analyses is data on one or several variables for many parasite taxa. The literature is full of information on certain parasite traits just waiting to be put to use, whereas information on other variables can sometimes be gathered firsthand for a subset of species. To date, the comparative approach has only been applied to some aspects of parasite biology, as will be obvious from this book. Sections in which no comparative studies are mentioned cover subjects that are yet to benefit from this approach. In particular, there is now information available for a sufficient number of species to allow population or community-level variables to be examined in the light of phylogeny.

The use of comparative studies alone is not sufficient to identify the forces shaping parasite biology over evolutionary time. For instance, a mechanistic approach should be used in parallel to comparative studies to identify phylogenetic constraints limiting the evolution of certain traits (McKitrick 1993). We need studies in developmental biology or functional morphology to confirm the existence of the constraints suggested by phylogenetic analyses. Perhaps the greatest weakness of the comparative approach used on its own is that it demonstrates an association between characters without demonstrating a causal link (Doughty 1996). This is where the comparative method connects with the experimental approach: we need to test causal hypotheses derived from comparative studies with experiments on extant taxa (Losos 1996). Model parasite species could be used in multigeneration experiments to test for the effect of particular selective regimes. An observed evolutionary change in the direction suggested by phylogenetically based comparative studies would validate the proposed causal mechanism.

So this is what I wish for the near future of parasite evolutionary ecology: a marriage between experimental studies and comparative studies based on robust phylogenies, linking large-scale or interspecific patterns with the action of selection within species and of ecological processes acting locally. This research strategy requires the collaboration of people with different expertise, and more importantly it requires that scientists in the different subdisciplines of parasitology adopt a holistic, evolutionary perspective of parasite biology. There are signs that this is happening: the past decade has witnessed an explosion in research activity on parasite evolutionary ecology. The field is in good health and its immediate future should be a bright one, as we prepare to confront some of the major outstanding questions still unanswered.

References

Adamson, M. L. 1986. Modes and transmission and evolution of life histories in zooparasitic nematodes. *Canadian Journal of Zoology* 64: 1375–84.

Adamson, M. L., and Noble, S. 1992. Structure of the pinworm (Oxyurida: Nematoda) guild in the hindgut of the American cockroach, *Periplaneta americana*. *Parasitology* 104: 497–507.

Adamson, M. L., and Noble, S. J. 1993. Interspecific and intraspecific competition among pinworms in the hindgut of *Periplaneta americana*. *Journal of Parasitology* 79: 50–56.

Adjei, E. L., Barnes, A., and Lester, R.J.G. 1986. A method for estimating possible parasite-related host mortality, illustrated using data from *Callitetrarhynchus gracilis* (Cestoda: Trypanorhyncha) in lizardfish (*Saurida* spp.). *Parasitology* 92: 227–43.

Adlard, R. D., and Lester, R.J.G. 1994. Dynamics of the interaction between the parasitic isopod, *Anilocra pomacentri*, and the coral reef fish, *Chromis nitida*. *Parasitology* 109: 311–24.

Aeby, G. S. 1992. The potential effect the ability of a coral intermediate host to regenerate has had on the evolution of its association with a marine parasite. *Proceedings of the Seventh International Coral Reef Symposium* 2: 809–15.

Aeby, G. S. 2002. Trade-offs for the butterflyfish, *Chaetodon multicinctus*, when feeding on coral prey infected with trematode metacercariae. *Behavioral Ecology and Sociobiology* 52: 158–65.

Agnew, P., and Koella, J. C. 1997. Virulence, parasite mode of transmission, and host fluctuating asymmetry. *Proceedings of the Royal Society of London B* 264: 9–15.

Aho, J. M. 1990. Helminth communities of amphibians and reptiles: comparative approaches to understanding patterns and processes. In *Parasite Communities: Patterns and Processes* (eds G. W. Esch, A. O. Bush, and J. M. Aho), Chapman & Hall, London, pp. 157–95.

Aho, J. M., and Kennedy, C. R. 1984. Seasonal population dynamics of the nematode *Cystidicoloides tenuissima* (Zeder) from the River Swincombe, England. *Journal of Fish Biology* 25: 473–89.

Aho, J. M., and Kennedy, C. R. 1987. Circulation pattern and transmission dynamics of the suprapopulation of the nematode *Cystidicoloides tenuissima* (Zeder) in the River Swincombe, England. *Journal of Fish Biology* 31: 123–41.

Albers, G.A.A., and Grey, G. D. 1986. Breeding for worm resistance: a perspective. In *Parasitology: Quo Vadit? Proceedings of the Sixth International Congress of Parasitology* (ed. M. J. Howell), Australian Academy of Science, Canberra, pp. 559–66.

Albon, S. D., Stien, A., Irvine, R. J., Langvatn, R., Ropstad, E., and Halvorsen, O. 2002. The role of parasites in the dynamics of a reindeer population. *Proceedings of the Royal Society of London B* 269: 1625–32.

Allen, A. P., Brown, J. H., and Gillooly, J. F. 2002. Global biodiversity, biochemical kinetics, and the energetic-equivalence rule. *Science* 297: 1545–48.

Alroy, J. 1998. Cope's rule and the dynamics of body mass evolution in North American fossil mammals. *Science* 280: 731–34.

Altizer, S., Harvell, D. and Friedle, E. 2003. Rapid evolutionary dynamics and disease threats to biodiversity. *Trends in Ecology and Evolution* 18: 589–96.

Amundsen, P.-A., Knudsen, R., Kuris, A. M., and Kristoffersen, R. 2003. Seasonal and ontogenetic dynamics in trophic transmission of parasites. *Oikos* 102: 285–93.

Andersen, K. I., and Valtonen, E. T. 1990. On the infracommunity structure of adult cestodes in freshwater fishes. *Parasitology* 101: 257–64.

Anderson, J. A., Blazek, K. J., Percival, T. J., and Janovy, J. Jr. 1993. The niche of the gill parasite *Dactylogyrus banghami* (Monogenea: Dactylogyridae) on *Notropis stramineus* (Pisces: Cyprinidae). *Journal of Parasitology* 79: 435–37.

Anderson, R. C. 1984. The origins of zooparasitic nematodes. *Canadian Journal of Zoology* 62: 317–28.

Anderson, R. C. 1996. Why do fish have so few roundworm (nematode) parasites? *Environmental Biology of Fishes* 46: 1–5.

Anderson, R. C. 2000. *Nematode Parasites of Vertebrates: Their Development and Transmission*, 2nd edition. CABI Publishing, Wallingford, UK.

Anderson, R. M. 1978. The regulation of host population growth by parasitic species. *Parasitology* 76: 119–57.

Anderson, R. M. 1982. Parasite dispersion patterns: generative mechanisms and dynamic consequences. In *Aspects of Parasitology* (ed. E. Meerovitch), McGill University Press, Montreal, pp. 1–40.

Anderson, R. M. 1993. Epidemiology. In *Modern Parasitology*, 2nd ed. (ed. F.E.G. Cox), Blackwell, Oxford, pp. 75–116.

Anderson, R. M. 1995. Evolutionary pressures in the spread and persistence of infectious agents in vertebrate populations. *Parasitology* 111: S15–S31.

Anderson, R. M., and Gordon, D. M. 1982. Processes influencing the distribution of parasite numbers within host populations with special emphasis on parasite-induced host mortalities. *Parasitology* 85: 373–98.

Anderson, R. M., and May, R. M. 1978. Regulation and stability of host-parasite population interactions. I. Regulatory processes. *Journal of Animal Ecology* 47: 219–47.

Anderson, R. M., and May, R. M. 1979. Population biology of infectious diseases: part I. *Nature* 280: 361–67.

Anderson, R. M., and May, R. M. 1982. Coevolution of hosts and parasites. *Parasitology* 85: 411–26.

Anderson, R. M., and May, R. M. 1991. *Infectious Diseases of Humans*. Oxford University Press, Oxford.

Anderson, T.J.C., Blouin, M. S., and Beech, R. N. 1998. Population biology of parasitic nematodes: applications of genetic markers. *Advances in Parasitology* 41: 219–83.

Anderson, T.J.C., Romero-Abal, M. E., and Jaenike, J. 1995. Mitochondrial DNA and *Ascaris* microepidemiology: the composition of parasite populations from individual hosts, families and villages. *Parasitology* 110: 221–29.

Andreassen, J., Ito, A., Ito, M., Nakao, M., and Nakaya, K. 2004. *Hymenolepis microstoma*: direct life cycle in immunodeficient mice. *Journal of Helminthology* 78: 1–5.

Anstensrud, M. 1990. Mating strategies of two parasitic copepods [*Lernaeocera branchialis* (L.) (Pennellidae) and *Lepeophtheirus pectoralis* (Müller) (Caligidae)] on flounder: polygamy, sex-specific age at maturity and sex ratio. *Journal of Experimental Marine Biology and Ecology* 136: 141–58.

Anstensrud, M., and Schram, T. A. 1988. Host and site selection by larval stages and adults of the parasitic copepod *Lernaeenicus sprattae* (Sowerby) (Copepoda, Pennellidae) in the Oslofjord. *Hydrobiologia* 167/168: 587–95.

Apakupakul, K., Siddall, M. E., and Burreson, E. M. 1999. Higher level relationships of leeches (Annelida: Clitellata: Euhirudinea) based on morphology and gene sequences. *Molecular Phylogenetics and Evolution* 12: 350–59.

Arneberg, P. 2001. An ecological law and its macroecological consequences as revealed by studies of relationships between host densities and parasite prevalence. *Ecography* 24: 352–58.

Arneberg, P. 2002. Host population density and body mass as determinants of species richness in parasite communities: comparative analyses of directly transmitted nematodes of mammals. *Ecography* 25: 88–94.

Arneberg, P., Skorping, A., Grenfell, B. T., and Read, A. F. 1998a. Host densities as determinants of abundance in parasite communities. *Proceedings of the Royal Society of London B* 265: 1283–89.

Arneberg, P., Skorping, A., and Read, A. F. 1997. Is population density a species character? Comparative analyses of the nematode parasites of mammals. *Oikos* 80: 289–300.

Arneberg, P., Skorping, A., and Read, A. F. 1998b. Parasite abundance, body size, life histories, and the energetic equivalence rule. *American Naturalist* 151: 497–513.

Arnold, A. J., Kelly, D. C., and Parker, W. C. 1995. Causality and Cope's rule: evidence from the planktonic foraminifera. *Journal of Paleontology* 69: 203–10.

Arthur, J. R. 1984. A survey of the parasites of walleye pollock (*Theragra chalcogramma*) from the northeastern Pacific Ocean off Canada and a zoogeographic analysis of the parasite fauna of this fish throughout its range. *Canadian Journal of Zoology* 62: 675–84.

Arundel, J. H., Beveridge, I., and Presidente, P. J. 1979. Parasites and pathological findings in enclosed and free-ranging populations of *Macropus rufus* (Demarest) (Marsupialia) at Menindee, New South Wales. *Australian Wildlife Research* 6: 361–79.

Ashworth, S. T., and Kennedy, C. R. 1999. Density-dependent effects on *Anguillicola crassus* (Nematoda) within its European eel definitive host. *Parasitology* 118: 289–96.

Athias-Binche, F., and Morand, S. 1993. From phoresy to parasitism: the example of mites and nematodes. *Research and Reviews in Parasitology* 53: 73–79.

Atkinson, D. 1994. Temperature and organism size: a biological law for ectotherms? *Advances in Ecological Research* 25: 1–58.

Avise, J. C. 2000. *Phylogeography: The History and Formation of Species*. Harvard University Press, Cambridge, Massachusetts.

Aznar, F. J., Bush, A. O., Balbuena, J. A., and Raga, J. A. 2001. *Corynosoma cetaceum* in the stomach of franciscanas, *Pontoporia blainvillei* (Cetacea): an exceptional case of habitat selection by an acanthocephalan. *Journal of Parasitology* 87: 536–41.

Babirat, C., Mouritsen, K. N., and Poulin, R. 2004. Equal partnership: two trematode species, not one, manipulate the burrowing behaviour of the New Zealand cockle, *Austrovenus stutchburyi*. *Journal of Helminthology* 78: 195–99.

Bagge, A. M., Poulin, R., and Valtonen, E. T. 2004. Fish population size, and not density, as the determining factor of parasite infection: a case study. *Parasitology* 128: 305–13.

Bagge, A. M., and Valtonen, E. T. 1999. Development of monogenean communities on the gills of roach fry (*Rutilus rutilus*). *Parasitology* 118: 479–87.

Baker, M. D., Vossbrinck, C. R., Becnel, J. J., and Andreadis, T. G. 1998. Phylogeny of *Amblyospora* (Microsporida: Amblyosporidae) and related genera based on small subunit ribosomal DNA data: a possible example of host-parasite cospeciation. *Journal of Invertebrate Pathology* 71: 199–206.

Ballabeni, P. 1995. Parasite-induced gigantism in a snail: a host adaptation? *Functional Ecology* 9: 887–93.

Ballabeni, P., and Ward, P. I. 1993. Local adaptation of the trematode *Diplostomum phoxini* to the European minnow *Phoxinus phoxinus*, its second intermediate host. *Functional Ecology* 7: 84–90.

Baltanás, A. 1992. On the use of some methods for the estimation of species richness. *Oikos* 65: 484–92.

Bandi, C., Dunn, A. M., Hurst, G.D.D., and Rigaud, T. 2001. Inherited microorganisms, sex-specific virulence and reproductive parasitism. *Trends in Parasitology* 17: 88–94.

Bandilla, M., Hakalahti, T., Hudson, P. J., and Valtonen, E. T. 2005. Aggregation of *Argulus coregoni* (Crustacea: Branchiura) on rainbow trout (*Oncorhynchus mykiss*): a consequence of host susceptibility or exposure? *Parasitology* 130: 169–76.

Barber, I., and Crompton, D.W.T. 1997. The distribution of the metacercariae of *Diplostomum phoxini* in the brain of minnows, *Phoxinus phoxinus*. *Folia Parasitologica* 44: 19–25.

Barber, I., Hoare, D., and Krause, J. 2000. Effects of parasites on fish behaviour: a review and evolutionary perspective. *Reviews in Fish Biology and Fisheries* 10: 131–65.

Barger, I. A. 1999. The role of epidemiological knowledge and grazing management for helminth control in small ruminants. *International Journal for Parasitology* 29: 41–47.

Barger, M. A., and Esch, G. W. 2000. *Plagioporus sinitsini* (Digenea: Opecoelidae): a one-host life cycle. *Journal of Parasitology* 86: 150–53.

Barger, M. A., and Esch, G. W. 2002. Host specificity and the distribution-abundance relationship in a community of parasites infecting fishes in streams of North Carolina. *Journal of Parasitology* 88: 446–53.

Barger, M. A., and Nickol, B. B. 1999. Effects of coinfection with *Pomphorhynchus bulbocolli* on development of *Leptorhynchoides thecatus* (Acanthocephala) in amphipods (*Hyalella azteca*). *Journal of Parasitology* 85: 60–63.

Barker, D. E., Marcogliese, D. J., and Cone, D. K. 1996. On the distribution and abundance of eel parasites in Nova Scotia: local versus regional patterns. *Journal of Parasitology* 82: 697–701.

Barker, S. C. 1991. Evolution of host-parasite associations among species of lice and rock-wallabies: coevolution? *International Journal for Parasitology* 21: 497–501.

Barker, S. C. 1994. Phylogeny and classification, origins, and evolution of host associations of lice. *International Journal for Parasitology* 24: 1285–91.

Barker, S. C. 1996. Lice, cospeciation and parasitism. *International Journal for Parasitology* 26: 219–22.

Barker, S. C., and Cribb, T. H. 1993. Sporocysts of *Mesostephanus haliasturis* (Digenea) produce miracidia. *International Journal for Parasitology* 23: 137–39.

Barker, S. C., Cribb, T. H., Bray, R. A., and Adlard, R. D. 1994. Host-parasite associations on a coral reef: pomacentrid fishes and digenean trematodes. *International Journal for Parasitology* 24: 643–47.

Barnard, C. J. 1990. Parasitic relationships. In *Parasitism and Host Behaviour* (eds. C. J. Barnard and J. M. Behnke), Taylor & Francis, London, pp. 1–33.

Barral, V., Morand, S., Pointier, J. P., and Théron, A. 1996. Distribution of schistosome genetic diversity within naturally infected *Rattus rattus* detected by RAPD markers. *Parasitology* 113: 511–17.

Barta, J. R. 1989. Phylogenetic analysis of the class Sporozoea (phylum Apicomplexa Levine, 1970): evidence for the independent evolution of heteroxenous life cycles. *Journal of Parasitology* 75: 195–206.

Basch, P. F. 1990. Why do schistosomes have separate sexes? *Parasitology Today* 6: 160–63.

Bascompte, J., Jordano, P., Melián, C. J., and Olesen, J. M. 2003. The nested assembly of plant-animal mutualistic networks. *Proceedings of the National Academy of Sciences of the USA* 100: 9383–87.

Baudoin, M. 1975. Host castration as a parasitic strategy. *Evolution* 29: 335–52.

Bauer, A., Trouvé, S., Grégoire, A., Bollache, L., and Cézilly, F. 2000. Differential influence of *Pomphorhynchus laevis* (Acanthocephala) on the behaviour of native and invader gammarid species. *International Journal for Parasitology* 30: 1453–57.

Bauer, G. 1994. The adaptive value of offspring size among freshwater mussels (Bivalvia; Unionoidea). *Journal of Animal Ecology* 63: 933–44.

Beck, C., and Forrester, D. J. 1988. Helminths of the Florida manatee, *Trichecus manatus latirostris*, with a discussion and summary of the parasites of sirenians. *Journal of Parasitology* 74: 628–37.

Begon, M., Townsend, C. R., and Harper, J. L. 2005. *Ecology*, 4th ed. Blackwell Science, Oxford.

Behnke, J. M., Gilbert, F. S., Abu-Madi, M. A., and Lewis, J. W. 2005. Do the helminth parasites of wood mice interact? *Journal of Animal Ecology* 74: 982–93.

Bell, A. S., Sommerville, C., and Gibson, D. I. 2002. Multivariate analyses of morphometrical features from *Apatemon gracilis* (Rudolphi, 1819) Szidat, 1928 and *A. annuligerum* (v. Nordmann, 1832) (Digenea: Strigeidae) metacercariae. *Systematic Parasitology* 51: 121–33.

Bell, G., and Burt, A. 1991. The comparative biology of parasite species diversity: internal helminths of freshwater fish. *Journal of Animal Ecology* 60: 1047–64.

Berdoy, M., Webster, J. P., and Macdonald, D. W. 2000. Fatal attraction in rats infected with *Toxoplasma gondii*. *Proceedings of the Royal Society of London B* 267: 1591–94.

Bethel, W. M., and Holmes, J. C. 1974. Correlation of development of altered evasive behavior in *Gammarus lacustris* (Amphipoda) harboring cystacanths of *Polymorphus paradoxus* (Acanthocephala) with infectivity to the definitive host. *Journal of Parasitology* 60: 272–74.

Beveridge, I., and Chilton, N. B. 2001. Co-evolutionary relationships between the nematode subfamily Cloacininae and its macropodid marsupial hosts. *International Journal for Parasitology* 31: 976–96.

Beveridge, I., Chilton, N. B., and Spratt, D. M. 2002. The occurrence of species flocks in the nematode genus *Cloacina* (Strongyloidea: Cloacininae), parasitic in the stomachs of kangaroos and wallabies. *Australian Journal of Zoology* 50: 597–620.

Billingsley, P. F., Snook, L. S., and Johnston, V. J. 2005. Malaria parasite growth is stimulated by mosquito probing. *Biology Letters* 1: 185–89.

Biron, D. G., Marché, L., Ponton, F., Loxdale, H. D., Galéotti, N., Renault, L., Joly, C., and Thomas, F. 2005. Behavioural manipulation in a grasshopper harbouring hairworm: a proteomics approach. *Proceedings of the Royal Society of London B* 272: 2117–26.

Bishop, C. A., and Threlfall, W. 1974. Helminth parasites of the common eider duck, *Somateria mollissima* (L.), in Newfoundland and Labrador. *Proceedings of the Helminthological Society of Washington* 41: 25–35.

Black, G. A. 1985. Reproductive output and population biology of *Cystidicola stigmatura* (Leidy) (Nematoda) in arctic char, *Salvelinus alpinus* (L.) (Salmonidae). *Canadian Journal of Zoology* 63: 617–22.

Black, G. A., and Lankester, M. W. 1980. Migration and development of swim-bladder nematodes, *Cystidicola* spp. (Habronematoidea), in their definitive hosts. *Canadian Journal of Zoology* 58: 1997–2005.

Black, G. A., and Lankester, M. W. 1981. The transmission, life span, and population biology of *Cystidicola cristivomeri* White, 1941 (Nematoda: Habronematoidea) in char, *Salvelinus* spp. *Canadian Journal of Zoology* 59: 498–509.

Blackburn, T. M., and Gaston, K. J. 1994a. Animal body size distributions: patterns, mechanisms and implications. *Trends in Ecology & Evolution* 9: 471–74.

Blackburn, T. M., and Gaston, K. J. 1994b. Animal body size distributions change as more species are described. *Proceedings of the Royal Society of London B* 257: 293–97.

Blackburn, T. M., and Gaston, K. J. 1997. A critical assessment of the form of the interspecific relationship between abundance and body size in animals. *Journal of Animal Ecology* 66: 233–49.

Blackburn, T. M., and Gaston, K. J. 2001. Linking patterns in macroecology. *Journal of Animal Ecology* 70: 338–52.

Blackburn, T. M., Gaston, K. J., and Loder, N. 1999. Geographic gradients in body size: a clarification of Bergmann's rule. *Diversity and Distributions* 5: 165–74.

Blackburn, T. M., and Lawton, J. H. 1994. Population abundance and body size in animal assemblages. *Philosophical Transactions of the Royal Society of London B* 343: 33–39.

Blaxter, M. L., De Ley, P., Garey, J. R., Liu, L. X., Scheldeman, P., Vierstraete, A., Vanfleteren, J. R., Mackey, L. Y., Dorris, M., Frisse, L. M., Vida, J. T., and Thomas, W. K. 1998. A molecular evolutionary framework for the phylum Nematoda. *Nature* 392: 71–75.

Bliss, C. I., and Fisher, R. A. 1953. Fitting the negative binomial distribution to biological data. *Biometrics* 9: 176–200.

Blouin, M. S. 2002. Molecular prospecting for cryptic species of nematodes: mitochondrial DNA versus internal transcribed spacer. *International Journal for Parasitology* 32: 527–31.

Blouin, M. S., Dame, J. B., Tarrant, C. A., and Courtney, C. H. 1992. Unusual population genetics of a parasitic nematode: mtDNA variation within and among populations. *Evolution* 46: 470–76.

Blouin, M. S., Yowell, C. A., Courtney, C. H., and Dame, J. B. 1995. Host movement and the genetic structure of populations of parasitic nematodes. *Genetics* 141: 1007–14.

Blower, S. M., and Roughgarden, J. 1989. Parasites detect host spatial pattern and density: a field experimental analysis. *Oecologia* 78: 138–41.

Boag, B., Lello, J., Fenton, A., Tompkins, D. M., and Hudson, P. J. 2001. Patterns of parasite aggregation in the wild European rabbit (*Oryctolagus cuniculus*). *International Journal for Parasitology* 31: 1421–28.

Boëte, C., Paul, R.E.L., and Koella, J. C. 2004. Direct and indirect immunosuppression by a malaria parasite in its mosquito vector. *Proceedings of the Royal Society of London B* 271: 1611–15.

Bohan, D. A. 2000. Spatial structuring and frequency distribution of the nematode *Steinernema feltiae* Filipjev. *Parasitology* 121: 417–25.

Boissier, J., Morand, S., and Moné, H. 1999. A review of performance and pathogenicity of male and female *Schistosoma mansoni* during the life cycle. *Parasitology* 119: 447–54.

Bonhoeffer, S., Lenski, R. E., and Ebert, D. 1996. The curse of the pharaoh: the evolution of virulence in pathogens with long living propagules. *Proceedings of the Royal Society of London B* 263: 715–21.

Bonner, J. T. 1988. *The Evolution of Complexity by Means of Natural Selection*. Princeton University Press, Princeton.

Bonner, J. T. 1993. *Life Cycles: Reflections of an Evolutionary Biologist*. Princeton University Press, Princeton.

Boone, E., Boettcher, A. A., Sherman, T. D., and O'Brien, J. J. 2004. What constrains the geographic and host range of the rhizocephalan *Loxothylacus texanus* in the wild? *Journal of Experimental Marine Biology and Ecology* 309: 129–39.

Boots, M., and Sasaki, A. 1999. "Small worlds" and the evolution of virulence: infection occurs locally and at a distance. *Proceedings of the Royal Society of London B* 266: 1933–38.

Borda, E., and Siddall, M. E. 2004. Arhynchobdellida (Annelida: Oligochaeta: Hirudinida): phylogenetic relationships and evolution. *Molecular Phylogenetics and Evolution* 30: 213–25.

Bottomley, C., Isham, V., and Basáñez, M.-G. 2005. Population biology of multispecies helminth infection: interspecific interactions and parasite distribution. *Parasitology* 131: 417–33.

Bouchet, P., and Perrine, D. 1996. More gastropods feeding at night on parrotfishes. *Bulletin of Marine Science* 59: 224–28.

Bouchon, D., Rigaud, T., and Juchault, P. 1998. Evidence for widespread *Wolbachia* infection in isopod crustaceans: molecular identification and host feminization. *Proceedings of the Royal Society of London B* 265: 1081–90.

Boulinier, T., Ives, A. R., and Danchin, E. 1996. Measuring aggregation of parasites at different host population levels. *Parasitology* 112: 581–87.

Bourtzis, K., and O'Neill, S. 1998. *Wolbachia* infections and arthropod reproduction. *BioScience* 48: 287–93.

Braisher, T. L., Gemmell, N. J., Grenfell, B. T., and Amos, W. 2004. Host isolation and patterns of genetic variability in three populations of *Teladorsagia* from sheep. *International Journal for Parasitology* 34: 1197–1204.

Brandle, M., and Brandl, R. 2001. Species richness of insects and mites on trees: expanding Southwood. *Journal of Animal Ecology* 70: 491–504.

Brandon, R. N. 1994. Theory and experiment in evolutionary biology. *Synthese* 99: 59–73.

Brant, S. V., and Gardner, S. L. 2000. Phylogeny of species of the genus *Litomosoides* (Nematoda: Onchocercidae): evidence of rampant host switching. *Journal of Parasitology* 86: 545–54.

Brant, S. V., and Orti, G. 2003. Evidence for gene flow in parasitic nematodes between two host species of shrews. *Molecular Ecology* 12: 2853–59.

Bray, R. A. 2004. The bathymetric distribution of the digenean parasites of deep-sea fishes. *Folia Parasitologica* 51: 268–74.

Bray, R. A., Littlewood, D.T.J., Herniou, E. A., Williams, B., and Henderson, R. E. 1999. Digenean parasites of deep-sea teleosts: a review and case studies of intrageneric phylogenies. *Parasitology* 119: S125–S144.

Brodeur, J., and McNeil, J. N. 1989. Seasonal microhabitat selection by an endoparasitoid through adaptive modification of host behavior. *Science* 244: 226–28.

Brodeur, J., and Vet, L.E.M. 1994. Usurpation of host behaviour by a parasitic wasp. *Animal Behaviour* 48: 187–92.

Brooks, D. R. 1980. Allopatric speciation and non-interactive parasite community structure. *Systematic Zoology* 29: 192–203.

Brooks, D. R. 1988. Macroevolutionary comparisons of host and parasite phylogenies. *Annual Review of Ecology and Systematics* 19: 235–59.

Brooks, D. R., Dowling, A.P.G., van Veller, M.G.P., and Hoberg, E. P. 2004. Ending a decade of deception: a valiant failure, a not-so-valiant failure, and a success story. *Cladistics* 20: 32–46.

Brooks, D. R., and McLennan, D. A. 1991. *Phylogeny, Ecology, and Behavior: A Research Program in Comparative Biology*. University of Chicago Press, Chicago.

Brooks, D. R., and McLennan, D. A. 1993a. *Parascript: Parasites and the Language of Evolution*. Smithsonian Institution Press, Washington, D.C.

Brooks, D. R., and McLennan, D. A. 1993b. Comparative study of adaptive radiations with an example using parasitic flatworms (Platyhelminthes: Cercomeria). *American Naturalist* 142: 755–78.

Brown, A. F. 1986. Evidence for density-dependent establishment and survival of *Pomphorhynchus laevis* (Müller, 1776) (Acanthocephala) in laboratory-infected *Salmo gairdneri* Richardson and its bearing on wild populations in *Leuciscus cephalus* (L.). *Journal of Fish Biology* 28: 659–69.

Brown, J. H. 1995. *Macroecology*. University of Chicago Press, Chicago.

Brown, J. H. 1999. Macroecology: progress and prospect. *Oikos* 87: 3–14.

Brown, S. P. 1999. Cooperation and conflict in host-manipulating parasites. *Proceedings of the Royal Society of London B* 266: 1899–1904.

Brown, S. P., De Lorgeril, J., Joly, C., and Thomas, F. 2003. Field evidence for density-dependent effects in the trematode *Microphallus papillorobustus* in its manipulated host, *Gammarus insensibilis*. *Journal of Parasitology* 89: 668–72.

Brown, S. P., Loot, G., Grenfell, B. T., and Guégan, J.-F. 2001b. Host manipulation by *Ligula intestinalis*: accident or adaptation? *Parasitology* 123: 519–29.

Brown, S. P., Renaud, F., Guégan, J.-F., and Thomas, F. 2001a. Evolution of trophic transmission in parasites: the need to reach a mating place? *Journal of Evolutionary Biology* 14: 815–20.

Brusca, R. C. 1981. A monograph on the Isopoda Cymothoidae (Crustacea) of the eastern Pacific. *Zoological Journal of the Linnean Society* 73: 117–99.

Buckling, A.G.J., Taylor, L. H., Carlton, J.M.R., and Read, A. F. 1997. Adaptive changes in *Plasmodium* transmission strategies following chloroquine chemotherapy. *Proceedings of the Royal Society of London B* 264: 553–59.

Bucknell, D., Hoste, H., Gasser, R. B., and Beveridge, I. 1996. The structure of the community of strongyloid nematodes of domestic equids. *Journal of Helminthology* 70: 185–92.

Bull, J. J., Molineux, I. J., and Rice, W. R. 1991. Selection of benevolence in a host-parasite system. *Evolution* 45: 875–82.

Bullini, L., Nascetti, G., Paggi, L., Orecchia, P., Mattiucci, S., and Berland, B. 1986. Genetic variation of ascaridoid worms with different life cycles. *Evolution* 40: 437–40.

Bulmer, M. 1994. *Theoretical Evolutionary Ecology*. Sinauer Associates, Sunderland, Massachusetts.

Bunkley-Williams, L., and Williams, E. H. Jr. 1998. Isopods associated with fishes: a synopsis and corrections. *Journal of Parasitology* 84: 893–96.

Burn, P. R. 1980. Density dependent regulation of a fish trematode population. *Journal of Parasitology* 66: 173–74.

Bush, A. O., Aho, J. M., and Kennedy, C. R. 1990. Ecological versus phylogenetic determinants of helminth parasite community richness. *Evolutionary Ecology* 4: 1–20.

Bush, A. O., Caira, J. N., Minchella, D. J., Nadler, S. A., and Seed, J. R. 1995. Parasitology year 2000. *Journal of Parasitology* 81: 835–42.

Bush, A. O., Fernández, J. C., Esch, G. W., and Seed, J. R. 2001. *Parasitism: The Diversity and Ecology of Animal Parasites*. Cambridge University Press, Cambridge.

Bush, A. O., and Forrester, D. J. 1976. Helminths of the white ibis in Florida. *Proceedings of the Helminthological Society of Washington* 43: 17–23.

Bush, A. O., Heard, R. W., and Overstreet, R. M. 1993. Intermediate hosts as source communities. *Canadian Journal of Zoology* 71: 1358–63.

Bush, A. O., and Holmes, J. C. 1986a. Intestinal helminths of lesser scaup ducks: patterns of association. *Canadian Journal of Zoology* 64: 132–41.

Bush, A. O., and Holmes, J. C. 1986b. Intestinal helminths of lesser scaup ducks: an interactive community. *Canadian Journal of Zoology* 64: 142–52.

Bush, A. O., and Kennedy, C. R. 1994. Host fragmentation and helminth parasites: hedging your bets against extinction. *International Journal for Parasitology* 24: 1333–43.

Bush, A. O., Lafferty, K. D., Lotz, J. M., and Shostak, A. W. 1997. Parasitology meets ecology on its own terms: Margolis et al. revisited. *Journal of Parasitology* 83: 575–83.

Butterworth, E. W., and Holmes, J. C. 1984. Character divergence in two species of trematodes (*Pharyngostomoides*: Strigeoidea). *Journal of Parasitology* 70: 315–16.

Caira, J. N., and Jensen, K. 2001. An investigation of the co-evolutionary relationships between onchobothriid tapeworms and their elasmobranch hosts. *International Journal for Parasitology* 31: 960–75.

Caira, J. N., Jensen, K., and Holsinger, K. E. 2003. On a new index of host specificity. In *Taxonomy, Ecology and Evolution of Metazoan Parasites*, vol. 1 (eds. C. Combes and J. Jourdane), Presses Universitaires de Perpignan, Perpignan, France, pp. 161–201.

Calero, C., Ortiz, P., and De Souza, L. 1950. Helminths in rats from Panama City and suburbs. *Journal of Parasitology* 36: 426.

Calero, C., Ortiz, P., and De Souza, L. 1951. Helminths in cats from Panama City and Balboa, C. Z. *Journal of Parasitology* 37: 326.

Calow, P. 1983. Pattern and paradox in parasite reproduction. *Parasitology* 86 (suppl.): 197–207.

Calvete, C., Blanco-Aguiar, J. A., Virgos, E., Cabezas-Diaz, S., and Villafuerte, R. 2004. Spatial variation in helminth community structure in the red-legged partridge (*Alectoris rufa* L.): effects of definitive host density. *Parasitology* 129: 101–13.

Campbell, R. A., Haedrich, R. L., and Munroe, T. A. 1980. Parasitism and ecological relationships among deep-sea benthic fishes. *Marine Biology* 57: 301–13.

Canning, E. U. 1982. An evaluation of protozoal characteristics in relation to biological control of pests. *Parasitology* 84: 119–49.

Canning, E. U., and Okamura, B. 2004. Biodiversity and evolution of the Myxozoa. *Advances in Parasitology* 56: 43–131.

Carney, J. P., and Brooks, D. R. 1991. Phylogenetic analysis of *Alloglossidium* Simer, 1929 (Digenea: Plagiorchiiformes: Macroderoididae) with discussion of the origin of truncated life cycle patterns in the genus. *Journal of Parasitology* 77: 890–900.

Carney, J. P., and Dick, T. A. 1999. Enteric helminths of perch (*Perca fluviatilis* L.) and yellow perch (*Perca flavescens* Mitchill): stochastic or predictable assemblages? *Journal of Parasitology* 85: 785–95.

Carney, J. P., and Dick, T. A. 2000a. The historical ecology of yellow perch (*Perca flavescens* [Mitchill]) and their parasites. *Journal of Biogeography* 27: 1337–47.

Carney, J. P., and Dick, T. A. 2000b. Helminth communities of yellow perch (*Perca flavescens* [Mitchill]): determinants of pattern. *Canadian Journal of Zoology* 78: 538–55.

Carney, W. P. 1969. Behavioral and morphological changes in carpenter ants harboring dicrocoelid metacercariae. *American Midland Naturalist* 82: 605–11.

Carter, N. P., Anderson, R. M., and Wilson, R. A. 1982. Transmission of *Schistosoma mansoni* from man to snail: laboratory studies on the influence of snail and miracidial densities on transmission success. *Parasitology* 85: 361–72.

Cavalier-Smith, T. 1985. *The Evolution of Genome Size*. Wiley, Chichester, New York.

Cézilly, F., Grégoire, A., and Bertin, A. 2000. Conflict between co-occurring manipulative parasites? An experimental study of the joint influence of two acanthocephalan parasites on the behaviour of *Gammarus pulex*. *Parasitology* 120: 625–30.

Charleston, M. A. 1998. Jungles: A new solution to the host-parasite phylogeny reconciliation problem. *Mathematical Biosciences* 149: 191–223.

Charleston, M. A. 2003. Recent results in cophylogeny mapping. *Advances in Parasitology* 54: 303–30.

Charlesworth, D., and Charlesworth, B. 1987. Inbreeding depression and its evolutionary consequences. *Annual Review of Ecology and Systematics* 18: 237–68.

Charnov, E. L. 1982. *The Theory of Sex Allocation*. Princeton University Press, Princeton.

Chehresa, A., Beech, R. N., and Scott, M. E. 1997. Life-history variation among lines isolated from a laboratory population of *Heligsomoides polygyrus bakeri*. *International Journal for Parasitology* 27: 541–51.

Chilton, N. B., Bao-Zhen, Q., Bøgh, H. O., and Nansen, P. 1999. An electrophoretic comparison of *Schistosoma japonicum* (Trematoda) from different provinces in the People's Republic of China suggests the existence of cryptic species. *Parasitology* 119: 375–83.

Chilton, N. B., Bull, C. M., and Andrews, R. H. 1992. Niche segregation in reptile ticks: attachment sites and reproductive success of females. *Oecologia* 90: 255–59.

Chisholm, L. A., and Whittington, I. D. 2000. Egg hatching in 3 species of monocotylid monogenean parasites from the shovelnose ray *Rhinobatos typus* at Heron Island, Australia. *Parasitology* 121: 303–13.

Choe, J. C., and Kim, K. C. 1988. Microhabitat preference and coexistence of ectoparasitic arthropods on Alaskan seabirds. *Canadian Journal of Zoology* 66: 987–97.

Choe, J. C., and Kim, K. C. 1989. Microhabitat selection and coexistence in feather mites (Acari: Analgoidea) on Alaskan seabirds. *Oecologia* 79: 10–14.

Choisy, M., Brown, S. P., Lafferty, K. D., and Thomas, F. 2003. Evolution of trophic transmission in parasites: why add intermediate hosts? *American Naturalist* 162: 172–81.

Choudhury, A., and Dick, T. A. 2000. Richness and diversity of helminth communities in tropical freshwater fishes: empirical evidence. *Journal of Biogeography* 27: 935–56.

Christen, M., Kurtz, J., and Milinski, M. 2002. Outcrossing increases infection success and competitive ability: experimental evidence from a hermaphrodite parasite. *Evolution* 56: 2243–51.

Churcher, T. S., Ferguson, N. M., and Basáñez, M.-G. 2005. Density dependence and overdispersion in the transmission of helminth parasites. *Parasitology* 131: 121–32.

Clarck, W. C. 1994. Origins of the parasitic habit in the Nematoda. *International Journal for Parasitology* 24: 1117–29.

Clayton, D. H., Bush, S. E., Goates, B. M., and Johnson, K. P. 2003. Host defense reinforces host-parasite cospeciation. *Proceedings of the National Academy of Sciences of the U.S.A.* 100: 15694–99.

Clayton, D. H., Bush, S. E., and Johnson, K. P. 2004. Ecology of congruence: past meets present. *Systematic Biology* 53: 165–73.

Clayton, D. H., and Johnson, K. P. 2001. Lice as probes. *Trends in Ecology and Evolution* 16: 433.

Clayton, D. H., and Johnson, K. P. 2003. Linking coevolutionary history to ecological process: doves and lice. *Evolution* 57: 2335–41.

Clayton, D. H., Price, R. D., and Page, R.D.M. 1996. Revision of *Dennyus* (*Collodennyus*) lice (Phthiraptera: Menoponidae) from swiftlets, with descriptions of new taxa and a comparison of host-parasite relationships. *Systematic Entomology* 21: 179–204.

Clayton, D. H., and Tompkins, D. M. 1994. Ectoparasite virulence is linked to mode of transmission. *Proceedings of the Royal Society of London B* 256: 211–17.

Clayton, D. H., and Walther, B. A. 2001. Influence of host ecology and morphology on the diversity of Neotropical bird lice. *Oikos* 94: 455–67.

Coats, D. W. 1999. Parasitic life styles of marine dinoflagellates. *Journal of Eukaryotic Microbiology* 46: 402–409.

Cockburn, A. 1991. *An Introduction to Evolutionary Ecology*. Blackwell Scientific Publications, Oxford.

Colwell, R. K., and Coddington, J. A. 1994. Estimating terrestrial biodiversity through extrapolation. *Philosophical Transactions of the Royal Society of London B* 345: 101–118.

Combes, C. 1990. Where do human schistosomes come from? An evolutionary approach. *Trends in Ecology and Evolution* 5: 334–37.

Combes, C. 1991a. Ethological aspects of parasite transmission. *American Naturalist* 138: 866–80.

Combes, C. 1991b. Evolution of parasite life cycles. In *Parasite-Host Associations: Coexistence or Conflict?* (eds. C. A. Toft, A. Aeschlimann, and L. Bolis), Oxford University Press, Oxford, pp. 62–82.

Combes, C. 1995. *Interactions Durables: Ecologie et Evolution du Parasitisme*. Masson, Paris.

Combes, C. 1996. Parasites, biodiversity and ecosystem stability. *Biodiversity and Conservation* 5: 953–62.

Combes, C. 1997. Fitness of parasites: pathology and selection. *International Journal for Parasitology* 27: 1–10.

Combes, C. 2001. *Parasitism: The Ecology and Evolution of Intimate Interactions*. University of Chicago Press, Chicago.

Combes, C., Fournier, A., Moné, H., and Théron, A. 1994. Behaviours in trematode cercariae that enhance parasite transmission: patterns and processes. *Parasitology* 109: S3–S13.

Combes, C., and Théron, A. 2000. Metazoan parasites and resource heterogeneity: constraints and benefits. *International Journal for Parasitology* 30: 299–304.

Conley, D. C., and Curtis, M. A. 1993. Effects of temperature and photoperiod on the duration of hatching, swimming, and copepodid survival of the parasitic copepod *Salmincola edwardsii*. *Canadian Journal of Zoology* 71: 972–76.

Connell, J. H. 1980. Diversity and the coevolution of competitors, or the ghost of competition past. *Oikos* 35: 131–38.

Conway Morris, S. 1981. Parasites and the fossil record. *Parasitology* 82: 489–509.

Cook, R. L., and Roberts, L. S. 1991. In vivo effects of putative crowding factors on development of *Hymenolepis diminuta*. *Journal of Parasitology* 77: 21–25.

Cook, R. R. 1995. The relationship between nested subsets, habitat subdivision, and species diversity. *Oecologia* 101: 204–10.

Cornell, H. V., and Lawton, J. H. 1992. Species interactions, local and regional processes, and limits to the richness of ecological communities: a theoretical perspective. *Journal of Animal Ecology* 61: 1–12.

Cornell, S. J., Isham, V. S., Smith, G., and Grenfell, B. T. 2003. Spatial parasite transmission, drug resistance, and the spread of rare genes. *Proceedings of the National Academy of Sciences of the U.S.A.* 100: 7401–05.

Cotgreave, P. 1993. The relationship between body size and population abundance in animals. *Trends in Ecology and Evolution* 8: 244–48.

Courtney, C. H., and Forrester, D. J. 1974. Helminth parasites of the brown pelican in Florida and Louisiana. *Proceedings of the Helminthological Society of Washington* 41: 89–93.

Coustau, C., Renaud, F., Delay, B., Robbins, I., and Mathieu, M. 1991. Mechanisms involved in parasitic castration: *in vitro* effects of the trematode *Prosorhynchus squamatus* on the gametogenesis and the nutrient storage metabolism of the marine bivalve mollusc *Mytilus edulis*. *Experimental Parasitology* 73: 36–43.

Cox, F.E.G. 1993. *Modern Parasitology*, 2nd ed. Blackwell Scientific Publications, Oxford.

Cribb, T. H., Bray, R. A., Olson, P. D., and Littlewood, D.T.J. 2003. Life cycle evolution in the Digenea: a new perspective from phylogeny. *Advances in Parasitology* 54: 197–254.

Cribb, T. H., Bray, R. A., Wright, T., and Pichelin, S. 2002. The trematodes of groupers (Serranidae: Epinephelinae): knowledge, nature and evolution. *Parasitology* 124: S23–S42.

Criscione, C. D., and Blouin, M. S. 2004. Life cycles shape parasite evolution: comparative population genetics of salmon trematodes. *Evolution* 58: 198–202.

Criscione, C. D., and Blouin, M. S. 2005. Effective sizes of macroparasite populations: a conceptual model. *Trends in Parasitology* 21: 212–17.

Criscione, C. D., Poulin, R., and Blouin, M. S. 2005. Molecular ecology of parasites: elucidating ecological and microevolutionary processes. *Molecular Ecology* 14: 2247–57.

Crofton, H. D. 1971. A quantitative approach to parasitism. *Parasitology* 62: 179–93.

Crompton, D.W.T. 1985. Reproduction. In *Biology of the Acanthocephala* (eds. D.W.T. Crompton, and B. B. Nickol), Cambridge University Press, Cambridge, pp. 213–71.

Crook, M., and Viney, M. E. 2005. The effect of non-immune stresses on the development of *Strongyloides ratti*. *Parasitology* 131: 383–92.

Cumming, G. S. 2000. Using habitat models to map diversity: pan-African species richness of ticks (Acari: Ixodida). *Journal of Biogeography* 27: 425–40.

Curtis, J., Sorensen, R. E., and Minchella, D. J. 2002. Schistosome genetic diversity: the implications of population structure as detected with microsatellite markers. *Parasitology* 125: S51–S59.

Curtis, L. A. 1987. Vertical distribution of an estuarine snail altered by a parasite. *Science* 235: 1509–11.

Curtis, L. A. 2003. Tenure of individual larval trematode infections in an estuarine gastropod. *Journal of the Marine Biological Association of the UK* 83: 1047–51.

Curtis, L. A., and Hubbard, K.M.K. 1993. Species relationships in a marine gastropod-trematode ecological system. *Biological Bulletin* 184: 25–35.

Curtis, M. A., and Rau, M. E. 1980. The geographical distribution of diplostomiasis (Trematoda: Strigeidae) in fishes from northern Quebec, Canada, in relation to the calcium ion concentrations of lakes. *Canadian Journal of Zoology* 58: 1390–94.

Dabert, J., Dabert, M., and Mironov, S. V. 2001. Phylogeny of feather mite subfamily Avenzoariinae (Acari: Analgoidea: Avenzoariidae) inferred from combined analyses of molecular and morphological data. *Molecular Phylogenetics and Evolution* 20: 124–35.

Dallas, J. F., Irvine, R. J., and Halvorsen, O. 2001. DNA evidence that *Marshallagia marshalli* Ransom, 1907 and *M. occidentalis* Ransom, 1907 (Nematoda: Ostertagiinae) from Svalbard reindeer are conspecific. *Systematic Parasitology* 50: 101–103.

Dash, K. M. 1981. Interaction between *Oesophagostomum columbianum* and *Oesophagostomum venulosum* in sheep. *International Journal for Parasitology* 11: 201–207.

Davies, C. M., Fairbrother, E., and Webster, J. P. 2002. Mixed strain schistosome infections of snails and the evolution of parasite virulence. *Parasitology* 124: 31–38.

Davies, C. M., Webster, J. P., and Woolhouse, M.E.J. 2001. Trade-offs in the evolution of virulence in an indirectly transmitted macroparasite. *Proceedings of the Royal Society of London B* 268: 251–57.

Davies, S. J., and McKerrow, J. H. 2003. Developmental plasticity in schistosomes and other helminths. *International Journal for Parasitology* 33: 1277–84.

Dawkins, R. 1982. *The Extended Phenotype*. Oxford University Press, Oxford.

Dawkins, R. 1990. Parasites, desiderata lists and the paradox of the organism. *Parasitology* 100: S63–S73.

Dawson, L.H.J., Renaud, F., Guégan, J.-F., and de Meeüs, T. 2000. Experimental evidence of asymmetrical competition between two species of parasitic copepods. *Proceedings of the Royal Society of London B* 267: 1973–78.

Day, J. F., and Edman, J. D. 1983. Malaria renders mice susceptible to mosquito feeding when gametocytes are most infective. *Journal of Parasitology* 69: 163–70.

Day, T. 2001. Parasite transmission modes and the evolution of virulence. *Evolution* 55: 2389–2400.

Day, T. 2003. Virulence evolution and the timing of disease life-history events. *Trends in Ecology and Evolution* 18: 113–18.

De Buron, I., and Morand, S. 2004. Deep-sea hydrothermal vent parasites: why do we not find more? *Parasitology* 128: 1–6.

De Meeüs, T., Hochberg, M. E., and Renaud, F. 1995. Maintenance of two genetic entities by habitat selection. *Evolutionary Ecology* 9: 131–38.

De Meeüs, T., Marin, R., and Renaud, F. 1992. Genetic heterogeneity within populations of *Lepeophtheirus europaensis* (Copepoda: Caligidae) parasitic on two host species. *International Journal for Parasitology* 22: 1179–81.

De Meeüs, T., and Renaud, F. 2002. Parasites within the new phylogeny of eukaryotes. *Trends in Parasitology* 18: 247–51.

De Meeüs, T., Renaud, F., and Gabrion, C. 1990. A model for studying isolation mechanisms in parasite populations: the genus *Lepeophtheirus* (Copepoda, Caligidae). *Journal of Experimental Zoology* 254: 207–14.

De Roode, J. C., Pansini, R., Cheesman, S. J., Helinski, M.E.H., Huijben, S., Wargo, A. R., Bell, A. S., Chan, B.H.K., Walliker, D., and Read, A. F. 2005. Virulence and competitive ability in genetically diverse malaria infections. *Proceedings of the National Academy of Sciences of the U.S.A.* 102: 7624–28.

Desdevises, Y., Morand, S., Jousson, O., and Legendre, P. 2002a. Coevolution between *Lamellodiscus* (Monogenea: Diplectanidae) and Sparidae (Teleostei): the study of a complex host-parasite system. *Evolution* 56: 2459–71.

Desdevises, Y., Morand, S., and Legendre, P. 2002b. Evolution and determinants of host specificity in the genus *Lamellodiscus* (Monogenea). *Biological Journal of the Linnean Society* 77: 431–43.

Despommier, D. D. 1990. *Trichinella spiralis*: the worm that would be virus. *Parasitology Today* 6: 193–96.

Despommier, D. D. 1993. *Trichinella spiralis* and the concept of niche. *Journal of Parasitology* 79: 472–82.

Dezfuli, B. S., Giari, L., De Biaggi, S., and Poulin, R. 2001b. Associations and interactions among intestinal helminths of the brown trout, *Salmo trutta*, in northern Italy. *Journal of Helminthology* 75: 331–36.

Dezfuli, B. S., Giari, L., and Poulin, R. 2000. Species associations among larval helminths in an amphipod intermediate host. *International Journal for Parasitology* 30: 1143–46.

Dezfuli, B. S., Giari, L., and Poulin, R. 2001a. Costs of intraspecific and interspecific host sharing in acanthocephalan parasites. *Parasitology* 122: 483–89.

Diekmann, O., and Heesterbeek, J.A.P. 2000. *Mathematical Epidemiology of Infectious Diseases: Model Building, Analysis and Interpretation.* John Wiley & Sons, Chichester, N.Y.

Dobson, A. P. 1985. The population dynamics of competition between parasites. *Parasitology* 91: 317–47.

Dobson, A. P. 1986. Inequalities in the individual reproductive success of parasites. *Parasitology* 92: 675–82.

Dobson, A. P. 1988. The population biology of parasite-induced changes in host behavior. *Quarterly Review of Biology* 63: 139–65.

Dobson, A. P. 1990. Models for multi-species parasite-host communities. In *Parasite Communities: Patterns and Processes* (eds. G. W. Esch, A. O. Bush, and J. M. Aho), Chapman & Hall, London, pp. 261–88.

Dobson, A. P. 2004. Population dynamics of pathogens with multiple host species. *American Naturalist* 164: S64–S78.

Dobson, A. P., and Keymer, A. E. 1985. Life-history models. In *Biology of the Acanthocephala* (eds. D.W.T. Crompton and B. B. Nickol), Cambridge University Press, Cambridge, pp. 347–84.

Dobson, A. P., and Merenlender, A. 1991. Coevolution of macroparasites and their hosts. In *Parasite-Host Associations: Coexistence or Conflict?* (eds. C. A. Toft, A. Aeschlimann, and L. Bolis), Oxford University Press, Oxford, pp. 83–101.

Dobson, A. P., and Roberts, M. 1994. The population dynamics of parasitic helminth communities. *Parasitology* 109: S97–S108.

Dobson, A. P., Waller, P. J., and Donald, A. D. (1990) Population dynamics of *Trichostrongylus colubriformis* in sheep: the effect of infection rate on the establishment of infective larvae and parasite fecundity. *International Journal for Parasitology* 20: 347–52.

Donald, K. M., Kennedy, M., Poulin, R., and Spencer, H. G. 2004. Host specificity and molecular phylogeny of larval Digenea isolated from New Zealand and Australian topshells (Gastropoda: Trochidae). *International Journal for Parasitology* 34: 557–68.

Donnelly, R. E., and Reynolds, J. D. 1994. Occurrence and distribution of the parasitic copepod *Lepeosphilus labrei* on corkwing wrasse (*Crenilabrus melops*) from Mulroy Bay, Ireland. *Journal of Parasitology* 80: 331–32.

Doughty, P. 1996. Statistical analysis of natural experiments in evolutionary biology: comments on recent criticisms of the use of comparative methods to study adaptation. *American Naturalist* 148: 943–56.

Dowling, A.P.G. 2002. Testing the accuracy of TreeMap and Brooks parsimony analyses of coevolutionary patterns using artificial associations. *Cladistics* 18: 416–35.

Dowling, A.P.G., van Veller, M.G.P., Hoberg, E. P., and Brooks, D. R. 2003. A priori and a posteriori methods in comparative evolutionary studies of host-parasite associations. *Cladistics* 19: 240–53.

Downes, B. J. 1989. Host specificity, host location and dispersal: experimental conclusions from freshwater mites (*Unionicola* spp.) parasitizing unionid mussels. *Parasitology* 98: 189–96.

Dreyer, H., and Wagele, J. W. 2001. Parasites of crustaceans (Isopoda: Bopyridae) evolved from fish parasites: molecular and morphological evidence. *Zoology—Analysis of Complex Systems* 103: 157–78.

Dufva, R. 1996. Sympatric and allopatric combinations of hen fleas and great tits: a test of the local adaptation hypothesis. *Journal of Evolutionary Biology* 9: 505–10.

Dunn, A. M., Adams, J., and Smith, J. E. 1993. Transovarial transmission and sex ratio distortion by a microsporidian parasite in a shrimp. *Journal of Invertebrate Pathology* 61: 248–52.

Dunn, A. M., and Hatcher, M. J. 1997. Prevalence, transmission and intensity of infection by a microsporidian sex ratio distorter in natural *Gammarus duebeni* populations. *Parasitology* 114: 231–36.

Dunn, A. M., Hatcher, M. J., Terry, R. S., and Tofts, C. 1995. Evolutionary ecology of vertically trans-mitted parasites: transovarial transmission of a microsporidian sex ratio distorter in *Gammarus duebeni*. *Parasitology* 111: S91–S109.

Dunn, A. M., Terry, R. S., and Smith, J. E. 2001. Transovarial transmission in the Microsporidia. *Advances in Parasitology* 48: 57–100.

Dupont, F., and Gabrion, C. 1987. The concept of specificity in the procercoid-copepod system: *Bothriocephalus claviceps* (Cestoda) a parasite of the eel (*Anguilla anguilla*). *Parasitology Research* 73: 151–58.

Durette-Desset, M.-C., Beveridge, I., and Spratt, D. M. 1994. The origins and evolutionary expansion of the Strongylida (Nematoda). *International Journal for Parasitology* 24: 1139–65.

Dybdahl, M. F., and Lively, C. M. 1996. The geography of coevolution: comparative population structures for a snail and its trematode parasite. *Evolution* 50: 2264–75.

Dybdahl, M. F., and Storfer, A. 2003. Parasite local adaptation: Red Queen versus Suicide King. *Trends in Ecology and Evolution* 18: 523–30.

Dyer, M., and Day, K. P. 2000. Commitment to gametocytogenesis in *Plasmodium falciparum*. *Parasitology Today* 16: 102–107.

Eberhard, W. G. 2000. Spider manipulation by a wasp larva. *Nature* 406: 255–56.

Ebert, D. 1994. Virulence and local adaptation of a horizontally transmitted parasite. *Science* 265: 1084–86.

Ebert, D. 1998. Experimental evolution of parasites. *Science* 282: 1432–35.

Ebert, D. 1999. The evolution and expression of parasite virulence. In *Evolution in Health and Disease* (ed. S. C. Stearns), Oxford University Press, Oxford, pp. 161–72.

Ebert, D. 2000. Experimental evidence for rapid parasite adaptation and its consequences for the evolution of virulence. In *Evolutionary Biology of Host-Parasite Relationships: Theory Meets Reality* (eds. R. Poulin, S. Morand, and A. Skorping), Elsevier, Amsterdam, pp. 163–84.

Ebert, D., Carius, H. J., Little, T., and Decaestecker, E. 2004. The evolution of virulence when parasites cause host castration and gigantism. *American Naturalist* 164: S19–S32.

Ebert, D., and Herre, E. A. 1996. The evolution of parasitic diseases. *Parasitology Today* 12: 96–101.

Ebert, D., Hottinger, J. W., and Pajunen, V. I. 2001. Temporal and spatial dynamics of parasite richness in a *Daphnia* metapopulation. *Ecology* 82: 3417–34.

Ebert, D., and Mangin, K. L. 1997. The influence of host demography on the evolution of virulence of a microsporidian gut parasite. *Evolution* 51: 1828–37.

Ebert, D., and Weisser, W. W. 1997. Optimal killing for obligate killers: the evolution of life histories and virulence of semelparous parasites. *Proceedings of the Royal Society of London B* 264: 985–91.

Edelaar, P., Drent, J., and de Goeij, P. 2003. A double test of the parasite manipulation hypothesis in a burrowing bivalve. *Oecologia* 134: 66–71.

Eisen, R. J., and Schall, J. J. 2000. Life history of a malaria parasite (*Plasmodium mexicanum*): independent traits and basis for variation. *Proceedings of the Royal Society of London B* 267: 793–99.

Elliot, S. L., Adler, F. R., and Sabelis, M. W. 2003. How virulent should a parasite be to its vector? *Ecology* 84: 2568–74.

Ellis, R. D., Pung, O. J., and Richardson, D. J. 1999. Site selection by intestinal helminths of the Virginia opossum (*Didelphis virginiana*). *Journal of Parasitology* 85: 1–5.

Elston, D. A., Moss, R., Boulinier, T., Arrowsmith, C., and Lambin, X. 2001. Analysis of aggregation, a worked example: numbers of ticks on red grouse chicks. *Parasitology* 122: 563–69.

Eppert, A., Lewis, F. A., Grzywacz, C., Coura-Filho, P., Caldas, I., and Minchella, D. J. 2002. Distribution of schistosome infections in molluscan hosts at different levels of parasite prevalence. *Journal of Parasitology* 88: 232–36.

Esch, G. W. 2004. *Parasites, People, and Places: Essays on Field Parasitology*. Cambridge University Press, Cambridge.

Esch, G. W., Bush, A. O., and Aho, J. M. 1990. *Parasite Communities: Patterns and Processes*. Chapman & Hall, London.

Esch, G. W., Curtis, L. A., and Barger, M. A. 2001. A perspective on the ecology of trematode communities in snails. *Parasitology* 123: S57–S75.

Esch, G. W., and Fernández, J. C. 1993. *A Functional Biology of Parasitism: Ecological and Evolutionary Implications*. Chapman & Hall, London.

Esch, G. W., Kennedy, C. R., Bush, A. O., and Aho, J. M. 1988. Patterns in helminth communities in freshwater fish in Great Britain: alternative strategies for colonization. *Parasitology* 96: 519–32.

Esch, G. W., Wetzel, E. J., Zelmer, D. A., and Schotthoefer, A. M. 1997. Long-term changes in parasite population and community structure: a case history. *American Midland Naturalist* 137: 369–87.

Euzet, L., and Combes, C. 1998. The selection of habitats among the Monogenea. *International Journal for Parasitology* 28: 1645–52.

Evans, N. A. 1985. The influence of environmental temperature upon transmission of the cercariae of *Echinostoma liei* (Digenea: Echinostomatidae). *Parasitology* 90: 269–75.

Ewald, P. W. 1983. Host-parasite relations, vectors, and the evolution of disease severity. *Annual Review of Ecology and Systematics* 14: 465–85.

Ewald, P. W. 1994. *Evolution of Infectious Disease*. Oxford University Press, Oxford.

Ewald, P. W. 1995. The evolution of virulence: a unifying link between parasitology and ecology. *Journal of Parasitology* 81: 659–69.

Fallon, S. M., Bermingham, E., and Ricklefs, R. E. 2005. Host specialization and geographic localization of avian malaria parasites: a regional analysis in the Lesser Antilles. *American Naturalist* 165: 466–80.

Fellis, K. J., and Esch, G. W. 2005. Autogenic-allogenic status affects interpond community similarity and species-area relationship of macroparasites in the bluegill sunfish, *Lepomis macrochirus*, from a series of freshwater ponds in the Piedmont area of North Carolina. *Journal of Parasitology* 91: 764–67.

Fenton, A., Fairbairn, J. P., Norman, R., and Hudson, P. J. 2002. Parasite transmission: reconciling theory and reality. *Journal of Animal Ecology* 71: 893–905.

Fenton, A., and Hudson, P. J. 2002. Optimal infection strategies: should macroparasites hedge their bets? *Oikos* 96: 92–101.

Fenton, A., Paterson, S., Viney, M. E., and Gardner, M. P. 2004. Determining the optimal developmental route of *Strongyloides ratti*: an evolutionarily stable strategy approach. *Evolution* 58: 989–1000.

Fenton, A., and Rands, S. A. 2004. Optimal parasite infection strategies: a state-dependent approach. *International Journal for Parasitology* 34: 813–21.

Ferguson, H. M., Mackinnon, M. J., Chan, B. H., and Read, A. F. 2003. Mosquito mortality and the evolution of malaria virulence. *Evolution* 57: 2792–2804.

Ferguson, H. M., and Read, A. F. 2002. Genetic and environmental determinants of malaria parasite virulence in mosquitoes. *Proceedings of the Royal Society of London B* 269: 1217–24.

Ferrari, N., Cattadori, I. M., Nespereira, J., Rizzoli, A., and Hudson, P. J. 2004. The role of host sex in parasite dynamics: field experiments on the yellow-necked mouse *Apodemus flavicollis*. *Ecology Letters* 7: 88–94.

Fichet-Calvet, E., Wang, J., Jomâa, I., Ismail, R. B., and Ashford, R. W. 2003. Patterns of the tapeworm *Raillietina trapezoides* infection in the fat sand rat *Psammomys obesus* in Tunisia: season, climatic conditions, host age and crowding effects. *Parasitology* 126: 481–92.

Fingerut, J. T., Zimmer, C. A., and Zimmer, R. K. 2003a. Patterns and processes of larval emergence in an estuarine parasite system. *Biological Bulletin* 205: 110–20.

Fingerut, J. T., Zimmer, C. A., and Zimmer, R. K. 2003b. Larval swimming overpowers turbulent mixing and facilitates transmission of a marine parasite. *Ecology* 84: 2502–15.

Fisher, M. C., and Viney, M. E. 1998. The population genetic structure of the facultatively sexual parasitic nematode *Strongyloides ratti* in wild rats. *Proceedings of the Royal Society of London B* 265: 703–709.

Floate, K. D., and Whitham, T. G. 1993. The "Hybrid Bridge" hypothesis: host shifting via plant hybrid swarms. *American Naturalist* 141: 651–62.

Font, W. F. 1998. Parasites in paradise: patterns of helminth distribution in Hawaiian stream fishes. *Journal of Helminthology* 72: 307–11.

Fox, C. W., Roff, D. A., and Fairbairn, D. J. 2001. *Evolutionary Ecology: Concepts and Case Studies.* Oxford University Press, Oxford.

Frank, A. C., Amiri, H., and Andersson, S.G.E. 2002. Genome deterioration: loss of repeated sequences and accumulation of junk DNA. *Genetica* 115: 1–12.

Frank, S. A. 1996. Models of parasite virulence. *Quarterly Review of Biology* 71: 37–78.

Fredensborg, B. L., and Poulin, R. 2005. Larval helminths in intermediate hosts: does competition early in life determine the fitness of adult parasites? *International Journal for Parasitology* 35: 1061–70.

Freeland, W. J. 1983. Parasites and the coexistence of animal host species. *American Naturalist* 121: 223–36.

Friggens, M. M., and Brown, J. H. 2005. Niche partitioning in the cestode communities of two elasmobranchs. *Oikos* 108: 76–84.

Fritz, R. S. 1982. Selection for host modification by insect parasitoids. *Evolution* 36: 283–88.

Fry, J. D. 1990. Trade-offs in fitness on different hosts: evidence from a selection experiment with a phytophagous mite. *American Naturalist* 136: 569–80.

Fulford, A.J.C. 1994. Dispersion and bias: can we trust geometric means? *Parasitology Today* 10: 446–48.

Gaba, S., Ginot, V., and Cabaret, J. 2005. Modelling macroparasite aggregation using a nematode-sheep system: the Weibull distribution as an alternative to the negative binomial distribution? *Parasitology* 131: 393–401.

Gage, J. D. 1996. Why are there so many species in deep-sea sediments? *Journal of Experimental Marine Biology and Ecology* 200: 257–86.

Galaktionov, K. V., and Dobrovolskij, A. A. 2003. *The Biology and Evolution of Trematodes.* Kluwer Academic Publishers, Dordrecht, The Netherlands.

Galvani, A. P. 2003a. Epidemiology meets evolutionary ecology. *Trends in Ecology and Evolution* 18: 132–39.

Galvani, A. P. 2003b. Immunity, antigenic heterogeneity, and aggregation of helminth parasites. *Journal of Parasitology* 89: 232–41.

Galvani, A. P., Coleman, R. M., and Ferguson, N. M. 2003. The maintenance of sex in parasites. *Proceedings of the Royal Society of London B* 270: 19–28.

Gandon, S. 2004. Evolution of multihost parasites. *Evolution* 58: 455–69.

Gandon, S., Mackinnon, M. J., Nee, S., and Read, A. F. 2001. Imperfect vaccines and the evolution of pathogen virulence. *Nature* 414: 751–56.

Gandon, S., and Michalakis, Y. 2000. Evolution of parasite virulence against qualitative or quantitative host resistance. *Proceedings of the Royal Society of London B* 267: 985–90.

Gandon, S., and Michalakis, Y. 2002. Multiple infection and its consequences for virulence management. In *Adaptive Dynamics of Infectious Diseases* (eds. U. Dieckmann, J.A.J. Metz, M. W. Sabelis, and K. Sigmund), Cambridge University Press, Cambridge, pp. 150–64.

Garcia-Varela, M., Pérez-Ponce de León, G., De la Torre, P., Cummings, M. P., Sarma, S.S.S., and Laclette, J. P. 2000. Phylogenetic relationships of Acanthocephala based on analysis of 18S ribosomal RNA gene sequences. *Journal of Molecular Evolution* 50: 532–40.

Garnick, E. 1992a. Niche breadth in parasites: an evolutionarily stable strategy model, with special reference to the protozoan parasite *Leishmania*. *Theoretical Population Biology* 42: 62–103.

Garnick, E. 1992b. Parasite virulence and parasite-host coevolution: a reappraisal. *Journal of Parasitology* 78: 381–86.

Gartner, J. V. Jr. and Zwerner, D. E. 1989. The parasite faunas of meso- and bathypelagic fishes of Norfolk Submarine Canyon, western North Atlantic. *Journal of Fish Biology* 34: 79–95.

Gaston, K. J. 1996. Species-range-size distributions: patterns, mechanisms and implications. *Trends in Ecology and Evolution* 11: 197–201.

Gaston, K. J., and Blackburn, T. M. 2000. *Pattern and Process in Macroecology*. Blackwell Science, Oxford.

Geets, A., Coene, H., and Ollevier, F. 1997. Ectoparasites of the whitespotted rabbitfish, *Siganus sutor* (Valenciennes, 1835) off the Kenyan coast: distribution within the host population and site selection on the gills. *Parasitology* 115: 69–79.

Gemmill, A. W., Skorping, A., and Read, A. F. 1999. Optimal timing of first reproduction in parasitic nematodes. *Journal of Evolutionary Biology* 12: 1148–56.

Gemmill, A. W., Viney, M. E., and Read, A. F. 1997. Host immune status determines sexuality in a parasitic nematode. *Evolution* 51: 393–401.

Gemmill, A. W., Viney, M. E., and Read, A. F. 2000. The evolutionary ecology of host-specificity: experimental studies with *Strongyloides ratti*. *Parasitology* 120: 429–37.

George-Nascimento, M., Muñoz, G., Marquet, P. A., and Poulin, R. 2004. Testing the energetic equivalence rule with helminth endoparasites of vertebrates. *Ecology Letters* 7: 527–31.

Gérard, C., Moné, H., and Théron, A. 1993. *Schistosoma mansoni—Biomphalaria glabrata*: dynamics of the sporocyst population in relation to the miracidial dose and the host size. *Canadian Journal of Zoology* 71: 1880–85.

Gibson, D. I., and Bray, R. A. 1994. The evolutionary expansion and host-parasite relationships of the Digenea. *International Journal for Parasitology* 24: 1213–26.

Goater, C. P., Baldwin, R. E., and Scrimgeour, G. J. 2005. Physico-chemical determinants of helminth component community structure in whitefish (*Coregonus clupeaformes*) from adjacent lakes in Northern Alberta, Canada. *Parasitology* 131: 713–22.

Goater, T. M., Esch G. W., and Bush, A. O. 1987. Helminth parasites of sympatric salamanders: ecological concepts at infracommunity, component and compound community levels. *American Midland Naturalist* 118: 289–300.

Godfray, H.C.J. 1987. The evolution of clutch size in invertebrates. *Oxford Surveys in Evolutionary Biology* 4: 117–54.

Godfray, H.C.J., and Werren, J. H. 1996. Recent developments in sex ratio studies. *Trends in Ecology and Evolution* 11: 59–63.

Goff, L. J., Ashen, J., and Moon, D. 1997. The evolution of parasites from their hosts: a case study in the parasitic red algae. *Evolution* 51: 1068–78.

Gorbushin, A. M., and Levakin, I. A. 1999. The effect of trematode parthenitae on the growth of *Onoba aculeus*, *Littorina saxatilis* and *L. obtusa* (Gastropoda: Prosobranchia). *Journal of the Marine Biological Association of the U.K.* 79: 273–79.

Gordon, D. M., and Rau, M. E. 1982. Possible evidence for mortality induced by the parasite *Apatemon gracilis* in a population of brook sticklebacks (*Culaea inconstans*). *Parasitology* 84: 41–47.

Gordon, D. M., and Whitfield, P. J. 1985. Interactions of the cysticercoids of *Hymenolepis diminuta* and *Raillietina cesticillus* in their intermediate host, *Tribolium confusum*. *Parasitology* 90: 421–31.

Gotelli, N. J., and Graves, G. R. 1996. *Null Models in Ecology*. Smithsonian Institution Press, Washington, D.C.

Gotelli, N. J., and Rohde, K. 2002. Co-occurrence of ectoparasites of marine fishes: a null model analysis. *Ecology Letters* 5: 86–94.

Gower, C. M., and Webster, J. P. 2004. Fitness of indirectly transmitted pathogens: restraint and constraint. *Evolution* 58: 1178–84.

Gower, C. M., and Webster, J. P. 2005. Intraspecific competition and the evolution of virulence in a parasitic trematode. *Evolution* 59: 544–53.

Grabda-Kazubska, B. 1976. Abbreviation of the life cycles in plagiorchid trematodes: general remarks. *Acta Parasitologica Polonica* 24: 125–41.

Grafen, A., and Woolhouse, M.E.J. 1993. Does the negative binomial distribution add up? *Parasitology Today* 9: 475–77.

Granath, W. O., and Esch, G. W. 1983a. Seasonal dynamics of *Bothriocephalus acheilognathi* in ambient and thermally altered areas of a North Carolina cooling reservoir. *Proceedings of the Helminthological Society of Washington* 50: 205–18.

Granath, W. O., and Esch, G. W. 1983b. Temperature and other factors that regulate the composition and infrapopulation densities of *Bothriocephalus acheiloghathi* (Cestoda) in *Gambusia affinis* (Pisces). *Journal of Parasitology* 69: 1116–24.

Gray, J. S. 1994. Is deep-sea species diversity really so high? Species diversity of the Norwegian continental shelf. *Marine Ecology Progress Series* 112: 205–209.

Gregory, R. D. 1990. Parasites and host geographic range as illustrated by waterfowl. *Functional Ecology* 4: 645–54.

Gregory, R. D., and Blackburn, T. M. 1991. Parasite prevalence and host sample size. *Parasitology Today* 7: 316–18.

Gregory, R. D., Keymer, A. E., and Harvey, P. H. 1991. Life history, ecology and parasite community structure in Soviet birds. *Biological Journal of the Linnean Society* 43: 249–62.

Gregory, R. D., Keymer, A. E., and Harvey, P. H. 1996. Helminth parasite richness among vertebrates. *Biodiversity and Conservation* 5: 985–97.

Gregory, R. D., and Woolhouse, M.E.J. 1993. Quantification of parasite aggregation: a simulation study. *Acta Tropica* 54: 131–39.

Grenfell, B. T., Wilson, K., Isham, V. S., Boyd, H.E.G., and Dietz, K. 1995. Modelling patterns of parasite aggregation in natural populations: trichostrongylid nematode—ruminant interactions as a case study. *Parasitology* 111: S135–S151.

Griffin, C. T., O'Callaghan, K. M., and Dix, I. 2001. A self-fertile species of *Steinernema* from Indonesia: further evidence of convergent evolution amongst entomopathogenic nematodes? *Parasitology* 122: 181–86.

Guégan, J.-F., and Agnèse, J.-F. 1991. Parasite evolutionary events inferred from host phylogeny: the case of *Labeo* species (Teleostei, Cyprinidae) and their dactylogyrid parasites (Monogenea, Dactylogyridae). *Canadian Journal of Zoology* 69: 595–603.

Guégan, J.-F., and Hugueny, B. 1994. A nested parasite species subset pattern in tropical fish: host as major determinant of parasite infracommunity structure. *Oecologia* 100: 184–89.

Guégan, J.-F., and Kennedy, C. R. 1993. Maximum local helminth parasite community richness in British freshwater fish: a test of the colonization time hypothesis. *Parasitology* 106: 91–100.

Guégan, J.-F., and Kennedy, C. R. 1996. Parasite richness/sampling effort/host range: the fancy three-piece jigsaw puzzle. *Parasitology Today* 12: 367–69.

Guégan, J.-F., Lambert, A., Lévêque, C., Combes, C., and Euzet, L. 1992. Can host body size explain the parasite species richness in tropical freshwater fishes? *Oecologia* 90: 197–204.

Guégan, J.-F., and Morand, S. 1996. Polyploid hosts: strange attractors for parasites? *Oikos* 77: 366–70.

Guinnee, M. A., Gemmill, A. W., Chan, B.H.K., Viney, M. E., and Read, A. F. 2003. Host immune status affects maturation time in two nematode species—but not as predicted by a simple life-history model. *Parasitology* 127: 507–12.

Guyatt, H. L., and Bundy, D.A.P. 1993. Estimation of intestinal nematode prevalence: influence of parasite mating patterns. *Parasitology* 107: 99–106.

Haag, K. L., Araujo, A. M., Gottstein, B., Siles-Lucas, M., Thompson, R.C.A., and Zaha, A. 1999. Breeding systems in *Echinococcus granulosus* (Cestoda: Taeniidae): selfing or outcrossing? *Parasitology* 118: 63–71.

Haas, W. 1994. Physiological analyses of host-finding behaviour in trematode cercariae: adaptations for transmission success. *Parasitology* 109: S15–S29.

Haas, W., Haberl, B., Kalbe, M., and Körner, M. 1995. Snail-host-finding by miracidia and cercariae: chemical host cues. *Parasitology Today* 11: 468–72.

Haas, W., Stiegeler, P., Keating, A., Kullmann, B., Rabenau, H., Schönamsgruber, E., and Haberl, B. 2002. *Diplostomum spathaceum* cercariae respond to a unique profile of cues during recognition of their fish host. *International Journal for Parasitology* 32: 1145–54.

Haberl, B., Körner, M., Spengler, Y., Hertel, J., Kalbe, M., and Haas, W. 2000. Host-finding in *Echinostoma caproni*: miracidia and cercariae use different signals to identify the same snail species. *Parasitology* 120: 479–86.

Hafner, M. S., and Nadler, S. A. 1988. Phylogenetic trees support the coevolution of parasites and their hosts. *Nature* 332: 258–59.

Hafner, M. S., and Nadler, S. A. 1990. Cospeciation in host-parasite assemblages: comparative analysis of rates of evolution and timing of cospeciation events. *Systematic Zoology* 39: 192–204.

Hafner, M. S., and Page, R. D.M. 1995. Molecular phylogenies and host-parasite cospeciation: gophers and lice as a model system. *Philosophical Transactions of the Royal Society of London B* 349: 77–83.

Hahn, D. C., Price, R. D., and Osenton, P. C. 2000. Use of lice to identify cowbird hosts. *Auk* 117: 943–51.

Haine, E. R., Brondani, E., Hume, K. D., Perrot-Minnot, M.-J., Gaillard, M., and Rigaud, T. 2004. Coexistence of three Microsporidia parasites in populations of the freshwater amphipod *Gammarus roeseli*: evidence for vertical transmission and positive effect on reproduction. *International Journal for Parasitology* 34: 1137–46.

Hairston, N. G. Jr., and Bohonak, A. J. 1998. Copepod reproductive strategies: life-history theory, phylogenetic pattern and invasion of inland waters. *Journal of Marine Systems* 15: 23–34.

Hakalahti, T., Häkkinen, H., and Valtonen, E. T. 2004. Ectoparasitic *Argulus coregoni* (Crustacea: Branchiura) hedge their bets—studies on egg hatching dynamics. *Oikos* 107: 295–302.

Hanelt, B., and Janovy, J. Jr. 2003. Spanning the gap: identification of natural paratenic hosts of horsehair worms (Nematomorpha: Gordioidea) by experimental determination of paratenic host specificity. *Invertebrate Biology* 122: 12–18.

Hanelt, B., and Janovy, J. Jr. 2004. Life cycle and paratenesis of American gordiids (Nematomorpha: Gordiida). *Journal of Parasitology* 90: 240–44.

Hanken, J., and Wake, D. B. 1993. Miniaturization of body size: organismal consequences and evolutionary significance. *Annual Review of Ecology and Systematics* 24: 501–19.

Hansen, F., Jeltsch, F., Tackmann, K., Staubach, C., and Thulke, H.-H. 2004. Processes leading to a spatial aggregation of *Echinococcus multilocularis* in its natural intermediate host *Microtus arvalis*. *International Journal for Parasitology* 34: 37–44.

Hanski, I. 1982. Dynamics of regional distribution: the core and satellite species hypothesis. *Oikos* 38: 210–21.

Hanski, I., and Gilpin, M. E. 1997. *Metapopulation Biology: Ecology, Genetics and Evolution*. Academic Press, London.

Hanski, I., and Simberloff, D. 1997. The metapopulation approach, its history, conceptual domain, and application to conservation. In *Metapopulation Biology: Ecology, Genetics and Evolution* (eds. I. A. Hanski and M. E. Gilpin), Academic Press, London, pp. 5–26.

Hartl, D. L., and Clark, A. G. 1997. *Principles of Population Genetics*, 3rd ed. Sinauer Associates, Sunderland, Massachusetts.

Hartvigsen, R., and Halvorsen, O. 1994. Spatial patterns in the abundance and distribution of parasites of freshwater fish. *Parasitology Today* 10: 28–31.

Harvell, C. D., Kim, K., Burkholder, J. M., Colwell, R. R., Epstein, P. R., Grimes, D. J., Hofmann, E. E., Lipp, E. K., Osterhaus, A.D.M.E., Overstreet, R. M., Porter, J. W., Smith, G. W., and

Vasta, G. R. 1999. Emerging marine diseases: climate links and anthropogenic factors. *Science* 285: 1505–10.

Harvell, C. D., Mitchell, C. E., Ward, J. R., Altizer, S., Dobson, A. P., Ostfeld, R. S., and Samuel, M. D. 2002. Climate warming and disease risks for terrestrial and marine biota. *Science* 296: 2158–62.

Harvey, P. H. 1996. Phylogenies for ecologists. *Journal of Animal Ecology* 65: 255–63.

Harvey, P. H., and Keymer, A. E. 1991. Comparing life histories using phylogenies. *Philosophical Transactions of the Royal Society of London B* 332: 31–39.

Harvey, P. H., and Pagel, M. D. 1991. *The Comparative Method in Evolutionary Biology*. Oxford University Press, Oxford.

Harvey, S. C., Gemmill, A. W., Read, A. F., and Viney, M. E. 2000. The control of morph development in the parasitic nematode *Strongyloides ratti*. *Proceedings of the Royal Society of London B* 267: 2057–63.

Harvey, S. C., Paterson, S., and Viney, M. E. 1999. Heterogeneity in the distribution of *Strongyloides ratti* infective stages among the faecal pellets of rats. *Parasitology* 119: 227–35.

Hassan, A.H.M., Haberl, B., Hertel, J., and Haas, W. 2003. Miracidia of an Egyptian strain of *Schistosoma mansoni* differentiate between sympatric snail species. *Journal of Parasitology* 89: 1248–50.

Hatcher, M. J., and Dunn, A. M. 1995. Evolutionary consequences of cytoplasmically inherited feminizing factors. *Philosophical Transactions of the Royal Society of London B* 348: 445–56.

Haukisalmi, V., Heino, M., and Kaitala, V. 1998. Body size variation in tapeworms (Cestoda): adaptation to intestinal gradients? *Oikos* 83: 152–60.

Haukisalmi, V., and Henttonen, H. 1993. Coexistence in helminths of the bank vole, *Clethrionomys glareolus*. I. Patterns of co-occurrence. *Journal of Animal Ecology* 62: 221–29.

Haukisalmi, V., and Henttonen, H. 1994. Distribution patterns and microhabitat segregation in gastrointestinal helminths of *Sorex* shrews. *Oecologia* 97: 236–42.

Haukisalmi, V., and Henttonen, H. 1998. Analysing interspecific associations in parasites: alternative methods and effects of sampling heterogeneity. *Oecologia* 116: 565–74.

Haukisalmi, V., and Henttonen, H. 1999. Determinants of helminth aggregation in natural host populations: individual differences or spatial heterogeneity? *Ecography* 22: 629–36.

Hay, K. B., Fredensborg, B. L., and Poulin, R. 2005. Trematode-induced alterations in shell shape of the mud snail *Zeacumantus subcarinatus* (Prosobranchia: Batillariidae). *Journal of the Marine Biological Association of the UK* 85: 989–92.

Hechtel, L. J., Johnson, C. L., and Juliano, S. A. 1993. Modification of antipredator behavior of *Caecidotea intermedius* by its parasite *Acanthocephalus dirus*. *Ecology* 74: 710–13.

Heins, D. C., Baker, J. A., and Martin, H. C. 2002. The "crowding effect" in the cestode *Schistocephalus solidus*: density-dependent effects on plerocercoid size and infectivity. *Journal of Parasitology* 88: 302–307.

Helluy, S., and Holmes, J. C. 1990. Serotonin, octopamine, and the clinging behavior induced by the parasite *Polymorphus paradoxus* (Acanthocephala) in *Gammarus lacustris* (Crustacea). *Canadian Journal of Zoology* 68: 1214–20.

Hendrickson, G. L., and Kingston, N. 1974. *Cercaria laramiensis* sp. n., a freshwater zygocercous cercaria from *Physa gyrina* Say, with a discussion of cercarial aggregation. *Journal of Parasitology* 60: 777–81.

Hendrickson, M. A., and Curtis, L. A. 2002. Infrapopulation sizes of co-occurring trematodes in the snail *Ilyanassa obsoleta*. *Journal of Parasitology* 88: 884–89.

Hengeveld, R. 1992. *Dynamic Biogeography*. Cambridge University Press, Cambridge.

Hengeveld, R., and Haeck, J. 1982. The distribution of abundance. I. Measurements. *Journal of Biogeography* 9: 303–16.

Herlyn, H., Piskurek, O., Schmitz, J., Ehlers, U., and Zischler, H. 2003. The syndermatan phylogeny and the evolution of acanthocephalan endoparasitism as inferred from 18S rDNA sequences. *Molecular Phylogenetics and Evolution* 26: 155–64.

Hernández-Alcántara, P., and Solis-Weiss, V. 1998. Parasitism among polychaetes: a rare case illustrated by a new species, *Labrorostratus zaragozensis* n. sp. (Oenonidae) found in the Gulf of California, Mexico. *Journal of Parasitology* 84: 978–82.

Herre, E. A. 1993. Population structure and the evolution of virulence in nematode parasites of fig wasps. *Science* 259: 1442–45.

Herre, E. A. 1995. Factors affecting the evolution of virulence: nematode parasites of fig wasps as a case study. *Parasitology* 111: S179–S191.

Herre, E. A., Knowlton, N., Mueller, U. G., and Rehner, S. A. 1999. The evolution of mutualisms: exploring the paths between conflict and cooperation. *Trends in Ecology and Evolution* 14: 49–53.

Hess, G. 1996. Disease in metapopulation models: implications for conservation. *Ecology* 77: 1617–32.

Hesselberg, C. A., and Andreassen, J. 1975. Some influences of population density on *Hymenolepis diminuta* in rats. *Parasitology* 71: 521–23.

Heuch, P. A., and Schram, T. A. 1996. Male mate choice in a natural population of the parasitic copepod *Lernaeocera branchialis* (Copepoda: Pennellidae). *Behaviour* 133: 221–39.

Hilburn, L. R., and Sattler, P. W. 1986. Electrophoretically detectable protein variation in natural populations of the lone star tick, *Amblyomma americanum* (Acari: Ixodidae). *Heredity* 56: 67–74.

Hillebrand, H. 2005. Regressions of local on regional diversity do not reflect the importance of local interactions or saturation of local diversity. *Oikos* 110: 195–98.

Hillebrand, H., and Azovsky, A. I. 2001. Body size determines the strength of the latitudinal diversity gradient. *Ecography* 24: 251–56.

Hillebrand, H., and Blenckner, T. 2002. Regional and local impact on species diversity: from pattern to processes. *Oecologia* 132: 479–91.

Hoberg, E. P., Alkire, N. L., de Queiroz, A., and Jones, A. 2001. Out of Africa: origins of the *Taenia* tapeworms in humans. *Proceedings of the Royal Society of London* B 268: 781–87.

Hoberg, E. P., Brooks, D. R., and Siegel-Causey, D. 1997. Host-parasite cospeciation: history, principles and prospects. In *Host-Parasite Evolution: General Principles and Avian Models* (eds. D. H. Clayton and J. Moore), Oxford University Press, Oxford, pp. 212–35.

Høeg, J. T. 1995. The biology and life cycle of the Rhizocephala (Cirripedia). *Journal of the Marine Biological Association of the U.K.* 75: 517–50.

Hogans, W. E., Brattey, J., Uhazy, L. S., and Hurley, P.C.F. 1983. Helminth parasites of swordfish (*Xiphias gladius* L.) from the northwest Atlantic Ocean. *Journal of Parasitology* 69: 1178–79.

Holland, C. 1984. Interactions between *Moniliformis* (Acanthocephala) and *Nippostrongylus* (Nematoda) in the small intestine of laboratory rats. *Parasitology* 88: 303–15.

Holmes, J. C. 1961. Effects of concurrent infections on *Hymenolepis diminuta* (Cestoda) and *Moniliformis dubius* (Acanthocephala). I. General effects and comparison with crowding. *Journal of Parasitology* 47: 209–16.

Holmes, J. C. 1973. Site segregation by parasitic helminths: interspecific interactions, site segregation, and their importance to the development of helminth communities. *Canadian Journal of Zoology* 51: 333–47.

Holmes, J. C., and Bethel, W. M. 1972. Modification of intermediate host behaviour by parasites. In *Behavioural Aspects of Parasite Transmission* (eds. E. U. Canning and C. A. Wright), Academic Press, London, pp. 123–49.

Holmes, J. C., and Price, P. W. 1980. Parasite communities: the roles of phylogeny and ecology. *Systematic Zoology* 29: 203–13.

Holmes, J. C., and Price, P. W. 1986. Communities of parasites. In *Community Ecology: Pattern and Process* (eds. D. J. Anderson and J. Kikkawa), Blackwell Scientific Publications, Oxford, pp. 187–213.

Holmes, J. C., and Zohar, S. 1990. Pathology and host behaviour. In *Parasitism and Host Behaviour* (eds. C. J. Barnard and J. M. Behnke), Taylor & Francis, London, pp. 34–64.

Holmstad, P. R., and Skorping, A. 1998. Covariation of parasite intensities in willow ptarmigan, *Lagopus lagopus* L. *Canadian Journal of Zoology* 76: 1581–88.

Hon, L. T., Forrester, D. J., and Williams, L. E. Jr. 1975. Helminths of wild turkeys in Florida. *Proceedings of the Helminthological Society of Washington* 42: 119–27.

Hoogenboom, I., and Dijkstra, C. 1987. *Sarcocystis cernae*: a parasite increasing the risk of predation of its intermediate host, *Microtus arvalis*. *Oecologia* 74: 86–92.

Howard, S. C., Donnelly, C. A., and Chan, M.-S. 2001. Methods for estimation of associations between multiple species parasite infections. *Parasitology* 122: 233–51.

Hubbell, S. P. 2001. *The Unified Neutral Theory of Biodiversity and Biogeography*. Princeton University Press, Princeton.

Hudson, P. J., and Dobson, A. P. 1997. Transmission dynamics and host-parasite interactions of *Trichostrongylus tenuis* in red grouse (*Lagopus lagopus scoticus*). *Journal of Parasitology* 83: 194–202.

Hudson, P. J., Dobson, A. P., and Newborn, D. 1992. Do parasites make prey vulnerable to predation? Red grouse and parasites. *Journal of Animal Ecology* 61: 681–92.

Hudson, P. J., Dobson, A. P., and Newborn, D. 1998. Prevention of population cycles by parasite removal. *Science* 282: 2256–58.

Hughes, D. P., Kathirithamby, J., Turillazzi, S., and Beani, L. 2004. Social wasps desert the colony and aggregate outside if parasitized: parasite manipulation? *Behavioral Ecology* 15: 1037–43.

Hulscher, J. B. 1973. Burying-depth and trematode infection in *Macoma balthica*. *Netherlands Journal of Sea Research* 6: 141–56.

Hung, G. C., Chilton, N. B., Beveridge, I., Zhu, X. Q., Lichtenfels, J. R., and Gasser, R. B. 1999. Molecular evidence for cryptic species within *Cylicostephanus minutus* (Nematoda: Strongylidae). *International Journal for Parasitology* 29: 285–91.

Hurd, H. 1990. Physiological and behavioural interactions between parasites and invertebrate hosts. *Advances in Parasitology* 29: 271–318.

Hurd, H. 2001. Host fecundity reduction: a strategy for damage limitation? *Trends in Parasitology* 17: 363–68.

Hurd, H., and Fogo, S. 1991. Changes induced by *Hymenolepis diminuta* (Cestoda) in the behaviour of the intermediate host *Tenebrio molitor* (Coleoptera). *Canadian Journal of Zoology* 69: 2291–94.

Hurst, L. D. 1993. The incidences, mechanisms and evolution of cytoplasmic sex ratio distorters in animals. *Biological Reviews* 68: 121–93.

Hutchinson, G. E. 1957. Concluding remarks. *Cold Spring Harbor Symposium on Quantitative Biology* 22: 415–27.

Huxham, M., Raffaelli, D., and Pike, A. 1993. The influence of *Cryptocotyle lingua* (Digenea: Platyhelminthes) infections on the survival and fecundity of *Littorina littorea* (Gastropoda: Prosobranchia): an ecological approach. *Journal of Experimental Marine Biology and Ecology* 168: 223–38.

Huyse, T., and Volckaert, F.A.M. 2005. Comparing host and parasite phylogenies: *Gyrodactylus* flatworms jumping from goby to goby. *Systematic Biology* 54: 710–18.

Imhoof, B., and Schmid-Hempel, P. 1998. Single-clone and mixed-clone infections versus host environment in *Crithidia bombi* infecting bumblebees. *Parasitology* 117: 331–36.

Irvine, R. J., Stien, A., Dallas, J. F., Halvorsen, O., Langvatn, R., and Albon, S. D. 2001. Contrasting regulation of fecundity in two abomasal nematodes of Svalbard reindeer (*Rangifer tarandus platyrhynchus*). *Parasitology* 122: 673–81.

Irvine, R. J., Stien, A., Halvorsen, O., Langvatn, R., and Albon, S. D. 2000. Life-history strategies and population dynamics of abomasal nematodes in Svalbard reindeer (*Rangifer tarandus platyrhynchus*). *Parasitology* 120: 297–311.

Jablonski, D. 1996. Body size and macroevolution. In *Evolutionary Paleobiology* (eds. D. Jablonski, D. H. Erwin, and J. H. Lipps), University of Chicago Press, Chicago, pp. 256–89.

Jablonski, D. 1997. Body-size evolution in Cretaceous molluscs and the status of Cope's rule. *Nature* 385: 250–52.

Jackson, C. J., Marcogliese, D. J., and Burt, M.D.B. 1997. Precociously developed *Ascarophis* sp. (Nematoda, Spirurata) and *Hemiurus levinseni* (Digenea, Hemiuridae) in their crustacean intermediate hosts. *Acta Parasitologica* 42: 31–35.

Jackson, J. A., and Tinsley, R. C. 2001. *Protopolystoma xenopodis* (Monogenea) primary and secondary infections in *Xenopus laevis*. *Parasitology* 123: 455–63.

Jackson, J. A., and Tinsley, R. C. 2003. Density dependence of postlarval survivorship in primary infections of *Protopolystoma xenopodis*. *Journal of Parasitology* 89: 958–60.

Jackson, J. A., Tinsley, R. C., and Hinkel, H. H. 1998. Mutual exclusion of congeneric monogenean species in a space-limited habitat. *Parasitology* 117: 563–69.

Jaenike, J. 1994. Aggregations of nematode parasites within *Drosophila*: proximate causes. *Parasitology* 108: 569–77.

Jaenike, J. 1996a. Suboptimal virulence of an insect-parasitic nematode. *Evolution* 50: 2241–47.

Jaenike, J. 1996c. Population-level consequences of parasite aggregation. *Oikos* 76: 155–60.

Jaenike, J., and Dombeck, I. 1998. General-purpose genotypes for host species utilization in a nematode parasite of *Drosophila*. *Evolution* 52: 832–40.

Jakes, K., O'Donoghue, P. J., and Cameron, S. L. 2003. Phylogenetic relationships of *Hepatozoon* (*Haemogregarina*) *boigae*, *Hepatozoon* sp., *Haemogregarina clelandi* and *Haemoproteus chelodina* from Australian reptiles to other Apicomplexa based on cladistic analyses of ultrastructural and life-cycle characters. *Parasitology* 126: 555–59.

James, E. R., and Green, D. R. 2004. Manipulation of apoptosis in the host-parasite interaction. *Trends in Parasitology* 20: 280–87.

Janovy, J. Jr., Clopton, R. E., Clopton, D. A., Snyder, S. D., Efting, A., and Krebs, L. 1995. Species density distributions as null models for ecologically significant interactions of parasite species in an assemblage. *Ecological Modelling* 77: 189–96.

Janovy, J. Jr., Ferdig, M. T., and McDowell, M. A. 1990. A model of dynamic behavior of a parasite species assemblage. *Journal of Theoretical Biology* 142: 517–29.

Janovy, J. Jr., and Kutish, G. W. 1988. A model of encounters between host and parasite populations. *Journal of Theoretical Biology* 134: 391–401.

Janovy, J. Jr., Snyder, S. D., and Clopton, R. E. 1997. Evolutionary constraints on population structure: the parasites of *Fundulus zebrinus* (Pisces: Cyprinodontidae) in the South Platte River of Nebraska. *Journal of Parasitology* 83: 584–92.

Jarne, P., and Charlesworth, D. 1993. The evolution of the selfing rate in functionally hermaphrodite plants and animals. *Annual Review of Ecology and Systematics* 24: 441–66.

Jarne, P., and Théron, A. 2001. Genetic structure in natural populations of flukes and snails: a practical approach and review. *Parasitology* 123: S27–S40.

Jennings, J. B., and Calow, P. 1975. The relationship between high fecundity and the evolution of entoparasitism. *Oecologia* 21: 109–15.

Johnson, K. P., Adams, R. J., and Clayton, D. H. 2002a. The phylogeny of the louse genus *Brueelia* does not reflect host phylogeny. *Biological Journal of the Linnean Society* 77: 233–47.

Johnson, K. P., Adams, R. J., Page, R.D.M., and Clayton, D. H. 2003. When do parasites fail to speciate in response to host speciation? *Systematic Biology* 52: 37–47.

Johnson, K. P., Bush, S. E., and Clayton, D. H. 2005. Correlated evolution of host and parasite body size: tests of Harrison's rule using birds and lice. *Evolution* 59: 1744–53.

Johnson, K. P., Williams, B. L., Drown, D. M., Adams, R. J., and Clayton, D. H. 2002b. The population genetics of host specificity: genetic differentiation in dove lice (Insecta: Phthiraptera). *Molecular Ecology* 11: 25–38.

Johnson, K. P., Yoshizawa, K., and Smith, V. S. 2004. Multiple origins of parasitism in lice. *Proceedings of the Royal Society of London B* 271: 1771–76.

Johst, K., and Brandl, R. 1997. Body size and extinction risk in a stochastic environment. *Oikos* 78: 612–17.

Jokela, J., Uotila, L., and Taskinen, J. 1993. Effect of the castrating trematode parasite *Rhipidocotyle fennica* on energy allocation of fresh-water clam *Anondota piscinalis*. *Functional Ecology* 7: 332–38.

Jones, J. T., Breeze, P., and Kusel, J. R. 1989. Schistosome fecundity: influence of host genotype and intensity of infection. *International Journal for Parasitology* 19: 769–77.

Jourdane, J., Imbert-Establet, D., and Tchuem Tchuenté, L. A. 1995. Parthenogenesis in Schistosomatidae. *Parasitology Today* 11: 427–30.

Jousson, O., Bartoli, P., and Pawlowski, J. 2000. Cryptic speciation among intestinal parasites (Trematoda: Digenea) infecting sympatric host fishes (Sparidae). *Journal of Evolutionary Biology* 13: 778–85.

Justine, J.-L. 1998. Non-monophyly of the monogeneans. *International Journal for Parasitology* 28: 1653–57.

Kain, D. E., Sperling, F.A.H., Daly, H. V., and Lane, R. S. 1999. Mitochondrial DNA sequence variation in *Ixodes pacificus* (Acari: Ixodidae). *Heredity* 83: 378–86.

Karban, R. 1989. Fine-scale adaptation of herbivorous thrips to individual host plants. *Nature* 340: 60–61.

Karvonen, A., Cheng, G.-H., and Valtonen, E. T. 2005. Within-lake dynamics in the similarity of parasite assemblages of perch (*Perca fluviatilis*). *Parasitology* 131: 817–23.

Karvonen, A., Hudson, P. J., Seppälä, O., and Valtonen, E. T. 2004a. Transmission dynamics of a trematode parasite: exposure, acquired resistance and parasite aggregation. *Parasitology Research* 92: 183–88.

Karvonen, A., Kirsi, S., Hudson, P. J., and Valtonen, P. J. 2004b. Patterns of cercarial production from *Diplostomum spathaceum*: terminal investment or bet hedging? *Parasitology* 129: 87–92.

Karvonen, A., Paukku, S., Valtonen, E. T., and Hudson, P. J. 2003. Transmission, infectivity and survival of *Diplostomum spathaceum* cercariae. *Parasitology* 127: 217–24.

Karvonen, A., and Valtonen, E. T. 2004. Helminth assemblages of whitefish (*Coregonus lavaretus*) in interconnected lakes: similarity as a function of species-specific parasites and geographical separation. *Journal of Parasitology* 90: 471–76.

Kathirithamby, J. 1989. Review of the order Strepsiptera. *Systematic Entomology* 14: 41–92.

Kawecki, T. J. 1997. Sympatric speciation via habitat specialization driven by deleterious mutations. *Evolution* 51: 1749–61.

Kawecki, T. J. 1998. Red Queen meets Santa Rosalia: arms races and the evolution of host specialization in organisms with parasitic lifestyles. *American Naturalist* 152: 635–51.

Kearn, G. C. 1986. The eggs of monogeneans. *Advances in Parasitology* 25: 175–273.

Kearn, G. C. 1998. *Parasitism and the Platyhelminths*. Chapman and Hall, London.

Kehr, A. I., Manly, B.F.J., and Hamann, M. I. 2000. Coexistence of helminth species in *Lysapsus limellus* (Anura: Pseudidae) from an Argentinean subtropical area: influence of biotic and abiotic factors. *Oecologia* 125: 549–58.

Kelly, C. K., and Southwood, T.R.E. 1999. Species richness and resource availability: a phylogenetic analysis of insects associated with trees. *Proceedings of the National Academy of Sciences of the U.S.A.* 96: 8013–16.

Kennedy, C. R. 1975. *Ecological Animal Parasitology*. Blackwell Scientific Publications, Oxford.

Kennedy, C. R. 1978. An analysis of the metazoan parasitocoenoses of brown trout *Salmo trutta* from British Lakes. *Journal of Fish Biology* 13: 255–63.

Kennedy, C. R. 1990. Helminth communities in freshwater fish: structured communities or stochastic assemblages? In *Parasite Communities: Patterns and Processes* (eds. G. W. Esch, A. O. Bush, and J. M. Aho), Chapman & Hall, London, pp. 131–56.

Kennedy, C. R. 1995. Richness and diversity of macroparasite communities in tropical eels *Anguilla reinhardtii* in Queensland, Australia. *Parasitology* 111: 233–45.

Kennedy, C. R. 2001. Metapopulation and community dynamics of helminth parasites of eels *Anguilla anguilla* in the River Exe system. *Parasitology* 122: 689–98.

Kennedy, C. R., and Bakke, T. A. 1989. Diversity patterns in helminth communities in common gulls, *Larus canus*. *Parasitology* 98: 439–45.

Kennedy, C. R., and Bush, A. O. 1992. Species richness in helminth communities: the importance of multiple congeners. *Parasitology* 104: 189–97.

Kennedy, C. R., and Bush, A. O. 1994. The relationship between pattern and scale in parasite communities: a stranger in a strange land. *Parasitology* 109: 187–96.

Kennedy, C. R., Bush, A. O., and Aho, J. M. 1986. Patterns in helminth communities: why are birds and fish so different? *Parasitology* 93: 205–15.

Kennedy, C. R., and Guégan, J.-F. 1994. Regional versus local helminth parasite richness in British freshwater fish: saturated or unsaturated parasite communities? *Parasitology* 109: 175–85.

Kennedy, C. R., and Guégan, J.-F. 1996. The number of niches in intestinal helminth communities of *Anguilla anguilla*: are there enough spaces for parasites? *Parasitology* 113: 293–302.

Kennedy, C. R., and Moriarty, C. 1987. Co-existence of congeneric species of Acanthocephala: *Acanthocephalus lucii* and *A. anguillae* in eels *Anguilla anguilla* in Ireland. *Parasitology* 95: 301–10.

Keymer, A. E. 1981. Population dynamics of *Hymenolepis diminuta* in the intermediate host. *Journal of Animal Ecology* 50: 941–50.

Keymer, A. E. 1982. Density-dependent mechanisms in the regulation of intestinal helminth populations. *Parasitology* 84: 573–87.

Keymer, A. E., and Anderson, R. M. 1979. The dynamics of infection of *Tribolium confusum* by *Hymenolepis diminuta*: the influence of infective-stage density and spatial distribution. *Parasitology* 79: 195–207.

Keymer, A. E., Crompton, D.W.T., and Singhvi, A. 1983. Mannose and the "crowding effect" of *Hymenolepis* in rats. *International Journal for Parasitology* 13: 561–70.

Keymer, A. E., and Hiorns, R. W. 1986. *Heligmosomoides polygyrus* (Nematoda): the dynamics of primary and repeated infection in outbred mice. *Proceedings of the Royal Society of London B* 229: 47–67.

Keymer, A. E., and Slater, A.F.G. 1987. Helminth fecundity: density dependence or statistical illusion? *Parasitology Today* 3: 56–58.

Khan, R. A., and Thulin, J. 1991. Influence of pollution on parasites of aquatic animals. *Advances in Parasitology* 30: 201–38.

Kirchner, T. B., Anderson, R. V., and Ingham, R. E. 1980. Natural selection and the distribution of nematode sizes. *Ecology* 61: 232–37.

Kirk, W.D.J. 1991. The size relationship between insects and their hosts. *Ecological Entomology* 16: 351–59.

Klompen, H., and Grimaldi, D. 2001. First Mesozoic record of a Parasitiform mite: a larval argasid tick in Cretaceous amber (Acari: Ixodida: Argasidae). *Annals of the Entomological Society of America* 94: 10–15.

Klompen, J.S.H., Black, W. C., Keirans, J. E., and Oliver, J. H. Jr. 1996. Evolution of ticks. *Annual Review of Entomology* 41: 141–61.

Knouft, J. H., and Page, L. M. 2003. The evolution of body size in extant groups of North American freshwater fishes: speciation, size distributions, and Cope's rule. *American Naturalist* 161: 413–21.

Knudsen, R., Curtis, M. A., and Kristoffersen, R. 2004. Aggregation of helminths: the role of feeding behavior of fish hosts. *Journal of Parasitology* 90: 1–7.

Knudsen, R., Gabler, H.-M., Kuris, A. M., and Amundsen, P.-A. 2001. Selective predation on parasitized prey—a comparison between two helminth species with different life-history strategies. *Journal of Parasitology* 87: 941–45.

Koella, J. C. 2000. Coevolution of parasite life cycles and host life-histories. In *Evolutionary Biology of Host-Parasite Relationships: Theory Meets Reality* (eds. R. Poulin, S. Morand, and A. Skorping), Elsevier, Amsterdam, pp. 185–200.

Koella, J. C., and Agnew, P. 1997. Blood-feeding success of the mosquito *Aedes aegypti* depends on the transmission route of its parasite *Edhazardia aedis*. *Oikos* 78: 311–16.

Koella, J. C., and Agnew, P. 1999. A correlated response of a parasite's virulence and life cycle to selection on its host's life history. *Journal of Evolutionary Biology* 12: 70–79.

Kostadinova, A., and Mavrodieva, R. S. 2005. Microphallids in *Gammarus insensibilis* Stock, 1966 from a Black Sea Lagoon: manipulation hypothesis going East? *Parasitology* 131: 337–46.

Kozlowski, J., and Gawelczyk, A. T. 2002. Why are species' body size distributions usually skewed to the right? *Functional Ecology* 16: 419–32.

Krasnov, B. R., Khokhlova, I. S., Burdelova, N. V., Mirzoyan, N. S., and Degen, A. A. 2004c. Fitness consequences of host selection in ectoparasites: testing reproductive patterns predicted by isodar theory in fleas parasitizing rodents. *Journal of Animal Ecology* 73: 815–20.

Krasnov, B. R., Mouillot, D., Shenbrot, G. I., Khokhlova, I. S., and Poulin, R. 2004a. Geographical variation in host specificity of fleas (Siphonaptera): the influence of phylogeny and local environmental conditions. *Ecography* 27: 787–97.

Krasnov, B. R., Poulin, R., Shenbrot, G. I., Mouillot, D., and Khokhlova, I. S. 2004b. Ectoparasitic "jacks-of-all-trades": relationship between abundance and host specificity in fleas (Siphonaptera) parasitic on small mammals. *American Naturalist* 164: 506–16.

Krasnov, B. R., Shenbrot, G. I., Khokhlova, I. S., and Poulin, R. 2004d. Relationships between parasite abundance and the taxonomic distance among a parasite's host species: an example with fleas parasitic on small mammals. *International Journal for Parasitology* 34: 1289–97.

Krasnov, B. R., Shenbrot, G. I., Medvedev, S. G., Vatschenok, V. S., and Khokhlova, I. S. 1997. Host-habitat relations as an important determinant of spatial distribution of flea assemblages (Siphonaptera) on rodents in the Negev Desert. *Parasitology* 114: 159–73.

Krasnov, B. R., Shenbrot, G. I., Mouillot, D., Khokhlova, I. S., and Poulin, R. 2005. Spatial variation in species diversity and composition of flea assemblages in small mammalian hosts: geographical distance or faunal similarity? *Journal of Biogeography* 32: 633–44.

Krist, A. C. 2000. Effect of the digenean parasite *Proterometra macrostoma* on host morphology in the freshwater snail *Elimia livescens*. *Journal of Parasitology* 86: 262–67.

Kuris, A. M. 1974. Trophic interactions: similarity of parasitic castrators to parasitoids. *Quarterly Review of Biology* 49: 129–48.

Kuris, A. M. 1997. Host behavior modification: an evolutionary perspective. In *Parasites and Pathogens: Effects on Host Hormones and Behavior* (ed. N. E. Beckage), Chapman and Hall, New York, pp. 293–315.

Kuris, A. M., and Blaustein, A. R. 1977. Ectoparasitic mites on rodents: application of the island biogeography theory? *Science* 195: 596–98.

Kuris, A. M., Blaustein, A. R., and Alió, J. J. 1980. Hosts as islands. *American Naturalist* 116: 570–86.

Kuris, A. M., and Lafferty, K. D. 1994. Community structure: larval trematodes in snail hosts. *Annual Review of Ecology and Systematics* 25: 189–217.

Lafferty, K. D. 1992. Foraging on prey that are modified by parasites. *American Naturalist* 140: 854–67.

Lafferty, K. D. 1997. Environmental parasitology: what can parasites tell us about human impacts on the environment? *Parasitology Today* 13: 251–55.

Lafferty, K. D. 1999. The evolution of trophic transmission. *Parasitology Today* 15: 111–15.

Lafferty, K. D., and Morris, A. K. 1996. Altered behavior of parasitized killifish increases susceptibility to predation by bird final hosts. *Ecology* 77: 1390–97.

Lafferty, K. D., Porter, J. W., and Ford, S. E. 2004. Are diseases increasing in the ocean? *Annual Review of Ecology, Evolution and Systematics* 35: 31–54.

Lafferty, K. D., Thomas, F., and Poulin, R. 2000. Evolution of host phenotype manipulation by parasites and its consequences. In *Evolutionary Biology of Host-Parasite Relationships: Theory Meets Reality* (eds. R. Poulin, S. Morand, and A. Skorping), Elsevier, Amsterdam, pp. 117–27.

Lajeunesse, M. J., and Forbes, M. R. 2002. Host range and local parasite adaptation. *Proceedings of the Royal Society of London* B 269: 703–10.

Lambert, A., and El Gharbi, S. 1995. Monogenean host specificity as a biological and taxonomic indicator for fish. *Biological Conservation* 72: 227–35.

Lampo, M., Rangel, Y., and Mata, A. 1998. Population genetic structure of a three-host tick, *Amblyomma dissimile*, in eastern Venezuela. *Journal of Parasitology* 84: 1137–42.

Latham, A.D.M., Fredensborg, B. L., McFarland, L. H., and Poulin, R. 2003. A gastropod scavenger serving as paratenic host for larval helminth communities in shore crabs. *Journal of Parasitology* 89: 862–64.

Lawlor, B. J., Read, A. F., Keymer, A. E., Parveen, G., and Crompton, D.W.T. 1990. Nonrandom mating in a parasitic worm: mate choice by males? *Animal Behaviour* 40: 870–76.

Lawton, J. H. 1999. Are there general laws in ecology? *Oikos* 84: 177–92.

Le Brun, N., Renaud, F., Berrebi, P., and Lambert, A. 1992. Hybrid zones and host-parasite relationships: effect on the evolution of parasitic specificity. *Evolution* 46: 56–61.

Lefebvre, F., and Poulin, R. 2005. Progenesis in digenean trematodes: a taxonomic and synthetic overview of species reproducing in their second intermediate hosts. *Parasitology* 130: 587–605.

Leignel, V., and Cabaret, J. 2001. Massive use of chemotherapy influences life traits of parasitic nematodes in domestic ruminants. *Functional Ecology* 15: 569–74.

Leignel, V., Cabaret, J., and Humbert, J. F. 2002. New molecular evidence that *Teladorsagia circumcincta* (Nematoda: Trichostrongylidea) is a species complex. *Journal of Parasitology* 88: 135–40.

Lello, J., Boag, B., Fenton, A., Stevenson, I. R., and Hudson, P. J. 2004. Competition and mutualism among the gut helminths of a mammalian host. *Nature* 428: 840–44.

Leong, T. S., and Holmes, J. C. 1981. Communities of metazoan parasites in open water fishes of Cold Lake, Alberta. *Journal of Fish Biology* 18: 693–713.

Leung, B. 1998. Aggregated parasite distributions on hosts in a homogeneous environment: examining the Poisson null model. *International Journal for Parasitology* 28: 1709–12.

Leung, B., and Forbes, M. R. 1998. The evolution of virulence: a stochastic simulation model examining parasitism at individual and population levels. *Evolutionary Ecology* 12: 165–77.

Levri, E. P. 1998. The influence of non-host predators on parasite-induced behavioral changes in a freshwater snail. *Oikos* 81: 531–37.

Levri, E. P. 1999. Parasite-induced change in host behavior of a freshwater snail: parasitic manipulation or byproduct of infection? *Behavioral Ecology* 10: 234–41.

Levsen, A., and Jakobsen, P. J. 2002. Selection pressure towards monoxeny in *Camallanus cotti* (Nematoda, Camallanidae) facing an intermediate host bottleneck situation. *Parasitology* 124: 625–29.

Light, J. E., and Siddall, M. E. 1999. Phylogeny of the leech family Glossiphoniidae based on mitochondrial gene sequences and morphological data. *Journal of Parasitology* 85: 815–23.

Lipsitch, M., Herre, E. A., and Nowak, M. A. 1995. Host population structure and the evolution of virulence: a "law of diminishing returns." *Evolution* 49: 743–48.

Lipsitch, M., Siller, S., and Nowak, M. A. 1996. The evolution of virulence in pathogens with vertical and horizontal transmission. *Evolution* 50: 1729–41.

Littlewood, D.T.J., Rohde, K., Bray, R. A., and Herniou, E. A. 1999a. Phylogeny of the Platyhelminthes and the evolution of parasitism. *Biological Journal of the Linnean Society* 68: 257–87.

Littlewood, D.T.J., Rohde, K., and Clough, K. A. 1999b. The interrelationships of all major groups of Platyhelminthes: phylogenetic evidence from morphology and molecules. *Biological Journal of the Linnean Society* 66: 75–114.

Lively, C. M. 1989. Adaptation by a parasitic trematode to local populations of its snail host. *Evolution* 43: 1663–71.

Lively, C. M., and Dybdahl, M. F. 2000. Parasite adaptation to locally common host genotypes. *Nature* 405: 679–81.

Lloyd, M. 1967. "Mean crowding." *Journal of Animal Ecology* 36: 1–30.

Lo, C. M. 1999. Mating rendezvous in monogenean gill parasites of the humbug *Dascyllus aruanus* (Pisces: Pomacentridae). *Journal of Parasitology* 85: 1178–80.

Lockhart, A. B., Thrall, P. H., and Antonovics, J. 1996. Sexually transmitted diseases in animals: ecological and evolutionary implications. *Biological Reviews* 71: 415–71.

Loker, E. S. 1983. A comparative study of the life-histories of mammalian schistosomes. *Parasitology* 87: 343–69.

Lopez, J. E., Gallinot, L. P., and Wade, M. J. 2005. Spread of parasites in metapopulations: an experimental study of the effects of host migration rate and local host population size. *Parasitology* 130: 323–32.

Losos, J. B. 1996. Phylogenies and comparative biology, stage II: testing causal hypotheses derived from phylogenies with data from extant taxa. *Systematic Biology* 45: 259–60.

Lotz, J. M., Bush, A. O., and Font, W. F. 1995. Recruitment-driven, spatially discontinuous communities: a null model for transferred patterns in target communities of intestinal helminths. *Journal of Parasitology* 81: 12–24.

Lotz, J. M., and Font, W. F. 1985. Structure of enteric helminth communities in two populations of *Eptesicus fuscus* (Chiroptera). *Canadian Journal of Zoology* 63: 2969–78.

Lotz, J. M., and Font, W. F. 1991. The role of positive and negative interspecific associations in the organization of communities of intestinal helminths of bats. *Parasitology* 103: 127–38.

Lotz, J. M., and Font, W. F. 1994. Excess positive associations in communities of intestinal helminths of bats: a refined null hypothesis and a test of the facilitation hypothesis. *Journal of Parasitology* 80: 398–413.

Lowenberger, C. A., and Rau, M. E. 1994. *Plagiorchis elegans*: emergence, longevity and infectivity of cercariae, and host behavioural modifications during cercarial emergence. *Parasitology* 109: 65–72.

Lowrie, F. M., Behnke, J. M., and Barnard, C. J. 2004. Density-dependent effects on the survival and growth of the rodent stomach worm *Protospirura muricola* in laboratory mice. *Journal of Helminthology* 78: 121–28.

Lüder, C.G.K., Gross, U., and Lopes, M. F. 2001. Intracellular protozoan parasites and apoptosis: diverse strategies to modulate parasite-host interactions. *Trends in Parasitology* 17: 480–86.

Ludwig, J. A., and Reynolds, J. F. 1988. *Statistical Ecology*. John Wiley & Sons, New York.

Lüscher, A., and Milinski, M. 2003. Simultaneous hermaphrodites reproducing in pairs self-fertilize some of their eggs: an experimental test of predictions of mixed-mating and hermaphrodite's dilemma theory. *Journal of Evolutionary Biology* 16: 1030–37.

Lüscher, A., and Wedekind, C. 2002. Size-dependent discrimination of mating partners in the simultaneous hermaphroditic cestode *Schistocephalus solidus*. *Behavioral Ecology* 13: 254–59.

Lydeard, C., Mulvey, M., and Davis, G. M. 1996. Molecular systematics and evolution of reproductive traits of North American freshwater unionacean mussels (Mollusca: Bivalvia) as inferred from 16S rRNA gene sequences. *Philosophical Transactions of the Royal Society of London B* 351: 1593–1603.

Lymbery, A. J. 1989. Host specificity, host range and host preference. *Parasitology Today* 5: 298.

Lymbery, A. J., Constantine, C. C., and Thompson, R.C.A. 1997. Self-fertilization without genomic or population structuring in a parasitic tapeworm. *Evolution* 51: 289–94.

Lymbery, A. J., and Thompson, R. C.A. 1989. Genetic differences between cysts of *Echinococcus granulosus* from the same host. *International Journal for Parasitology* 19: 961–64.

Lyndon, A. R., and Kennedy, C. R. 2001. Colonisation and extinction in relation to competition and resource partitioning in acanthocephalans of freshwater fishes of the British Isles. *Folia Parasitologica* 48: 37–46.

Lysne, D. A., Hemmingsen, W., and Skorping, A. 1997. Regulation of infrapopulations of *Cryptocotyle lingua* on cod. *Parasitology* 114: 145–50.

Lysne, D. A., and Skorping, A. 2002. The parasite *Lernaeocera branchialis* on caged cod: infection pattern is caused by differences in host susceptibility. *Parasitology* 124: 69–76.

MacArthur, R. H., and Wilson, E. O. 1967. *The Theory of Island Biogeography*. Princeton University Press, Princeton.

Mackinnon, M. J., and Read, A. F. 1999a. Genetic relationships between parasite virulence and transmission in the rodent malaria *Plasmodium chabaudi*. *Evolution* 53: 689–703.

Mackinnon, M. J., and Read, A. F. 1999b. Selection for high and low virulence in the malaria parasite *Plasmodium chabaudi*. *Proceedings of the Royal Society of London B* 266: 741–48.

Mackinnon, M. J., and Read, A. F. 2004. Virulence in malaria: an evolutionary viewpoint. *Philosophical Transactions of the Royal Society of London B* 359: 965–86.

Maitland, D. P. 1994. A parasitic fungus infecting yellow dungflies manipulates host perching behaviour. *Proceedings of the Royal Society of London B* 258: 187–93.

Mangin, K. L., Lipsitch, M., and Ebert, D. 1995. Virulence and transmission modes of two microsporidia in *Daphnia magna*. *Parasitology* 111: 133–42.

Marcogliese, D. J. 1995. The role of zooplankton in the transmission of helminth parasites to fish. *Reviews in Fish Biology and Fisheries* 5: 336–71.

Marcogliese, D. J. 1997. Fecundity of sealworm (*Pseudoterranova decipiens*) infecting grey seals (*Halichoerus grypus*) in the Gulf of St. Lawrence, Canada: lack of density-dependent effects. *International Journal for Parasitology* 27: 1401–09.

Marcogliese, D. J. 2001. Implications of climate change for parasitism of animals in the aquatic environment. *Canadian Journal of Zoology* 79: 1331–52.

Marcogliese, D. J. 2002. Food webs and the transmission of parasites to marine fish. *Parasitology* 124: S83–S99.

Marcogliese, D. J. 2005. Parasites and the superorganism: are they indicators of ecosystem health? *International Journal for Parasitology* 35: 705–16.

Marcogliese, D. J., and Cone, D. K. 1991. Importance of lake characteristics in structuring parasite communities of salmonids from insular Newfoundland. *Canadian Journal of Zoology* 69: 2962–67.

Marcogliese, D. J., and Cone, D. K. 1996. On the distribution and abundance of eel parasites in Nova Scotia: influence of pH. *Journal of Parasitology* 82: 389–99.

Marcogliese, D. J., and Cone, D. K. 1998. Comparison of richness and diversity of macroparasite communities among eels from Nova Scotia, the United Kingdom and Australia. *Parasitology* 116: 73–83.

Mariaux, J. 1998. A molecular phylogeny of the Eucestoda. *Journal of Parasitology* 84: 114–24.

May, R. M., and Anderson, R. M. 1978. Regulation and stability of host-parasite population interactions. II. Destabilizing processes. *Journal of Animal Ecology* 47: 249–67.

May, R. M., and Anderson, R. M. 1979. Population biology of infectious diseases: part II. *Nature* 280: 455–61.

May, R. M., and Nowak, M. A. 1995. Coinfection and the evolution of parasite virulence. *Proceedings of the Royal Society of London B* 261: 209–15.

May, R. M., and Woolhouse, M.E.J. 1993. Biased sex ratios and parasite mating probabilities. *Parasitology* 107: 287–95.

Mayhew, P. J., and Blackburn, T. M. 1999. Does development mode organize life-history traits in the parasitoid Hymenoptera? *Journal of Animal Ecology* 68: 906–16.

Maynard, B. J., DeMartini, L., and Wright, W. G. 1996. *Gammarus lacustris* harboring *Polymorphus paradoxus* show altered patterns of serotonin-like immunoreactivity. *Journal of Parasitology* 82: 663–66.

Maynard Smith, J. 1978. *The Evolution of Sex*. Cambridge University Press, Cambridge.

McCallum, H. 2000. *Population Parameters: Estimation for Ecological Models*. Blackwell Science, Oxford.

McCarthy, A. M. 1990. The influence of second intermediate host dispersion pattern upon the transmission of cercariae of *Echinoparyphium recurvatum* (Digenea: Echinostomatidae). *Parasitology* 101: 43–47.

McCarthy, H. O., Irwin, S.W.B., and Fitzpatrick, S. M. 1999. *Nucella lapillus* as a paratenic host for *Maritrema arenaria*. *Journal of Helminthology* 73: 281–82.

McCarthy, H. O., Fitzpatrick, S. M., and Irwin, S.W.B. 2002. Life history and life cycles: production and behavior of trematode cercariae in relation to host exploitation and next-host characteristics. *Journal of Parasitology* 88: 910–18.

McCarthy, H. O., Fitzpatrick, S. M., and Irwin, S.W.B. 2004. Parasite alteration of host shape: a quantitative approach to gigantism helps elucidate evolutionary advantages. *Parasitology* 128: 7–14.

McCoy, K. D., Boulinier, T., and Tirard, C. 2005b. Comparative host-parasite population structures: disentangling prospecting and dispersal in the black-legged kittiwake *Rissa tridactyla*. *Molecular Ecology* 14: 2825–38.

McCoy, K. D., Boulinier, T., Tirard, C., and Michalakis, Y. 2001. Host specificity of a generalist parasite: genetic evidence of sympatric host races in the seabird tick *Ixodes uriae*. *Journal of Evolutionary Biology* 14: 395–405.

McCoy, K. D., Boulinier, T., Tirard, C., and Michalakis, Y. 2003. Host-dependent genetic structure of parasite populations: differential dispersal of seabird tick host races. *Evolution* 57: 288–96.

McCoy, K. D., Chapuis, E., Tirard, C., Boulinier, T., Michalakis, Y., Le Bohec, C., Le Maho, Y., and Gauthier-Clerc, M. 2005a. Recurrent evolution of host-specialized races in a globally distributed parasite. *Proceedings of the Royal Society of London B* 272: 2389–95.

McFarland, L. H., Mouritsen, K. N., and Poulin, R. 2003. From first to second and back to first intermediate host: the complex transmission routes of *Curtuteria australis* (Digenea: Echinostomatidae). *Journal of Parasitology* 89: 625–28.

McKenzie, C. E., and Welch, H. E. 1979. Parasite fauna of the muskrat, *Ondatra zibethica* (Linnaeus, 1766), in Manitoba, Canada. *Canadian Journal of Zoology* 57: 640–46.

McKitrick, M. C. 1993. Phylogenetic constraint in evolutionary theory: has it any explanatory power? *Annual Review of Ecology and Systematics* 24: 307–30.

McLachlan, A., Ladle, R., and Bleay, C. 1999. Is infestation the result of adaptive choice behaviour by the parasite? A study of mites and midges. *Animal Behaviour* 58: 615–20.

Medica, D. L., and Sukhdeo, M.V.K. 2001. Estimating transmission potential in gastrointestinal nematodes (Order: Strongylida). *Journal of Parasitology* 87: 442–45.

Merrett, N. R., and Haedrich, R. L. 1997. *Deep-sea Demersal Fish and Fisheries*. Chapman and Hall, London.

Mettrick, D. F. 1987. Parasitology: today and tomorrow. *Canadian Journal of Zoology* 65: 812–22.

Michel, J. F. 1974. Arrested development of nematodes and some related phenomena. *Advances in Parasitology* 12: 279–366.

Minchella, D. J. 1985. Host life-history variation in response to parasitism. *Parasitology* 90: 205–16.

Minchella, D. J., Eddings, A. R., and Neel, S. T. 1994. Genetic, phenotypic, and behavioral variation in North American sylvatic isolates of *Trichinella*. *Journal of Parasitology* 80: 696–704.

Minchella, D. J., and Scott, M. E. 1991. Parasitism: a cryptic determinant of animal community structure. *Trends in Ecology and Evolution* 6: 250–54.

Minchella, D. J., Sollenberger, K. M., and Pereira de Souza, C. 1995. Distribution of schistosome genetic diversity within molluscan intermediate hosts. *Parasitology* 111: 217–20.

Mitchell, G. F., Garcia, E. G., Wood, S. M., Diasanta, R., Almonte, R., Calica, E., Davern, K. M., and Tiu, W. U. 1990. Studies on the sex ratio of worms in schistosome infections. *Parasitology* 101: 27–34.

Mohandas, A. 1975. Further studies on the sporocyst capable of producing miracidia. *Journal of Helminthology* 49: 167–71.

Møller, A. P. 1996. Effects of host sexual selection on the population biology of parasites. *Oikos* 75: 340–44.

Møller, A. P., Christe, P., and Garamszegi, L. Z. 2005. Coevolutionary arms races: increased host immune defense promotes specialization by avian fleas. *Journal of Evolutionary Biology* 18: 46–59.

Moné, H. 1997. Change in schistosome sex ratio under the influence of a biotic environmental-related factor. *Journal of Parasitology* 83: 220–23.

Moné, H., and Boissier, J. 2004. Sexual biology of schistosomes. *Advances in Parasitology* 57: 89–189.

Moore, J. 1981. Asexual reproduction and environmental predictability in cestodes (Cyclophyllidea: Taeniidae). *Evolution* 35: 723–41.

Moore, J. 1983. Responses of an avian predator and its isopod prey to an acanthocephalan parasite. *Ecology* 64: 1000–15.

Moore, J. 1984. Altered behavioral responses in intermediate hosts: an acanthocephalan parasite strategy. *American Naturalist* 123: 572–77.

Moore, J. 1993. Parasites and the behavior of biting flies. *Journal of Parasitology* 79: 1–16.

Moore, J. 1995. The behavior of parasitized animals. *BioScience* 45: 89–96.

Moore, J. 2002. *Parasites and the Behavior of Animals*. Oxford University Press, Oxford.

Moore, J., and Brooks, D. R. 1987. Asexual reproduction in cestodes (Cyclophyllidea: Taeniidae): ecological and phylogenetic influences. *Evolution* 41: 882–91.

Moore, J., and Freehling, M. 2002. Cockroach hosts in thermal gradients suppress parasite development. *Oecologia* 133: 261–66.

Moore, J., Freehling, M., and Gotelli, N. J. 1994. Altered behavior in two species of blattid cockroaches infected with *Moniliformis moniliformis* (Acanthocephala). *Journal of Parasitology* 80: 220–23.

Moore, J., and Gotelli, N. J. 1990. A phylogenetic perspective on the evolution of altered host behaviours: a critical look at the manipulation hypothesis. In *Parasitism and Host Behaviour* (eds. C. J. Barnard and J. M. Behnke), Taylor & Francis, London, pp. 193–233.

Moore, J., and Gotelli, N. J. 1996. Evolutionary patterns of altered behavior and susceptibility in parasitized hosts. *Evolution* 50: 807–19.

Moore, J., and Simberloff, D. 1990. Gastrointestinal helminth communities of bobwhite quail. *Ecology* 71: 344–59.

Mopper, S., and Strauss, S. Y. 1998. *Genetic Structure and Local Adaptation in Natural Insect Populations*. Chapman and Hall, London.

Moqbel, R., and Wakelin, D. 1979. *Trichinella spiralis* and *Strongyloides ratti*: immune interaction in adult rats. *Experimental Parasitology* 47: 65–72.

Moran, N. A., and Wernegreen, J. J. 2000. Lifestyle evolution in symbiotic bacteria: insights from genomics. *Trends in Ecology and Evolution* 15: 321–26.

Morand, S. 1993. Sexual transmission of a nematode: study of a model. *Oikos* 66: 48–54.

Morand, S. 1996a. Biodiversity of parasites in relation to their life-cycles. In *Aspects of the Genesis and Maintenance of Biological Diversity* (eds. M. E. Hochberg, J. Clobert, and R. Barbault), Oxford University Press, Oxford, pp. 243–60.

Morand, S. 1996b. Life-history traits in parasitic nematodes: a comparative approach for the search of invariants. *Functional Ecology* 10: 210–18.

Morand, S. 2000. Wormy world: comparative tests of theoretical hypotheses on parasite species richness. In *Evolutionary Biology of Host-Parasite Relationships: Theory Meets Reality* (eds. R. Poulin, S. Morand, and A. Skorping), Elsevier Science, Amsterdam, pp. 63–79.

Morand, S., Cribb, T. H., Kulbicki, M., Chauvet, C., Dufour, V., Faliex, E., Galzin, R., Lo, C., Lo-Yat, A., Pichelin, S. P., Rigby, M. C., and Sasal, P. 2000b. Determinants of endoparasite species richness of New Caledonian Chaetodontidae. *Parasitology* 121: 65–73.

Morand, S., Hafner, M. S., Page, R.D.M., and Reed, D. L. 2000a. Comparative body size relationships in pocket gophers and their chewing lice. *Biological Journal of the Linnean Society* 70: 239–49.

Morand, S., and Harvey, P. H. 2000. Mammalian metabolism, longevity and parasite species richness. *Proceedings of the Royal Society of London B* 267: 1999–2003.

Morand, S., and Hugot, J.-P. 1998. Sexual size dimorphism in parasitic oxyurid nematodes. *Biological Journal of the Linnean Society* 63: 397–410.

Morand, S., Legendre, P., Gardner, S. L., and Hugot, J.-P. 1996a. Body size evolution of oxyurid (Nematoda) parasites: the role of hosts. *Oecologia* 107: 274–82.

Morand, S., Manning, S. D., and Woolhouse, M.E.J. 1996b. Parasite-host coevolution and geographic patterns of parasite infectivity and host susceptibility. *Proceedings of the Royal Society of London B* 263: 119–28.

Morand, S., and Müller-Graf, C.D.M. 2000. Muscles or testes? Comparative evidence for sexual competition among dioecious blood parasites (Schistosomatidae) of vertebrates. *Parasitology* 120: 45–56.

Morand, S., Pointier, J.-P., Borel, G., and Théron, A. 1993. Pairing probability of schistosomes related to their distribution among the host population. *Ecology* 74: 2444–49.

Morand, S., Pointier, J.-P., and Théron, A. 1999a. Population biology of *Schistosoma mansoni* in the black rat: host regulation and basic transmission rate. *International Journal for Parasitology* 29: 673–84.

Morand, S., and Poulin, R. 1998. Density, body mass and parasite species richness of terrestrial mammals. *Evolutionary Ecology* 12: 717–27.

Morand, S., and Poulin, R. 2000. Optimal time to patency in parasitic nematodes: host mortality matters. *Ecology Letters* 3: 186–90.

Morand, S., and Poulin, R. 2002. Body size—density relationships and species diversity in parasitic nematodes: patterns and likely processes. *Evolutionary Ecology Research* 4: 951–61.

Morand, S., and Poulin, R. 2003. Phylogenies, the comparative method and parasite evolutionary ecology. *Advances in Parasitology* 54: 281–302.

Morand, S., Poulin, R., Rohde, K., and Hayward, C. 1999b. Aggregation and species coexistence of ectoparasites of marine fishes. *International Journal for Parasitology* 29: 663–72.

Morand, S., and Rivault, C. 1992. Infestation dynamics of *Blatticola blattae* Graeffe (Nematoda: Thelastomatidae), a parasite of *Blattella germanica* L. (Dictyoptera: Blattellidae). *International Journal for Parasitology* 22: 983–89.

Morand, S., Robert, F., and Connors, V. A. 1995. Complexity in parasite life cycles: population biology of cestodes in fish. *Journal of Animal Ecology* 64: 256–64.

Morand, S., and Sorci, G. 1998. Determinants of life-history evolution in nematodes. *Parasitology Today* 14: 193–96.

Moravec, F. 1994. *Parasitic Nematodes of Freshwater Fishes of Europe*. Academia, Prague.

Moreira, D., and López-Garcia, P. 2003. Are hydrothermal vents oases for parasitic protists? *Trends in Parasitology* 19: 556–58.

Morgan, J.A.T., DeJong, R. J., Jung, Y., Khallaayoune, K., Kock, S., Mkoji, G. M., and Loker, E. S. 2002. A phylogeny of planorbid snails, with implications for the evolution of *Schistosoma* parasites. *Molecular Phylogenetics and Evolution* 25: 477–88.

Mouillot, D., George-Nascimento, M., and Poulin, R. 2003. How parasites divide resources: a test of the niche apportionment hypothesis. *Journal of Animal Ecology* 72: 757–64.

Mouillot, D., George-Nascimento, M., and Poulin, R. 2005. Richness, structure and functioning in metazoan parasite communities. *Oikos* 109: 447–60.

Mouillot, D., and Poulin, R. 2004. Taxonomic partitioning shedding light on the diversification of parasite communities. *Oikos* 104: 205–207.

Moulia, C., Le Brun, N., Dallas, J., Orth, A., and Renaud, F. 1993. Experimental evidence of genetic determinism in high susceptibility to intestinal pinworm infection in mice: a hybrid zone model. *Parasitology* 106: 387–93.

Mouritsen, K. N. 1997. Crawling behaviour in the bivalve *Macoma balthica*: the parasite manipulation hypothesis revisited. *Oikos* 79: 513–20.

Mouritsen, K. N. 2002a. The *Hydrobia ulvae—Maritrema subdolum* association: influence of temperature, salinity, light, water-pressure and secondary host exudates on cercarial emergence and longevity. *Journal of Helminthology* 76: 341–47.

Mouritsen, K. N. 2002b. The parasite-induced surfacing behaviour in the cockle *Austrovenus stutchburyi*: a test of an alternative hypothesis and identification of potential mechanisms. *Parasitology* 124: 521–28.

Mouritsen, K. N., and Jensen, K. T. 1994. The enigma of gigantism: effect of larval trematodes on growth, fecundity, egestion and locomotion in *Hydrobia ulvae* (Pennant) (Gastropoda: Prosobranchia). *Journal of Experimental Marine Biology and Ecology* 181: 53–66.

Mouritsen, K. N., and Poulin, R. 2002a. Parasitism, climate oscillations and the structure of natural communities. *Oikos* 97: 462–68.

Mouritsen, K. N., and Poulin, R. 2002b. Parasitism, community structure and biodiversity in intertidal ecosystems. *Parasitology* 124: S101–S117.

Mouritsen, K. N., and Poulin, R. 2003. Parasite-induced trophic facilitation exploited by a non-host predator: a manipulator's nightmare. *International Journal for Parasitology* 33: 1043–50.

Mulcahy, G., O'Neill, S., Fanning, J., McCarthy, E., and Sekiya, M. 2005. Tissue migration by parasitic helminths: an immunoevasive strategy? *Trends in Parasitology* 21: 273–77.

Mulvey, M., Aho, J. M., Lydeard, C., Leberg, P. L., and Smith, M. H. 1991. Comparative population genetic structure of a parasite (*Fascioloides magna*) and its definitive host. *Evolution* 45: 1628–40.

Munger, J. C., Karasov, W. H., and Chang, D. 1989. Host genetics as a cause of overdispersion of parasites among hosts: how general a phenomenon? *Journal of Parasitology* 75: 707–10.

Muñoz, G., and Cribb, T. H. 2005. Infracommunity structure of parasites of *Hemigymnus melapterus* (Pisces: Labridae) from Lizard Island, Australia: the importance of habitat and parasite body size. *Journal of Parasitology* 91: 38–44.

Muñoz, G., Mouillot, D., and Poulin, R. 2006. Testing the niche apportionment hypothesis with parasite communities: is random assortment always the rule? *Parasitology* 132: 717–24.

Murrell, A., and Barker, S. C. 2005. Multiple origins of parasitism in lice: phylogenetic analysis of SSU rDNA indicates that the Phthiraptera and Psocoptera are not monophyletic. *Parasitology Research* 97: 274–80.

Nadler, S. A. 1990. Molecular approaches to studying helminth population genetics and phylogeny. *International Journal for Parasitology* 20: 11–29.

Nadler, S. A. 1995. Microevolution and the genetic structure of parasite populations. *Journal of Parasitology* 81: 395–403.

Nadler, S. A., Lindquist, R. L., and Near, T. J. 1995. Genetic structure of midwestern *Ascaris suum* populations: a comparison of isoenzyme and RAPD markers. *Journal of Parasitology* 81: 385–94.

Nascetti, G., Cianchi, R., Mattiucci, S., D'Amelio, S., Orecchia, P., Paggi, L., Brattey, J., Berland, B., Smith, J. W., and Bullini, L. 1993. Three sibling species within *Contracaecum osculatum* (Nematoda, Ascaridida, Ascaridoidea) from the Atlantic Arctic—Boreal region: reproductive isolation and host preferences. *International Journal for Parasitology* 23: 105–20.

Navarro, P., Lluch, J., and Font, E. 2005. The component helminth community in six sympatric species of Ardeidae. *Journal of Parasitology* 91: 775–79.

Near, T. J., Garey, J. R., and Nadler, S. A. 1998. Phylogenetic relationships of the Acanthocephala inferred from 18S ribosomal DNA sequences. *Molecular Phylogenetics and Evolution* 10: 287–98.

Nee, S., West, S. A., and Read, A. F. 2002. Inbreeding and parasite sex ratios. *Proceedings of the Royal Society of London B* 269: 755–60.

Nelson, P. A., and Dick, T. A. 2002. Factors shaping the parasite communities of trout-perch, *Percopsis omiscomaycus* Walbaum (Osteichthyes: Percopsidae), and the importance of scale. *Canadian Journal of Zoology* 80: 1986–99.

Nevo, E. 1978. Genetic variation in natural populations: patterns and theory. *Theoretical Population Biology* 13: 121–77.

Nie, P., and Kennedy, C. R. 1993. Infection dynamics of larval *Bothriocephalus claviceps* in *Cyclops vicinus*. *Parasitology* 106: 503–509.

Niebel, A., Gheysen, G., and Van Montagu, M. 1994. Plant-cyst nematode and plant-root-knot nematode interactions. *Parasitology Today* 10: 424–30.

Nieberding, C., Morand, S., Libois, R., and Michaux, J. R. 2004. A parasite reveals the cryptic phylogeographic history of its host. *Proceedings of the Royal Society of London B* 271: 2559–68.

Noble, E. R. 1973. Parasites and fishes in a deep-sea environment. *Advances in Marine Biology* 11: 121–95.

Noble, E. R., Noble, G. A., Schad, G. A., and MacInnes, A. J. 1989. *Parasitology: The Biology of Animal Parasites*, 6th ed. Lea & Febiger, Philadelphia.

Noble, G. A. 1967. A forty-foot fluke. *Journal of Parasitology* 53: 645.

Norton, J., Lewis, J. W., and Rollinson, D. 2003. Parasite infracommunity diversity in eels: a reflection of local component community diversity. *Parasitology* 127: 475–82.

Norton, J., Lewis, J. W., and Rollinson, D. 2004b. Temporal and spatial patterns of nestedness in eel macroparasite communities. *Parasitology* 129: 203–11.

Norton, J., Rollinson, D., and Lewis, J. W. 2004a. Patterns of infracommunity species richness in eels, *Anguilla anguilla*. *Journal of Helminthology* 78: 141–46.

Norval, R.A.I., Andrew, H. R., and Yunker, C. E. 1989. Pheromone-mediation of host-selection in bont ticks (*Amblyomma hebraeum* Koch). *Science* 243: 364–65.

Nosil, P. 2002. Transition rates between specialization and generalization in phytophagous insects. *Evolution* 56: 1701–1706.

Nosil, P., and Mooers, A. Ø. 2005. Testing hypotheses about ecological specialization using phylogenetic trees. *Evolution* 59: 2256–63.

Nunn, C. L., Altizer, S., Jones, K. E., and Sechrest, W. 2003. Comparative tests of parasite species richness in primates. *American Naturalist* 162: 597–614.

Obrebski, S. 1975. Parasite reproductive strategy and evolution of castration of hosts by parasites. *Science* 188: 1314–16.

Okamura, B., and Canning, E. U. 2003. Orphan worms and homeless parasites enhance bilaterian diversity. *Trends in Ecology and Evolution* 18: 633–39.

Oldroyd, B. P. 1999. Coevolution while you wait: *Varroa jacobsoni*, a new parasite of western honeybees. *Trends in Ecology and Evolution* 14: 312–15.

Oliva, M. E., and González, M. T. 2005. The decay of similarity over geographical distance in parasite communities of marine fishes. *Journal of Biogeography* 32: 1327–32.

Oliver, J. H. Jr. 1989. Biology and systematics of ticks (Acari: Ixodida). *Annual Review of Ecology and Systematics* 20: 397–430.

Olson, P. D., Littlewood, D.T.J., Bray, R. A., and Mariaux, J. 2001. Interrelationships and evolution of the tapeworms (Platyhelminthes: Cestoda). *Molecular Phylogenetics and Evolution* 19: 443–67.

Orme, C.D.L., Quicke, D.L.J., Cook, J. M., and Purvis, A. 2002. Body size does not predict species richness among the metazoan phyla. *Journal of Evolutionary Biology* 15: 235–47.

Osgood, S. M., Eisen, R. J., Wargo, A. R., and Schall, J. J. 2003. Manipulation of the vertebrate host's testosterone does not affect gametocyte sex ratio of a malaria parasite. *Journal of Parasitology* 89: 190–92.

Osgood, S. M., and Schall, J. J. 2004. Gametocyte sex ratio of a malaria parasite: response to experimental manipulation of parasite clonal diversity. *Parasitology* 128: 23–29.

Outreman, Y., Bollache, L., Plaistow, S., and Cézilly, F. 2002. Patterns of intermediate host use and levels of association between two conflicting manipulative parasites. *International Journal for Parasitology* 32: 15–20.

Øverli, Ø., Páll, M., Borg, B., Jobling, M., and Winberg, S. 2001. Effects of *Schistocephalus solidus* infection on brain monoaminergic activity in female three-spined sticklebacks *Gasterosteus aculeatus*. *Proceedings of the Royal Society of London B* 268: 1411–15.

Page, L. K., Swihart, R. K., and Kazacos, K. R. 2001. Changes in transmission of *Baylisascaris procyonis* to intermediate hosts as a function of spatial scale. *Oikos* 93: 213–20.

Page, R.D.M. 1990. Component analysis: a valiant failure? *Cladistics* 6: 119–36.

Page, R.D.M. 1993. Parasites, phylogeny and cospeciation. *International Journal for Parasitology* 23: 499–506.

Page, R.D.M. 1994. Parallel phylogenies: reconstructing the history of host-parasite assemblages. *Cladistics* 10: 155–73.

Page, R.D.M. 2003. *Tangled Trees: Phylogeny, Cospeciation, and Coevolution.* University of Chicago Press, Chicago.

Page, R.D.M., Clayton, D. H., and Paterson, A. M. 1996. Lice and cospeciation: a response to Barker. *International Journal for Parasitology* 26: 213–18.

Page, R.D.M., Lee, P.L.M., Becher, S. A., Griffiths, R., and Clayton, D. H. 1998. A different tempo of mitochondrial DNA evolution in birds and their parasitic lice. *Molecular Phylogenetics and Evolution* 9: 276–93.

Paggi, L., Nascetti, G., Cianchi, R., Orecchia, P., Mattiucci, S., D'Amelio, S., Berland, B., Brattey, J., Smith, J. W., and Bullini, L. 1991. Genetic evidence for three species within *Pseudoterranova decipiens* (Nematoda, Ascaridida, Ascaridoidea) in the North Atlantic and Norwegian and Barents Seas. *International Journal for Parasitology* 21: 195–212.

Palmer, M. W. 1990. The estimation of species richness by extrapolation. *Ecology* 71: 1195–98.

Palmieri, J. R., Thurman, J. B., and Andersen, F. L. 1978. Helminth parasites of dogs in Utah. *Journal of Parasitology* 64: 1149–50.

Palumbi, S. R. 2001. Humans as the world's greatest evolutionary force. *Science* 293: 1786–90.

Pampoulie, C., and Morand, S. 2002. Nonrandom association patterns in parasite infections caused by the host life cycle: empirical evidence from *Kudoa camarguensis* (Myxosporea) and *Aphalloides coelomicola* (Trematoda). *Journal of Parasitology* 88: 817–19.

Park, T. 1948. Experimental studies of interspecies competition. I. Competition between populations of the flour beetles, *Tribolium confusum* Duval and *Tribolium castaneum* Herbst. *Ecological Monographs* 18: 267–307.

Parker, G. A., Chubb, J. C., Ball, M. A., and Roberts, G. N. 2003a. Evolution of complex life cycles in helminth parasites. *Nature* 425: 480–84.

Parker, G. A., Chubb, J. C., Roberts, G. N., Michaud, M., and Milinski, M. 2003b. Optimal growth strategies of larval helminths in their intermediate hosts. *Journal of Evolutionary Biology* 16: 47–54.

Parmentier, E., and Das, K. 2004. Commensal vs. parasitic relationship between Carapini fish and their hosts: some further insight through δ^{13}C and δ^{15}N measurements. *Journal of Experimental Marine Biology and Ecology* 310: 45–58.

Partridge, L., and Harvey, P. H. 1988. The ecological context of life history evolution. *Science* 241: 1449–55.

Pasternak, A. F., Huntingford, F. A., and Crompton, D.W.T. 1995. Changes in metabolism and behaviour of the freshwater copepod *Cyclops strenuus abyssorum* infected with *Diphyllobothrium* spp. *Parasitology* 110: 395–99.

Pasternak, Z., Blasius, B. and Abelson, A. 2004. Host location by larvae of a parasitic barnacle: larval chemotaxis and plume tracking in flow. *Journal of Plankton Research* 26: 487–93.

Paterson, A. M., and Banks, J. 2001. Analytical approaches to measuring cospeciation of host and parasites: through a glass, darkly. *International Journal for Parasitology* 31: 1012–22.

Paterson, A. M., and Gray, R. D. 1997. Host-parasite cospeciation, host switching and missing the boat. In *Host-Parasite Evolution: General Principles and Avian Models* (eds. D. H. Clayton and J. Moore), Oxford University Press, Oxford, pp. 236–50.

Paterson, A. M., Gray, R. D., and Wallis, G. P. 1993. Parasites, petrels and penguins: does louse presence reflect seabird phylogeny? *International Journal for Parasitology* 23: 515–26.

Paterson, A. M., Palma, R. L., and Gray, R. D. 1999. How frequently do avian lice miss the boat? Implications for coevolutionary studies. *Systematic Biology* 48: 214–23.

Paterson, A. M., and Poulin, R. 1999. Have chondracanthid copepods co-speciated with their teleost hosts? *Systematic Parasitology* 44: 79–85.

Paterson, S. 2005. No evidence for specificity between host and parasite genotypes in experimental *Strongyloides ratti* (Nematoda) infections. *International Journal for Parasitology* 35: 1539–45.

Paterson, S., Fisher, M. C., and Viney, M. E. 2000. Inferring infection processes of a parasitic nematode using population genetics. *Parasitology* 120: 185–94.

Paterson, S., and Viney, M. E. 2002. Host immune responses are necessary for density dependence in nematode infections. *Parasitology* 125: 283–92.

Patrick, M. J. 1991. Distribution of enteric helminths in *Glaucomys volans* L. (Sciuridae): a test for competition. *Ecology* 72: 755–58.

Patterson, B. D., and Atmar, W. 1986. Nested subsets and the structure of insular mammalian faunas and archipelagos. *Biological Journal of the Linnean Society* 28: 65–82.

Patz, J. A., Graczyk, T. K., Geller, N., and Vittor, A. Y. 2000. Effect of environmental change on emerging parasitic diseases. *International Journal for Parasitology* 30: 1395–1405.

Paul, R.E.L., Nu, V.A.T., Krettli, A. U., and Brey, P. T. 2002. Interspecific competition during transmission of two sympatric malaria parasite species to the mosquito vector. *Proceedings of the Royal Society of London B* 269: 2551–57.

Pedersen, A. B., Altizer, S., Poss, M., Cunningham, A. A., and Nunn, C. L. 2005. Patterns of host specificity and transmission among parasites of wild primates. *International Journal for Parasitology* 35: 647–57.

Perkins, S. E., Cattadori, I. M., Tagliapietra, V., Rizzoli, A. P., and Hudson, P. J. 2003. Empirical evidence for key hosts in persistence of a tick-borne disease. *International Journal for Parasitology* 33: 909–17.

Perlman, S. J., and Jaenike, J. 2003. Infection success in novel hosts: an experimental and phylogenetic study of *Drosophila*-parasitic nematodes. *Evolution* 57: 544–57.

Perlman, S. J., Spicer, G. S., Shoemaker, D. D., and Jaenike, J. 2003. Associations between mycophagous *Drosophila* and their *Howardula* nematode parasites: a worldwide phylogenetic shuffle. *Molecular Ecology* 12: 237–49.

Perrot-Minnot, M.-J. 2004. Larval morphology, genetic divergence, and contrasting levels of host manipulation between forms of *Pomphorhynchus laevis* (Acanthocephala). *International Journal for Parasitology* 34: 45–54.

Perry, R. N. 1999. Desiccation survival of parasitic nematodes. *Parasitology* 119: S19–S30.

Peters, R. H. 1983. *The Ecological Implications of Body Size*. Cambridge University Press, Cambridge.

Petney, T. N. 2001. Environmental, cultural and social changes and their influence on parasite infections. *International Journal for Parasitology* 31: 919–32.

Pettibone, M. H. 1957. Endoparasitic polychaetous annelids of the family Arabellidae with descriptions of new species. *Biological Bulletin* 113: 170–87.

Phares, K. 1996. An unusual host-parasite relationship: the growth hormone-like factor from plerocercoids of spirometrid tapeworms. *International Journal for Parasitology* 26: 575–88.

Pianka, E. R. 1970. On r- and K- selection. *American Naturalist* 104: 592–97.

Pianka, E. R. 1994. *Evolutionary Ecology*, 5th ed. Harper Collins, New York.

Pica-Mattoccia, L., Moroni, R., Tchuem Tchuenté, L. A., Southgate, V. R., and Cioli, D. 2000. Changes of mate occur in *Schistosoma mansoni. Parasitology* 120: 495–500.

Pietrock, M., and Marcogliese, D. J. 2003. Free-living endohelminth stages: at the mercy of environmental conditions. *Trends in Parasitology* 19: 293–99.

Platt, T. R., and Brooks, D. R. 1997. Evolution of the schistosomes (Digenea: Schistosomatoidea): the origin of dioecy and colonization of the venous system. *Journal of Parasitology* 83: 1035–44.

Poddubnaya, L. G., Mackiewicz, J. S., and Kuperman, B. I. 2003. Ultrastructure of *Archigetes sieboldi* (Cestoda: Caryophyllidea): relationship between progenesis, development and evolution. *Folia Parasitologica* 50: 275–92.

Poinar, G. O. Jr. 1983. *The Natural History of Nematodes*. Prentice Hall, Englewood Cliffs, New Jersey.

Poinar, G. O. Jr. 1993. Origins and phylogenetic relationships of the entomophilic rhabditids *Heterorhabditis* and *Steinernema. Fundamental and Applied Nematology* 16: 333–38.

Poinar, G. O. Jr. 1999. *Paleochordodes protus* n.g., n.sp. (Nematomorpha, Chordodidae), parasites of a fossil cockroach, with a critical examination of other fossil hairworms and helminths of extant cockroaches (Insecta: Blattaria). *Invertebrate Biology* 118: 109–15.

Poinar, G. O. Jr., Krantz, G. W., Boucot, A. J., and Pike, T. M. 1997. A unique Mesozoic parasitic association. *Naturwissenschaften* 84: 321–22.

Poulin, R. 1992a. Determinants of host-specificity in parasites of freshwater fishes. *International Journal for Parasitology* 22: 753–58.

Poulin, R. 1992b. Toxic pollution and parasitism in freshwater fish. *Parasitology Today* 8: 58–61.

Poulin, R. 1993. The disparity between observed and uniform distributions: a new look at parasite aggregation. *International Journal for Parasitology* 23: 937–44.

Poulin, R. 1994a. The evolution of parasite manipulation of host behaviour: a theoretical analysis. *Parasitology* 109: S109–S118.

Poulin, R. 1994b. Meta-analysis of parasite-induced behavioural changes. *Animal Behaviour* 48: 137–46.

Poulin, R. 1995a. Evolutionary and ecological parasitology: a changing of the guard? *International Journal for Parasitology* 25: 861–62.

Poulin, R. 1995b. Clutch size and egg size in free-living and parasitic copepods: a comparative analysis. *Evolution* 49: 325–36.

Poulin, R. 1995c. Evolutionary influences on body size in free-living and parasitic isopods. *Biological Journal of the Linnean Society* 54: 231–44.

Poulin, R. 1995d. Evolution of parasite life history traits: myths and reality. *Parasitology Today* 11: 342–45.

Poulin, R. 1995e. "Adaptive" changes in the behaviour of parasitized animals: a critical review. *International Journal for Parasitology* 25: 1371–83.

Poulin, R. 1995f. Phylogeny, ecology, and the richness of parasite communities in vertebrates. *Ecological Monographs* 65: 283–302.

Poulin, R. 1996a. How many parasite species are there: are we close to answers? *International Journal for Parasitology* 26: 1127–29.

Poulin, R. 1996b. The evolution of life history strategies in parasitic animals. *Advances in Parasitology* 37: 107–34.

Poulin, R. 1996c. The evolution of body size in the Monogenea: the role of host size and latitude. *Canadian Journal of Zoology* 74: 726–32.

Poulin, R. 1996d. Sexual size dimorphism and transition to parasitism in copepods. *Evolution* 50: 2520–23.

Poulin, R. 1996e. Observations on the free-living adult stage of *Gordius dimorphus* (Nematomorpha: Gordioidea). *Journal of Parasitology* 82: 845–46.

Poulin, R. 1996f. Measuring parasite aggregation: defending the index of discrepancy. *International Journal for Parasitology* 26: 227–29.

Poulin, R. 1996g. Richness, nestedness, and randomness in parasite infracommunity structure. *Oecologia* 105: 545–51.

Poulin, R. 1997a. Parasite faunas of freshwater fish: the relationship between richness and the specificity of parasites. *International Journal for Parasitology* 27: 1091–98.

Poulin, R. 1997b. Egg production in adult trematodes: adaptation or constraint? *Parasitology* 114: 195–204.

Poulin, R. 1997c. Covariation of sexual size dimorphism and adult sex ratio in parasitic nematodes. *Biological Journal of the Linnean Society* 62: 567–80.

Poulin, R. 1997d. Population abundance and sex ratio in dioecious helminth parasites. *Oecologia* 111: 375–80.

Poulin, R. 1997e. Species richness of parasite assemblages: evolution and patterns. *Annual Review of Ecology and Systematics* 28: 341–58.

Poulin, R. 1998a. Large-scale patterns of host use by parasites of freshwater fishes. *Ecology Letters* 1: 118–28.

Poulin, R. 1998b. Host and environmental correlates of body size in ticks (Acari: Argasidae and Ixodidae). *Canadian Journal of Zoology* 76: 925–30.

Poulin, R. 1998c. Comparison of three estimators of species richness in parasite component communities. *Journal of Parasitology* 84: 485–90.

Poulin, R. 1999a. The intra- and interspecific relationships between abundance and distribution in helminth parasites of birds. *Journal of Animal Ecology* 68: 719–25.

Poulin, R. 1999b. Body size vs. abundance among parasite species: positive relationships? *Ecography* 22: 246–50.

Poulin, R. 1999c. Speciation and diversification of parasite lineages: an analysis of congeneric parasite species in vertebrates. *Evolutionary Ecology* 13: 455–67.

Poulin, R. 2000. Manipulation of host behaviour by parasites: a weakening paradigm? *Proceedings of the Royal Society of London B* 267: 787–92.

Poulin, R. 2001a. Body size and segmentation patterns in free-living and parasitic polychaetes. *Canadian Journal of Zoology* 79: 741–45.

Poulin, R. 2001b. Interactions between species and the structure of helminth communities. *Parasitology* 122: S3–S11.

Poulin, R. 2001c. Another look at the richness of helminth communities in tropical freshwater fish. *Journal of Biogeography* 28: 737–43.

Poulin, R. 2002. The evolution of monogenean diversity. *International Journal for Parasitology* 32: 245–54.

Poulin, R. 2003a. Information about transmission opportunities triggers a life history switch in a parasite. *Evolution* 57: 2899–903.

Poulin, R. 2003b. The decay of similarity with geographical distance in parasite communities of vertebrate hosts. *Journal of Biogeography* 30: 1609–15.

Poulin, R. 2004. Parasites and the neutral theory of biodiversity. *Ecography* 27: 119–23.

Poulin, R. 2005a. Relative infection levels and taxonomic distances among the host species used by a parasite: insights into parasite specialization. *Parasitology* 130: 109–15.

Poulin, R. 2005b. Evolutionary trends in body size of parasitic flatworms. *Biological Journal of the Linnean Society* 85: 181–89.

Poulin, R. 2005c. Detection of interspecific competition in parasite communities. *Journal of Parasitology* 91: 1232–35.

Poulin, R., and Combes, C. 1999. The concept of virulence: interpretations and implications. *Parasitology Today* 15: 474–75.

Poulin, R., and Cribb, T. H. 2002. Trematode life cycles: short is sweet? *Trends in Parasitology* 18: 176–83.

Poulin, R., Curtis, M. A., and Rau, M. E. 1992. Effects of *Eubothrium salvelini* (Cestoda) on the behaviour of *Cyclops vernalis* (Copepoda) and its susceptibility to fish predators. *Parasitology* 105: 265–71.

Poulin, R., Fredensborg, B. L., Hansen, E., and Leung, T.L.F. 2005. The true cost of host manipulation by parasites. *Behavioural Processes* 68: 241–44.

Poulin, R., Giari, L., Simoni, E., and Dezfuli, B. S. 2003c. Effects of conspecifics and heterospecifics on individual worm mass in four helminth species parasitic in fish. *Parasitology Research* 90: 143–47.

Poulin, R., and Guégan, J.-F. 2000. Nestedness, anti-nestedness, and the relationship between prevalence and intensity in ectoparasite assemblages of marine fish: a spatial model of species coexistence. *International Journal for Parasitology* 30: 1147–52.

Poulin, R., and Hamilton, W. J. 1995. Ecological determinants of body size and clutch size in amphipods: a comparative approach. *Functional Ecology* 9: 364–70.

Poulin, R., and Hamilton, W. J. 1997. Ecological correlates of body size and egg size in parasitic Ascothoracida and Rhizocephala (Crustacea). *Acta Oecologica* 18: 621–35.

Poulin, R., and Hamilton, W. J. 2000. Egg size variation as a function of environmental variability in parasitic trematodes. *Canadian Journal of Zoology* 78: 564–69.

Poulin, R., Krasnov, B. R., Shenbrot, G. I., Mouillot, D., and Khokhlova, I. S. 2006. Evolution of host specificity in fleas: is it directional and irreversible? *International Journal for Parasitology* 36: 185–91.

Poulin, R., and Latham, A.D.M. 2002. Parasitism and the burrowing depth of the beach hopper *Talorchestia quoyana* (Amphipoda: Talitridae). *Animal Behaviour* 63: 269–75.

Poulin, R., and Latham, A.D.M. 2003. Effects of initial (larval) size and host body temperature on growth in trematodes. *Canadian Journal of Zoology* 81: 574–81.

Poulin, R., and Luque, J. L. 2003. A general test of the interactive-isolationist continuum in gastrointestinal parasite communities of fish. *International Journal for Parasitology* 33: 1623–30.

Poulin, R., Marshall, L. J., and Spencer, H. G. 2000. Metazoan parasite species richness and genetic variation among freshwater fish species: cause or consequence? *International Journal for Parasitology* 30: 697–703.

Poulin, R., and Morand, S. 1997. Parasite body size distributions: interpreting patterns of skewness. *International Journal for Parasitology* 27: 959–64.

Poulin, R., and Morand, S. 1999. Geographical distances and the similarity among parasite communities of conspecific host populations. *Parasitology* 119: 369–74.

Poulin, R., and Morand, S. 2000a. The diversity of parasites. *Quarterly Review of Biology* 75: 277–93.

Poulin, R., and Morand, S. 2000b. Testes size, body size and male-male competition in acanthocephalan parasites. *Journal of Zoology* 250: 551–58.

Poulin, R., and Morand, S. 2000c. Parasite body size and interspecific variation in levels of aggregation among nematodes. *Journal of Parasitology* 86: 642–47.

Poulin, R., and Morand, S. 2004. *Parasite Biodiversity*. Smithsonian Institution Press, Washington, D.C.

Poulin, R., and Mouillot, D. 2003a. Parasite specialization from a phylogenetic perspective: a new index of host specificity. *Parasitology* 126: 473–80.

Poulin, R., and Mouillot, D. 2003b. Host introductions and the geography of parasite taxonomic diversity. *Journal of Biogeography* 30: 837–45.

Poulin, R., and Mouillot, D. 2004. The relationship between specialization and local abundance: the case of helminth parasites of birds. *Oecologia* 140: 372–78.

Poulin, R., and Mouillot, D. 2005. Combining phylogenetic and ecological information into a new index of host specificity. *Journal of Parasitology* 91: 511–14.

Poulin, R., Nichol, K., and Latham, A.D.M. 2003b. Host sharing and host manipulation by larval helminths in shore crabs: cooperation or conflict? *International Journal for Parasitology* 33: 425–33.

Poulin, R., Rau, M. E., and Curtis, M. A. 1991. Infection of brook trout fry, *Salvelinus fontinalis*, by ectoparasitic copepods: the role of host behaviour and initial parasite load. *Animal Behaviour* 41: 467–76.

Poulin, R., and Rohde, K. 1997. Comparing the richness of metazoan ectoparasite communities of marine fishes: controlling for host phylogeny. *Oecologia* 110: 278–83.

Poulin, R., and Thomas, F. 1999. Phenotypic variability induced by parasites: extent and evolutionary implications. *Parasitology Today* 15: 28–32.

Poulin, R., and Valtonen, E. T. 2001a. Nested assemblages resulting from host size variation: the case of endoparasite communities in fish hosts. *International Journal for Parasitology* 31: 1194–1204.

Poulin, R., and Valtonen, E. T. 2001b. Interspecific associations among larval helminths in fish. *International Journal for Parasitology* 31: 1589–96.

Poulin, R., and Valtonen, E. T. 2002. The predictability of helminth community structure in space: a comparison of fish populations from adjacent lakes. *International Journal for Parasitology* 32: 1235–43.

Poulin, R., Wise, M., and Moore, J. 2003a. A comparative analysis of adult body size and its correlates in acanthocephalan parasites. *International Journal for Parasitology* 33: 799–805.

Price, P. W. 1974. Strategies for egg production. *Evolution* 28: 76–84.

Price, P. W. 1980. *Evolutionary Biology of Parasites.* Princeton University Press, Princeton.

Price, P. W. 1987. Evolution in parasite communities. *International Journal for Parasitology* 17: 209–14.

Price, P. W., and Clancy, K. M. 1983. Patterns in number of helminth parasite species in freshwater fishes. *Journal of Parasitology* 69: 449–54.

Prugnolle, F., Liu, H., de Meeûs, T., and Balloux, F. 2005a. Population genetics of complex life-cycle parasites: an illustration with trematodes. *International Journal for Parasitology* 35: 255–63.

Prugnolle, F., Théron, A., Pointier, J.-P., Jabbour-Zahab, R., Jarne, P., Durand, P., and de Meeûs, T. 2005b. Dispersal in a parasitic worm and its two hosts: consequence for local adaptation. *Evolution* 59: 296–303.

Pulkkinen, K., Pasternak, A. F., Hasu, T., and Valtonen, E. T. 2000. Effect of *Triaenophorus crassus* (Cestoda) infection on behavior and susceptibility to predation of the first intermediate host *Cyclops strenuus* (Copepoda). *Journal of Parasitology* 86: 664–70.

Quicke, D.L.J., and Belshaw, R. 1999. Incongruence between morphological data sets: an example from the evolution of endoparasitism among parasitic wasps (Hymenoptera: Braconidae). *Systematic Biology* 48: 436–54.

Quinnell, R. J. 2003. Genetics of susceptibility to human helminth infection. *International Journal for Parasitology* 33: 1219–31.

Quinnell, R. J., Medley, G. F., and Keymer, A. E. 1990. The regulation of gastrointestinal helminth populations. *Philosophical Transactions of the Royal Society of London B* 330: 191–201.

Radovsky, F. J., Krantz, G. W., and Whitaker, J. O. Jr. 1997. A remarkable example of predation in the parasitic mite family Macronyssidae. *International Journal of Acarology* 23: 3–6.

Radtke, A., McLennan, D. A., and Brooks, D. R. 2002. Resource tracking in North American *Telorchis* spp. (Digenea: Plagiorchiformes: Telorchidae). *Journal of Parasitology* 88: 874–79.

Raibaut, A., and Trilles, J. P. 1993. The sexuality of parasitic crustaceans. *Advances in Parasitology* 32: 367–444.

Raikova, E. V. 1994. Life cycle, cytology, and morphology of *Polypodium hydriforme*, a coelenterate parasite of the eggs of Acipenseriform fishes. *Journal of Parasitology* 80: 1–22.

Rauch, G., Kalbe, M., and Reusch, T.B.H. 2005. How a complex life cycle can improve a parasite's sex life. *Journal of Evolutionary Biology* 18: 1069–75.

Rauque, C. A., Semenas, L. G., and Viozzi, G. P. 2002. Post-cyclic transmission in *Acanthocephalus tumescens* (Acanthocephala, Echinorhynchidae). *Folia Parasitologica* 49: 127–30.

Rauque, C. A., Viozzi, G. P., and Semenas, L. G. 2003. Component population study of *Acanthocephalus tumescens* (Acanthocephala) in fishes from Lake Moreno, Argentina. *Folia Parasitologica* 50: 72–78.

Read, A. F., Anwar, M., Shutler, D., and Nee, S. 1995. Sex allocation and population structure in malaria and related parasitic protozoa. *Proceedings of the Royal Society of London B* 260: 359–63.

Read, A. F., Gemmill, A. W., and Skorping, A. 2000. Evolution of nematode life histories: theory meets reality? In *Evolutionary Biology of Host-Parasite Relationships: Theory Meets Reality* (eds. R. Poulin, S. Morand, and A. Skorping), Elsevier, Amsterdam, pp. 237–46.

Read, A. F., Narara, A., Nee, S., Keymer, A. E., and Day, K. P. 1992. Gametocyte sex ratios as indirect measures of outcrossing rates in malaria. *Parasitology* 104: 387–95.

Read, A. F., and Skorping, A. 1995. The evolution of tissue migration by parasitic nematode larvae. *Parasitology* 111: 359–71.

Read, A. F., and Taylor, L. H. 2001. The ecology of genetically diverse infections. *Science* 292: 1099–1102.

Read, C. P. 1951. The "crowding effect" in tapeworm infections. *Journal of Parasitology* 37: 174–78.

Rékási, J., Rózsa, L., and Kiss, B. J. 1997. Patterns in the distribution of avian lice (Phthiraptera: Amblycera, Ischnocera). *Journal of Avian Biology* 28: 150–56.

Reversat, J., Silan, P., and Maillard, C. 1992. Structure of monogenean populations, ectoparasites of the gilthead sea bream *Sparus aurata*. *Marine Biology* 112: 43–47.

Reyda, F. B., and Nickol, B. B. 2001. A comparison of biological performance among a laboratory-isolated population and two wild populations of *Moniliformis moniliformis*. *Journal of Parasitology* 87: 330–38.

Richardson, D. J., and Nickol, B. B. 2000. Experimental investigation of physiological factors that may influence microhabitat specificity exhibited by *Leptorhynchoides thecatus* (Acanthocephala) in green sunfish (*Lepomis cyanellus*). *Journal of Parasitology* 86: 685–90.

Ricklefs, R. E., and Fallon, S. M. 2002. Diversification and host switching in avian malaria parasites. *Proceedings of the Royal Society of London B* 269: 885–92.

Ricklefs, R. E., and Schluter, D. 1993. *Species Diversity in Ecological Communities: Historical and Geographical Perspectives*. University of Chicago Press, Chicago.

Rigaud, T., and Moret, Y. 2003. Differential phenoloxidase activity between native and invasive gammarids infected by local acanthocephalans: differential immunosuppression? *Parasitology* 127: 571–77.

Riggs, M. R., and Esch, G. W. 1987. The suprapopulation dynamics of *Bothriocephalus acheilognathi* in a North Carolina reservoir: abundance, dispersion, and prevalence. *Journal of Parasitology* 73: 877–92.

Riggs, M. R., Lemly, A. D., and Esch, G. W. 1987. The growth, biomass, and fecundity of *Bothriocephalus acheilognathi* in a North Carolina cooling reservoir. *Journal of Parasitology* 73: 893–900.

Ritchie, M. E., and Olff, H. 1999. Spatial scaling laws yield a synthetic theory of biodiversity. *Nature* 440: 557–60.

Robb, T., and Reid, M. L. 1996. Parasite-induced changes in the behaviour of cestode-infected beetles: adaptation or simple pathology? *Canadian Journal of Zoology* 74: 1268–74.

Robert, F., Renaud, F., Mathieu, E., and Gabrion, C. 1988. Importance of the paratenic host in the biology of *Bothriocephalus gregarius* (Cestoda, Pseudophyllidea), a parasite of the turbot. *International Journal for Parasitology* 18: 611–21.

Roberts, L. S. 2000. The crowding effect revisited. *Journal of Parasitology* 86: 209–11.

Roberts, L. S., and Janovy, J. Jr. 1996. *Foundations of Parasitology*, 5th ed. W. C. Brown Publishers, Dubuque, Iowa.

Roberts, M. G., Dobson, A. P., Arneberg, P., de Leo, G. A., Krecek, R. C., Manfredi, M. T., Lanfranchi, P., and Zaffaroni, E. 2002. Parasite community ecology and biodiversity. In *The Ecology of Wildlife Diseases* (eds. P. J. Hudson, A. Rizzoli, B. T. Grenfell, H. Heesterbeek and A. P. Dobson), Oxford University Press, Oxford, pp. 63–82.

Roche, M., and Patrzek, D. 1966. The female to male ratio (FMR) in hookworm. *Journal of Parasitology* 52: 117–21.

Rodgers-Gray, T. P., Smith, J. E., Ashcroft, A. E., Isaac, R. E., and Dunn, A. M. 2004. Mechanisms of parasite-induced sex reversal in *Gammarus duebeni*. *International Journal for Parasitology* 34: 747–53.

Roff, D. A. 1992. *The Evolution of Life Histories: Theory and Analysis*. Chapman & Hall, New York.

Rohde, K. 1979. A critical evaluation of intrinsic and extrinsic factors responsible for niche restriction in parasites. *American Naturalist* 114: 648–71.

Rohde, K. 1980. Host specificity indices of parasites and their application. *Experientia* 36: 1369–71.

Rohde, K. 1989. At least eight types of sense receptors in an endoparasitic flatworm: a countertrend to sacculinization. *Naturwissenschaften* 76: 383–85.

Rohde, K. 1991. Intra- and interspecific interactions in low density populations in resource-rich habitats. *Oikos* 60: 91–104.

Rohde, K. 1992. Latitudinal gradients in species diversity: the search for the primary cause. *Oikos* 65: 514–27.

Rohde, K. 1993. *Ecology of Marine Parasites*, 2nd ed. CAB International, Wallingford, UK.

Rohde, K. 1994a. The origins of parasitism in the Platyhelminthes. *International Journal for Parasitology* 24: 1099–1115.

Rohde, K. 1994b. Niche restriction in parasites: proximate and ultimate causes. *Parasitology* 109: S69–S84.

Rohde, K. 1996. Robust phylogenies and adaptive radiations: a critical examination of methods used to identify key innovations. *American Naturalist* 148: 481–500.

Rohde, K. 1998. Is there a fixed number of niches for endoparasites of fish? *International Journal for Parasitology* 28: 1861–65.

Rohde, K. 1999. Latitudinal gradients in species diversity and Rapoport's rule revisited: a review of recent work and what can parasites teach us about the causes of the gradients? *Ecography* 22: 593–613.

Rohde, K. 2001. Spatial scaling laws may not apply to most animal species. *Oikos* 93: 499–504.

Rohde, K. 2002. Ecology and biogeography of marine parasites. *Advances in Marine Biology* 43: 1–86.

Rohde, K., and Heap, M. 1998. Latitudinal differences in species and community richness and in community structure of metazoan endo- and ectoparasites of marine teleost fish. *International Journal for Parasitology* 28: 461–74.

Rohde, K., and Rohde, P. P. 2005. The ecological niches of parasites. In *Marine Parasitology* (ed. K. Rohde), CSIRO Publishing, Melbourne, Australia, pp. 286–93.

Rohde, K., Worthen, W. B., Heap, M., Hugueny, B., and Guégan, J.-F. 1998. Nestedness in assemblages of metazoan ecto- and endoparasites of marine fish. *International Journal for Parasitology* 28: 543–49.

Rolff, J. 2000. Water mite parasitism in damselflies during emergence: two hosts, one pattern. *Ecography* 23: 273–82.

Rondelaud, D., and Barthe, D. 1987. *Fasciola hepatica* L.: étude de la productivité d'un sporocyste en fonction de la taille de *Lymnaea truncatula* Müller. *Parasitology Research* 74: 155–60.

Ronquist, F. 1994. Evolution of parasitism among closely related species: phylogenetic relationships and the origin of inquilinism in gall wasps (Hymenoptera, Cynipidae). *Evolution* 48: 241–66.

Rosen, R., and Dick, T. A. 1983. Development and infectivity of the procercoid of *Triaenophorus crassus* Forel and mortality of the first intermediate host. *Canadian Journal of Zoology* 61: 2120–28.

Rosen, R., and Dick, T. A. 1984. Growth and migration of plerocercoids of *Triaenophorus crassus* Forel and pathology in experimentally infected whitefish, *Coregonus clupeaformis*. *Canadian Journal of Zoology* 62: 203–11.

Rosenzweig, M. L. 1995. *Species Diversity in Space and Time*. Cambridge University Press, Cambridge.

Rothschild, M., and Clay, T. 1952. *Fleas, Flukes and Cuckoos*. Collins, London.

Rothsey, S., and Rohde, K. 2002. The responses of larval copepods and monogeneans to light, gravity and magnetic fields. *Acta Parasitologica* 47: 167–72.

Roubal, F. R., and Quartararo, N. 1992. Observations on the pigmentation of the monogeneans, *Anoplodiscus* spp. (family Anoplodiscidae) in different microhabitats on their sparid teleost hosts. *International Journal for Parasitology* 22: 459–64.

Rousset, F., Thomas, F., De Meeûs, T., and Renaud, F. 1996. Inference of parasite induced host mortality from distributions of parasite loads. *Ecology* 77: 2203–11.

Rózsa, L. 1992. Endangered parasite species. *International Journal for Parasitology* 22: 265–66.

Rózsa, L. 1993. Speciation patterns of ectoparasites and "straggling" lice. *International Journal for Parasitology* 23: 859–64.

Rózsa, L. 1997. Adaptive sex ratio manipulation in *Pediculus humanus capitis*: possible interpretations of Buxton's data. *Journal of Parasitology* 83: 543–44.

Rózsa, L., Reiczigel, J., and Majoros, G. 2000. Quantifying parasites in samples of hosts. *Journal of Parasitology* 86: 228–32.

Rózsa, L., Rékási, J., and Reiczigel, J. 1996. Relationship of host coloniality to the population ecology of avian lice (Insecta: Phthiraptera). *Journal of Animal Ecology* 65: 242–48.

Sage, R. D., Heyneman, D., Lim, K.-C., and Wilson, A. C. 1986. Wormy mice in a hybrid zone. *Nature* 324: 60–63.

Sakanari, J. A., and Moser, M. 1990. Adaptation of an introduced host to an indigenous parasite. *Journal of Parasitology* 76: 420–23.

Saladin, K. S. 1979. Behavioral parasitology and perspectives on miracidial host-finding. *Zeitschrift für Parasitenkunde* 60: 197–210.

Salewski, V. 2003. Satellite species in lampreys: a worldwide trend for ecological speciation in sympatry? *Journal of Fish Biology* 63: 267–79.

Sandland, G. J., and Goater, C. P. 2000. Development and intensity dependence of *Ornithodiplostomum ptychocheilus* metacercariae in fathead minnows (*Pimephales promelas*). *Journal of Parasitology* 86: 1056–60.

Sangster, N. C. 1999. Anthelmintic resistance: past, present and future. *International Journal for Parasitology* 29: 115–24.

Sapp, K. K., and Loker, E. S. 2000. Mechanisms underlying digenean-snail specificity: role of miracidial attachment and host plasma factors. *Journal of Parasitology* 86: 1012–19.

Sasal, P., Trouvé, S., Müller-Graf, C., and Morand, S. 1999. Specificity and host predictability: a comparative analysis among monogenean parasites of fish. *Journal of Animal Ecology* 68: 437–44.

Schad, G. A. 1963. Niche diversification in a parasite species flock. *Nature* 198: 404–406.

Schad, G. A., and Anderson, R. M. 1985. Predisposition to hookworm infection in humans. *Science* 228: 1537–40.

Schall, J. J. 1989. The sex ratio of *Plasmodium* gametocytes. *Parasitology* 98: 343–50.

Schall, J. J. 2002. Parasite virulence. In *The Behavioural Ecology of Parasites* (eds. E. E. Lewis, J. F. Campbell, and M.V.K. Sukhdeo), CAB International, Wallingford, UK, pp. 283–313.

Schallig, H.D.F.H., Sassen, M.J.M., Hordijk, P. L., and De Jong-Brink, M. 1991. *Trichobilharzia ocellata*: influence of infection on the fecundity of its intermediate snail host *Lymnaea stagnalis* and cercarial induction of the release of schistosomin, a snail neuropeptide antagonizing female gonadotropic hormones. *Parasitology* 102: 85–91.

Schärer, L., and Wedekind, C. 1999. Lifetime reproductive output in a hermaphrodite cestode when reproducing alone or in pairs: a time cost of pairing. *Evolutionary Ecology* 13: 381–94.

Schärer, L., and Wedekind, C. 2001. Social situation, sperm competition and sex allocation in a simultaneous hermaphrodite parasite, the cestode *Schistocephalus solidus*. *Journal of Evolutionary Biology* 14: 942–53.

Schjørring, S. 2004. Delayed selfing in relation to the availability of a mating partner in the cestode *Schistocephalus solidus*. *Evolution* 58: 2591–96.

Schjørring, S., and Koella, J. C. 2003. Sub-lethal effects of pathogens can lead to the evolution of lower virulence in multiple infections. *Proceedings of the Royal Society of London B* 270: 189–93.

Schluter, D. 1984. A variance test for detecting species associations, with some example applications. *Ecology* 65: 998–1005.

Schluter, D. 2000. Ecological character displacement in adaptive radiation. *American Naturalist* 156: S4–S16.

Schmidt, G. D., and Roberts, L. S. 1989. *Foundations of Parasitology*, 4th ed. Mosby, St. Louis.

Scott, M. E. 1987. Temporal changes in aggregation: a laboratory study. *Parasitology* 94: 583–95.

Scott, M. E., and Anderson, R. M. 1984. The population dynamics of *Gyrodactylus bullatarudis* (Monogenea) within laboratory populations of the fish host *Poecilia reticulata*. *Parasitology* 89: 159–94.

Searcy, D. G., and MacInnis, A. J. 1970. Measurements by DNA renaturation of the genetic basis of parasitic reduction. *Evolution* 24: 796–806.

Seed, J. R., and Sechelski, J. B. 1996. The individual host, a unique evolutionary island for rapidly dividing parasites: a theoretical approach. *Journal of Parasitology* 82: 263–67.

Shaw, D. J., and Dobson, A. P. 1995. Patterns of macroparasite abundance and aggregation in wildlife populations: a quantitative review. *Parasitology* 111: S111–S133.

Shaw, D. J., Grenfell, B. T., and Dobson, A. P. 1998. Patterns of macroparasite aggregation in wildlife host populations. *Parasitology* 117: 597–610.

Shine, R. 1989. Ecological causes for the evolution of sexual dimorphism: a review of the evidence. *Quarterly Review of Biology* 64: 419–61.

Shirakashi, S., and Goater, C. P. 2005. Chronology of parasite-induced alteration of fish behaviour: effects of parasite maturation and host experience. *Parasitology* 130: 177–83.

Shoop, W. L. 1988. Trematode transmission patterns. *Journal of Parasitology* 74: 46–59.

Shostak, A. W., and Dick, T. A. 1987. Individual variability in reproductive success of *Triaenophorus crassus* Forel (Cestoda: Pseudophyllidea), with comments on use of the Lorenz curve and Gini coefficient. *Canadian Journal of Zoology* 65: 2878–85.

Shostak, A. W., and Dick, T. A. 1989. Variability in timing of egg hatch of *Triaenophorus crassus* Forel (Cestoda: Pseudophyllidea) as a mechanism increasing temporal dispersion of coracidia. *Canadian Journal of Zoology* 67: 1462–70.

Shostak, A. W., and Esch, G. W. 1990. Photocycle-dependent emergence by cercariae of *Halipegus occidualis* from *Helisoma anceps*, with special reference to cercarial emergence patterns as adaptations for transmission. *Journal of Parasitology* 76: 790–95.

Shostak, A. W., and Scott, M. E. 1993. Detection of density-dependent growth and fecundity of helminths in natural infections. *Parasitology* 106: 527–39.

Shutler, D., Bennett, G. F., and Mullie, A. 1995. Sex proportions of *Haemoproteus* blood parasites and local mate competition. *Proceedings of the National Academy of Sciences of the U.S.A.* 92: 6748–52.

Shutler, D., and Read, A. F. 1998. Local mate competition, and extraordinary and ordinary blood parasite sex ratio. *Oikos* 82: 417–24.

Shutler, D., Reece, S. E., Mullie, A., Billingsley, P. F., and Read, A. F. 2005. Rodent malaria parasites *Plasmodium chabaudi* and *P. vinckei* do not increase their rates of gametocytogenesis in response to mosquito probing. *Proceedings of the Royal Society of London B* 272: 2397–2402.

Sibly, R. M., and Calow, P. 1986. *Physiological Ecology of Animals: An Evolutionary Approach*. Blackwell Scientific Publications, Oxford.

Siddall, M. E. 2004. Fallacies of false attribution: the defense of BPA by Brooks, Dowling, van Veller, and Hoberg. *Cladistics* 20: 376–77.

Siddall, M. E. 2005. Bracing for another decade of deception: the promise of Secondary Brooks Parsimony Analysis. *Cladistics* 21: 90–99.

Siddall, M. E., Brooks, D. R., and Desser, S. S. 1993. Phylogeny and the reversibility of parasitism. *Evolution* 47: 308–13.

Siddall, M. E., and Perkins, S. L. 2003. Brooks Parsimony Analysis: a valiant failure. *Cladistics* 19: 554–64.

Silver, B. B., Dick, T. A., and Welch, H. E. 1980. Concurrent infections of *Hymenolepis diminuta* and *Trichinella spiralis* in the rat intestine. *Journal of Parasitology* 66: 786–91.

Simberloff, D. 1990. Free-living communities and alimentary tract helminths: hypotheses and pattern analyses. In *Parasite Communities: Patterns and Processes* (eds. G. W. Esch, A. O. Bush, and J. M. Aho), Chapman & Hall, London, pp. 289–19.

Simberloff, D., and Moore, J. 1997. Community ecology of parasites and free-living animals. In *Host-Parasite Evolution: General Principles and Avian Models* (eds. D. H. Clayton and J. Moore), Oxford University Press, Oxford, pp. 174–97.

Simkova, A., Kadlec, D., Gelnar, M., and Morand, S. 2002. Abundance-prevalence relationship of gill congeneric ectoparasites: testing for the core-satellite hypothesis and ecological specialisation. *Parasitology Research* 88: 682–86.

Simkova, A., Morand, S., Jobet, E., Gelnar, M., and Verneau, O. 2004. Molecular phylogeny of congeneric monogenean parasites (*Dactylogyrus*): a case of intrahost speciation. *Evolution* 58: 1001–18.

Sire, C., Durand, P., Pointier, J.-P., and Théron, A. 1999. Genetic diversity and recruitment pattern of *Schistosoma mansoni* in a *Biomphalaria glabrata* snail population: a field study using random-amplified polymorphic DNA markers. *Journal of Parasitology* 85: 436–41.

Sire, C., Durand, P., Pointier, J.-P., and Théron, A. 2001a. Genetic diversity of *Schistosoma mansoni* within and among individual hosts (*Rattus rattus*): infrapopulation differentiation at microspatial scale. *International Journal for Parasitology* 31: 1609–16.

Sire, C., Langand, J., Barral, V., and Théron, A. 2001b. Parasite (*Schistosoma mansoni*) and host (*Biomphalaria glabrata*) genetic diversity: population structure in a fragmented landscape. *Parasitology* 122: 545–54.

Skerikova, A., Hypsa, V., and Scholz, T. 2001. Phylogenetic analysis of European species of *Proteocephalus* (Cestoda: Proteocephalidea): compatibility of molecular and morphological data, and parasite-host coevolution. *International Journal for Parasitology* 31: 1121–28.

Skorping, A., and Read, A. F. 1998. Drugs and parasites: global experiments in life history evolution? *Ecology Letters* 1: 10–12.

Skorping, A., Read., A. F., and Keymer, A. E. 1991. Life history covariation in intestinal nematodes of mammals. *Oikos* 60: 365–72.

Slapeta, J. R., Modry, D., Votypka, J., Jirku, M., Koudela, B., and Lukes, J. 2001. Multiple origin of the dihomoxenous life cycle in sarcosporidia. *International Journal for Parasitology* 31: 413–17.

Smith, B. P. 1998. Loss of larval parasitism in parasitengonine mites. *Experimental and Applied Acarology* 22: 187–99.

Smith, E. N. 1984. Alteration of behavior by parasites: a problem for evolutionists. *Creation Research Society Quarterly* 21: 124.

Smith, G., and Grenfell, B. T. 1985. The population biology of *Ostertagia ostertagi*. *Parasitology Today* 1: 76–81.

Smith, T. G., Kim, B., Hong, H., and Desser, S. S. 2000. Intraerythrocytic development of species of *Hepatozoon* infecting ranid frogs: evidence for convergence of life cycle characteristics among apicomplexans. *Journal of Parasitology* 86: 451–58.

Smith Trail, D. R. 1980. Behavioral interactions between parasites and hosts: host suicide and the evolution of complex life cycles. *American Naturalist* 116: 77–91.

Smithers, S. R., and Terry, R. J. 1969. The immunology of schistosomiasis. *Advances in Parasitology* 7: 41–93.

Smythe, A. B., and Font, W. F. 2001. Phylogenetic analysis of *Alloglossidium* (Digenea: Macroderoididae) and related genera: life-cycle evolution and taxonomic revision. *Journal of Parasitology* 87: 386–91.

Snyder, D. E., and Fitzgerald, P. R. 1985. Helminth parasites from Illinois raccoons (*Procyon lotor*). *Journal of Parasitology* 71: 274–78.

Snyder, S. D., and Janovy, J. Jr. 1996. Behavioral basis of second intermediate host specificity among four species of *Haematoloechus* (Digenea: Haematoloechidae). *Journal of Parasitology* 82: 94–99.

Snyder, S. D., and Loker, E. S. 2000. Evolutionary relationships among the Schistosomatidae (Platyhelminthes: Digenea) and an Asian origin for *Schistosoma*. *Journal of Parasitology* 86: 283–88.

Sokal, R. R., and Rohlf, F. J. 1995. *Biometry*, 3rd ed. W. H. Freeman & Co., New York.

Sorci, G., Morand, S., and Hugot, J.-P. 1997. Host-parasite coevolution: comparative evidence for covariation of life history traits in primates and oxyurid parasites. *Proceedings of the Royal Society of London* B 264: 285–89.

Sorci, G., Skarstein, F., Morand, S., and Hugot, J.-P. 2003. Correlated evolution between host immunity and parasite life histories in primates and oxyurid parasites. *Proceedings of the Royal Society of London* B 270: 2481–84.

Sorensen, R. E., and Minchella, D. J. 1998. Parasite influences on host life history: *Echinostoma revolutum* parasitism of *Lymnaea elodes* snails. *Oecologia* 115: 188–95.

Sorensen, R. E., and Minchella, D. J. 2001. Snail-trematode life history interactions: past trends and future directions. *Parasitology* 123: S3–S18.

Sousa, W. P. 1983. Host life history and the effect of parasitic castration on growth: a field study of *Cerithidea californica* Haldemann (Gastropoda: Prosobranchia) and its trematode parasites. *Journal of Experimental Marine Biology and Ecology* 73: 273–96.

Sousa, W. P. 1992. Interspecific interactions among larval trematode parasites of freshwater and marine snails. *American Zoologist* 32: 583–92.

Sousa, W. P. 1993. Interspecific antagonism and species coexistence in a diverse guild of larval trematode parasites. *Ecological Monographs* 63: 103–28.

Southwood, T.R.E. 1978. *Ecological Methods, with Particular Reference to the Study of Insect Populations*. John Wiley & Sons, New York.

Southwood, T.R.E. 1988. Tactics, strategies and templets. *Oikos* 52: 3–18.

Sparkes, T. C., Wright, V. M., Renwick, D. T., Weil, K. A., Talkington, J. A., and Milhalyov, M. 2004. Intra-specific host sharing in the manipulative parasite *Acanthocephalus dirus*: does conflict occur over host modification? *Parasitology* 129: 335–40.

Sprent, J.F.A. 1992. Parasites lost. *International Journal for Parasitology* 22: 139–51.

Srivastava, D. S. 1999. Using local—regional richness plots to test for species saturation: pitfalls and potentials. *Journal of Animal Ecology* 68: 1–16.

Stankiewicz, M., Cowan, P. E., and Heath, D. D. 1997a. Endoparasites of brushtail possums (*Trichosurus vulpecula*) from the South Island, New Zealand. *New Zealand Veterinary Journal* 45: 257–60.

Stankiewicz, M., Heath, D. D., and Cowan, P. E. 1997b. Internal parasites of possums (*Trichosurus vulpecula*) from Kawau Island, Chatham Island and Stewart Island. *New Zealand Veterinary Journal* 45: 247–50.

Stankiewicz, M., Jowett, G. H., Roberts, M. G., Heath, D. D., Cowan, P., Clark, J. M., Jowett, J., and Charleston, W.A.G. 1996. Internal and external parasites of possums (*Trichosurus vulpecula*) from forest and farmland, Wanganui, New Zealand. *New Zealand Journal of Zoology* 23: 345–53.

Stanko, M., Miklisova, D., Goüy de Bellocq, J., and Morand, S. 2002. Mammal density and patterns of ectoparasite species richness and abundance. *Oecologia* 131: 289–95.

Stanley, S. M. 1973. An explanation for Cope's rule. *Evolution* 27: 1–26.

Stear, M. J., Bairden, K., Duncan, J. L., Holmes, P. H., McKellar, Q. A., Park, M., Strain, S., Murray, M., Bishop., S. C., and Gettinby, G. 1997. How hosts control worms. *Nature* 389: 27.

Stearns, S. C. 1989. Tradeoffs in life history evolution. *Functional Ecology* 3: 259–68.

Stearns, S. C. 1992. *The Evolution of Life Histories*. Oxford University Press, Oxford.

Stearns, S. C. 1999. *Evolution in Health and Disease*. Oxford University Press, Oxford.

Steinauer, M. L., and Nickol, B. B. 2003. Effect of cystacanth body size on adult success. *Journal of Parasitology* 89: 251–54.

Stepien, C. A., and Brusca, R. C. 1985. Nocturnal attacks on nearshore fishes in southern California by crustacean zooplankton. *Marine Ecology—Progress Series* 25: 91–105.

Stevens, J. R. 2003. The evolution of myiasis in blowflies (Calliphoridae). *International Journal for Parasitology* 33: 1105–13.

Stien, A., Dallimer, M., Irvine, R. J., Halvorsen, O., Langvatn, R., Albon, S. D., and Dallas, J. F. 2005. Sex ratio variation in gastrointestinal nematodes of Svalbard reindeer: density dependence and implications for estimates of species composition. *Parasitology* 130: 99–107.

Stien, A., Halvorsen, O., and Leinaas, H.-P. 1996. Density-dependent sex ratio in *Echinomermella matsi* (Nematoda), a parasite of the sea urchin *Strongylocentrotus droebachiensis*. *Parasitology* 112: 105–12.

Stireman, J. O. III. 2005. The evolution of generalization? Parasitoid flies and the perils of inferring host range evolution from phylogenies. *Journal of Evolutionary Biology* 18: 325–36.

Stock, T. M., and Holmes, J. C. 1987. Host specificity and exchange of intestinal helminths among four species of grebes (Podicipedidae). *Canadian Journal of Zoology* 65: 669–76.

Stock, T. M., and Holmes, J. C. 1988. Functional relationships and microhabitat distributions of enteric helminths of grebes (Podicipedidae): the evidence for interactive communities. *Journal of Parasitology* 74: 214–27.

Stohler, R. A., Curtis, J., and Minchella, D. J. 2004. A comparison of microsatellite polymorphism and heterozygosity among field and laboratory populations of *Schistosoma mansoni*. *International Journal for Parasitology* 34: 595–601.

Stromberg, B. E., and Averbeck, G. A. 1999. The role of parasite epidemiology in the management of grazing cattle. *International Journal for Parasitology* 29: 33–39.

Stuart, R. J., and Gaugler, R. 1996. Genetic adaptation and founder effect in laboratory populations of the entomopathogenic nematode *Steinernema glaseri*. *Canadian Journal of Zoology* 74: 164–70.

Su, C., Evans, D., Cole, R. H., Kissinger, J. C., Ajioka., J. W., and Sibley, L. D. 2003. Recent expansion of *Toxoplasma* through enhanced oral transmission. *Science* 299: 414–16.

Sugihara, G. 1980. Minimal community structure: an explanation of species abundance patterns. *American Naturalist* 116: 770–87.

Sukhdeo, M.V.K. 1990a. The relationship between intestinal location and fecundity in adult *Trichinella spiralis*. *International Journal for Parasitology* 21: 855–58.

Sukhdeo, M.V.K. 1990b. Habitat selection by helminths: a hypothesis. *Parasitology Today* 6: 234–37.

Sukhdeo, M.V.K. 1997. Earth's third environment: the worm's eye view. *BioScience* 47: 141–49.

Sukhdeo, M.V.K., and Mettrick, D. F. 1987. Parasite behaviour: understanding platyhelminth responses. *Advances in Parasitology* 26: 73–144.

Sukhdeo, M.V.K., and Sukhdeo, S. C. 1994. Optimal habitat selection by helminths within the host environment. *Parasitology* 109: S41–S55.

Sukhdeo, M.V.K., and Sukhdeo, S. C. 2002. Fixed behaviours and migration in parasitic flatworms. *International Journal for Parasitology* 32: 329–42.

Sukhdeo, S. C., Sukhdeo, M.V.K., Black., M. B., and Vrijenhoek, R. C. 1997. The evolution of tissue migration in parasitic nematodes (Nematoda: Strongylida) inferred from a protein-coding mitochondrial gene. *Biological Journal of the Linnean Society* 61: 281–98.

Sun, P., and Hughes, G. 1994. On a statistical test to detect spatial pattern. *Biometrical Journal* 36: 951–58.

Sures, B. 2004. Environmental parasitology: relevancy of parasites in monitoring environmental pollution. *Trends in Parasitology* 20: 170–77.

Sutherst, R. W. 2001. The vulnerability of animal and human health to parasites under global change. *International Journal for Parasitology* 31: 933–48.

Swennen, C. 1969. Crawling-tracks of trematode infected *Macoma balthica* (L.). *Netherlands Journal of Sea Research* 4: 376–79.

Szalai, A. J., and Dick, T. A. 1989. Differences in numbers and inequalities in mass and fecundity during the egg-producing period for *Raphidascaris acus* (Nematoda: Anisakidae). *Parasitology* 98: 489–95.

Takemoto, R. M., Pavanelli, G. C., Lizama, M.A.P., Luque, J. L., and Poulin, R. 2005. Host population density as the major determinant of endoparasite species richness in floodplain fishes of the upper Paraná River, Brazil. *Journal of Helminthology* 79: 75–84.

Tanguay, G. V., and Scott, M. E. 1992. Factors generating aggregation of *Heligmosomoides polygyrus* (Nematoda) in laboratory mice. *Parasitology* 104: 519–29.

Taskinen, J. 1998. Influence of trematode parasitism on the growth of a bivalve host in the field. *International Journal for Parasitology* 28: 599–602.

Taylor, J., and Purvis, A. 2003. Have mammals and their chewing lice diversified in parallel? In *Tangled Trees: Phylogeny, Cospeciation, and Coevolution* (ed. R.D.M. Page), University of Chicago Press, Chicago, pp. 240–61.

Taylor, L. H., Mackinnon, M. J., and Read, A. F. 1998. Virulence of mixed-clone and single-clone infections of the rodent malaria *Plasmodium chabaudi*. *Evolution* 52: 583–91.

Taylor, L. H., and Read, A. F. 1997. Why so few transmission stages? Reproductive restraint by malaria parasites. *Parasitology Today* 13: 135–40.

Taylor, L. H., Walliker, D., and Read, A. F. 1997. Mixed-genotype infections of the rodent malaria *Plasmodium chabaudi* are more infectious to mosquitoes than single-genotype infections. *Parasitology* 115: 121–32.

Terry, R. J. 1994. Human immunity to schistosomes: concomitant immunity? *Parasitology Today* 10: 377–78.

Théron, A. 1984. Early and late shedding patterns of *Schistosoma mansoni* cercariae: ecological significance in transmission to human and murine hosts. *Journal of Parasitology* 70: 652–55.

Théron, A., Gérard, C., and Moné, H. 1992. Early enhanced growth of the digestive gland of *Biomphalaria glabrata* infected with *Schistosoma mansoni*: side effect or parasite manipulation? *Parasitology Research* 78: 445–50.

Théron, A., Sire, C., Rognon, A., Prugnolle, F., and Durand, P. 2004. Molecular ecology of *Schistosoma mansoni* transmission inferred from the genetic composition of larval and adult infrapopulations within intermediate and definitive hosts. *Parasitology* 129: 571–85.

Thomas, F., Adamo, S., and Moore, J. 2005. Parasitic manipulation: where are we and where should we go? *Behavioural Processes* 68: 185–99.

Thomas, F., Brown, S. P., Sukhdeo, M., and Renaud, F. 2002a. Understanding parasite strategies: a state-dependent approach? *Trends in Parasitology* 18: 387–90.

Thomas, F., Fauchier, J., and Lafferty, K. D. 2002c. Conflict of interest between a nematode and a trematode in an amphipod host: test of the "sabotage" hypothesis. *Behavioral Ecology and Sociobiology* 51: 296–301.

Thomas, F., Guldner, E., and Renaud, F. 2000. Differential parasite (Trematoda) encapsulation in *Gammarus aequicauda* (Amphipoda). *Journal of Parasitology* 86: 650–54.

Thomas, F., Mete, K., Helluy, S., Santalla, F., Verneau, O., De Meeüs, T., Cézilly, F., and Renaud, F. 1997. Hitch-hiker parasites or how to benefit from the strategy of another parasite. *Evolution* 51: 1316–18.

Thomas, F., and Poulin, R. 1998. Manipulation of a mollusc by a trophically transmitted parasite: convergent evolution or phylogenetic inheritance? *Parasitology* 116: 431–36.

Thomas, F., Poulin, R., and Renaud, F. 1998b. Nonmanipulative parasites in manipulated hosts: "hitch-hikers" or simply "lucky passengers"? *Journal of Parasitology* 84: 1059–61.

Thomas, F., Renaud, F., and Guégan, J.-F. 2005. *Parasitism and Ecosystems*. Oxford University Press, Oxford.

Thomas, F., Renaud, F., and Poulin, R. 1998a. Exploitation of manipulators: "hitch-hiking" as a parasite transmission strategy. *Animal Behaviour* 56: 199–206.

Thomas, F., Renaud, F., Rousset, F., Cézilly, F., and De Meeüs, T. 1995. Differential mortality of two closely related host species induced by one parasite. *Proceedings of the Royal Society of London B* 260: 349–52.

Thomas, F., Schmidt-Rhaesa, A., Martin, G., Manu, C., Durand, P., and Renaud, F. 2002b. Do hairworms (Nematomorpha) manipulate the water seeking behaviour of their terrestrial hosts? *Journal of Evolutionary Biology* 15: 356–61.

Thomas, F., Ulitsky, P., Augier, R., Dusticier, N., Samuel, D., Strambi, C., Biron, D. G., and Cayre, M. 2003. Biochemical and histological changes in the brain of the cricket *Nemobius sylvestris* infected by the manipulative parasite *Paragordius tricuspidatus* (Nematomorpha). *International Journal for Parasitology* 33: 435–43.

Thompson, J. N. 1994. *The Coevolutionary Process*. University of Chicago Press, Chicago.

Thompson, J. N. 2005. *The Geographic Mosaic of Coevolution*. University of Chicago Press, Chicago.

Thompson, R.C.A. 2001. The future impact of societal and cultural factors on parasitic disease: some emerging issues. *International Journal for Parasitology* 31: 949–59.

Thompson, R.C.A., and Lymbery, A. J. 1996 Genetic variability in parasites and host-parasite interactions. *Parasitology* 112: S7–S22.

Thompson, S. N., and Kavaliers, M. 1994. Physiological bases for parasite-induced alterations of host behaviour. *Parasitology* 109: S119–S138.

Thomson, J. D. 1980. Implications of different sorts of evidence for competition. *American Naturalist* 116: 719–26.

Thul, J. E., Forrester, D. J., and Abercrombie, C. L. 1985. Ecology of parasitic helminths of wood ducks, *Aix sponsa*, in the Atlantic flyway. *Proceedings of the Helminthological Society of Washington* 52: 297–310.

Timi, J. T. 2003. Habitat selection by *Lernanthropus cynoscicola* (Copepoda: Lernanthropidae): host as physical environment, a major determinant of niche restriction. *Parasitology* 127: 155–63.

Timi, J. T., and Poulin, R. 2003. Parasite community structure within and across host populations of a marine pelagic fish: how repeatable is it? *International Journal for Parasitology* 33: 1353–62.

Tingley, G. A., and Anderson, R. M. 1986. Environmental sex determination and density-dependent population regulation in the entomogenous nematode *Romanomermis culicivorax*. *Parasitology* 92: 431–49.

Tinsley, M. C., and Reilly, S. D. 2002. Reproductive ecology of the saltmarsh-dwelling marine ectoparasite *Paragnathia formica* (Crustacea: Isopoda). *Journal of the Marine Biological Association of the UK* 82: 79–84.

Tinsley, R. C. 1990. Host behaviour and opportunism in parasite life cycles. In *Parasitism and Host Behaviour* (eds. C. J. Barnard and J. M. Behnke), Taylor & Francis, London, pp. 158–92.

Tinsley, R. C. 1995. Parasitic disease in amphibians: control by the regulation of worm burden. *Parasitology* 111: S153–S178.

Tinsley, R. C. 1999. Parasite adaptation to extreme conditions in a desert environment. *Parasitology* 119: S31–S36.

Tinsley, R. C. 2004. Platyhelminth parasite reproduction: some general principles derived from monogeneans. *Canadian Journal of Zoology* 82: 270–91.

Toft, C. A., and Karter, A. J. 1990. Parasite-host coevolution. *Trends in Ecology and Evolution* 5: 326–29.

Tokeshi, M. 1999. *Species Coexistence: Ecological and Evolutionary Perspectives*. Blackwell Science, Oxford.

Tompkins, D. M., and Clayton, D. H. 1999. Host resources govern the specificity of swiftlet lice: size matters. *Journal of Animal Ecology* 68: 489–500.

Tompkins, D. M., and Hudson, P. J. 1999. Regulation of nematode fecundity in the ring-necked pheasant (*Phasianus colchicus*): not just density dependence. *Parasitology* 118: 417–23.

Torchin, M. E., Lafferty, K. D., Dobson, A. P., McKenzie, V. J., and Kuris, A. M. 2003. Introduced species and their missing parasites. *Nature* 421: 628–30.

Touassem, R., and Théron, A. 1989. *Schistosoma rodhaini*: dynamics and cercarial production for mono- and pluri-miracidial infections of *Biomphalaria glabrata*. *Journal of Helminthology* 63: 79–83.

Tripet, F., Christe, P. and Møller, A. P. 2002. The importance of host spatial distribution for parasite specialization and speciation: a comparative study of bird fleas (Siphonaptera: Ceratophyllidae). *Journal of Animal Ecology* 71: 735–48.

Trouvé, S., Jourdane, J., Renaud, F., Durand, P., and Morand, S. 1999. Adaptive sex allocation in a simultaneous hermaphrodite. *Evolution* 53: 1599–1604.

Trouvé, S., and Morand, S. 1998. Evolution of parasites' fecundity. *International Journal for Parasitology* 28: 1817–19.

Trouvé, S., Sasal, P., Jourdane, J., Renaud, F., and Morand, S. 1998. The evolution of life-history traits in parasitic and free-living Platyhelminthes: a new perspective. *Oecologia* 115: 370–78.

Tsai, M.-L., Li, J.-J., and Dai, C.-F. 1999. Why selection favors protandrous sex change for the parasitic isopod, *Ichthyoxenus fushanensis* (Isopoda: Cymothoidae). *Evolutionary Ecology* 13: 327–38.

Tsai, M.-L., Li, J.-J., and Dai, C.-F. 2001. How host size may constrain the evolution of parasite body size and clutch size: the parasitic isopod *Ichthyoxenus fushanensis* and its host fish, *Varicorhinus bacbatulus*, as an example. *Oikos* 92: 13–19.

Turner, C.M.R., Aslam, N., and Dye, C. 1995. Replication, differentiation, growth and the virulence of *Trypanosoma brucei* infections. *Parasitology* 111: 289–300.

Uebelacker, J. M. 1978. A new parasitic polychaetous annelid (Arabellidae) from the Bahamas. *Journal of Parasitology* 64: 151–54.

Urdal, K., Tierney, J. F., and Jakobsen, P. J. 1995. The tapeworm *Schistocephalus solidus* alters the activity and response, but not the predation susceptibility of infected copepods. *Journal of Parasitology* 81: 330–33.

Uznanski, R. L., and Nickol, B. B. 1982. Site selection, growth and survival of *Leptorhynchoides thecatus* (Acanthocephala) during the prepatent period in *Lepomis cyanellus*. *Journal of Parasitology* 68: 686–90.

Valera, F., Hoi, H., Darolova, A., and Kristofik, J. 2004. Size versus health as a cue for host choice: a test of the tasty chick hypothesis. *Parasitology* 129: 59–68.

van Baalen, M., and Sabelis, M. W. 1995. The dynamics of multiple infection and the evolution of virulence. *American Naturalist* 146: 881–910.

Vance, S. A. 1996. Morphological and behavioural sex reversal in mermithid-infected mayflies. *Proceedings of the Royal Society of London B* 263: 907–12.

Vaughn, C. C., and Taylor, C. M. 2000. Macroecology of a host-parasite relationship. *Ecography* 23: 11–20.

Vázquez, D. P., and Aizen, M. A. 2003. Null model analyses of specialization in plant-pollinator interactions. *Ecology* 84: 2493–2501.

Vázquez, D. P., Poulin, R., Krasnov, B. R., and Shenbrot, G. I. 2005. Species abundance and the distribution of specialization in host-parasite interaction networks. *Journal of Animal Ecology* 74: 946–55.

Verneau, O., Renaud, F., and Catzeflis, F. M. 1991. DNA reassociation kinetics and genome complexity of a fish (*Psetta maxima*: Teleostei) and its gut parasite (*Bothriocephalus gregarius*: Cestoda). *Comparative Biochemistry and Physiology* 99B: 883–86.

Vicente, J., Fierro, Y., Martinez, M., and Gortazar, C. 2004. Long-term epidemiology, effect on body condition and interspecific interactions of concomitant infection by nasopharyngeal bot fly larvae (*Cephenemyia auribarbis* and *Pharyngomyia picta*, Oestridae) in a population of Iberian red deer (*Cervus elaphus hispanicus*). *Parasitology* 129: 349–61.

Vickery, W. L., and Poulin, R. 1998. Parasite extinction and colonization and the evolution of parasite communities: a simulation study. *International Journal for Parasitology* 28: 727–37.

Vickery, W. L., and Poulin, R. 2002. Can helminth community patterns be amplified when transferred by predation from intermediate to definitive hosts? *Journal of Parasitology* 88: 650–56.

Vidal-Martinez, V. M., and Kennedy, C. R. 2000. Potential interactions between the intestinal helminths of the cichlid fish *Cichlasoma synspilum* from southeastern Mexico. *Journal of Parasitology* 86: 691–95.

Vidal-Martinez, V. M., Kennedy, C. R., and Aguirre-Macedo, M. L. 1998. The structuring process of the macroparasite community of an experimental population of *Cichlasoma urophthalmus* through time. *Journal of Helminthology* 72: 199–207.

Vidal-Martinez, V. M., and Poulin, R. 2003. Spatial and temporal repeatability in parasite community structure of tropical fish hosts. *Parasitology* 127: 387–98.

Vignoles, P., Ménard, A., Rondelaud, D., Agoulon, A., and Dreyfuss, G. 2004. *Fasciola hepatica*: the growth and larval productivity of redial generations in *Galba truncatula* subjected to miracidia differing in their mammalian origin. *Journal of Parasitology* 90: 430–33.

Vilas, R., Paniagua, E., and Sanmartin, M. L. 2003. Genetic variation within and among infrapopulations of the marine digenetic trematode *Lecithochirium fusiforme*. *Parasitology* 126: 465–72.

Viney, M. E. 1996. Developmental switching in the parasitic nematode *Strongyloides ratti*. *Proceedings of the Royal Society of London B* 263: 201–208.

Vizoso, D. B., and Ebert, D. 2005. Phenotypic plasticity of host-parasite interactions in response to the route of infection. *Journal of Evolutionary Biology* 18: 911–21.

Wakelin, D. 1978. Genetic control of susceptibility and resistance to parasitic infection. *Advances in Parasitology* 16: 219–308.

Wakelin, D. 1985. Genetic control of immunity to helminth infections. *Parasitology Today* 1: 17–23.

Waller, P. J. 1993. Nematophagous fungi: prospective biological control agents of animal parasitic nematodes? *Parasitology Today* 9: 429–31.

Walter, D., and Proctor, H. 1999. *Mites: Ecology, Evolution and Behaviour*. CABI Publishing, Wallingford, UK.

Walther, B. A., Cotgreave, P., Price, R. D., Gregory, R. D., and Clayton, D. H. 1995. Sampling effort and parasite species richness. *Parasitology Today* 11: 306–10.

Walther, B. A., and Ewald, P. W. 2004. Pathogen survival in the external environment and the evolution of virulence. *Biological Reviews* 79: 849–69.

Walther, B. A., and Morand, S. 1998. Comparative performance of species richness estimation methods. *Parasitology* 116: 395–405.

Ward, S. A. 1992. Assessing functional explanations of host-specificity. *American Naturalist* 139: 883–91.

Washburn, J. O., Gross, M. E., Mercer, D. R., and Anderson, J. R. 1988. Predator-induced trophic shift of a free-living ciliate: parasitism of mosquito larvae by their prey. *Science* 240: 1193–95.

Wassom, D. L., Dick, T. A., Arnason, N., Strickland, D., and Grundmann, A. W. 1986. Host genetics: a key factor in regulating the distribution of parasites in natural host populations. *Journal of Parasitology* 72: 334–37.

Waters, A. P., Higgins, D. G., and McCutchan, T. F. 1991. *Plasmodium falciparum* appears to have arisen as a result of lateral transfer between avian and human hosts. *Proceedings of the National Academy of Sciences of the U.S.A.* 88: 3140–44.

Watters, G. T. 1992. Unionids, fishes, and the species-area curve. *Journal of Biogeography* 19: 481–90.

Watve, M. G., and Sukumar, R. 1995. Parasite abundance and diversity in mammals: correlates with host ecology. *Proceedings of the National Academy of Sciences of the USA* 92: 8945–49.

Webb, T. J., and Hurd, H. 1999. Direct manipulation of insect reproduction by agents of parasite origin. *Proceedings of the Royal Society of London B* 266: 1537–41.

Webster, J. P., Gowtage-Sequeira, S., Berdoy, M. and Hurd, H. 2000. Predation of beetles (*Tenebrio molitor*) infected with tapeworms (*Hymenolepis diminuta*): a note of caution for the manipulation hypothesis. *Parasitology* 120: 313–18.

Weckstein, J. D. Biogeography explains cophylogenetic patterns in toucan chewing lice. *Systematic Biology* 53: 154–64.

Wedekind, C., and Rüetschi, A. 2000. Parasite heterogeneity affects infection success and the occurrence of within-host competition: an experimental study with a cestode. *Evolutionary Ecology Research* 2: 1031–43.

Wedekind, C., Strahm, D., and Schärer, L. 1998. Evidence for strategic egg production in a hermaphroditic cestode. *Parasitology* 117: 373–82.

Werren, J. H., Nur, U., and Wu, C.-I. 1988. Selfish genetic elements. *Trends in Ecology and Evolution* 3: 297–302.

Wesenburg-Lund, C. 1931. Contributions to the development of the Trematoda Digenea. Part I. The biology of *Leucochloridium paradoxum*. *Mémoires de l'Académie Royale des Sciences et des Lettres de Danemark, Section des Sciences (series 9)* 4: 90–142.

West, S. A., Gemmill, A. W., Graham, A., Viney, M. E., and Read, A. F. 2001a. Immune stress and facultative sex in a parasitic nematode. *Journal of Evolutionary Biology* 14: 333–37.

West, S. A., Herre, E. A., and Sheldon, B. C. 2000a. The benefits of allocating sex. *Science* 290: 288–90.

West, S. A., Reece, S. E., and Read, A. F. 2001b. Evolution of gametocyte sex ratios in malaria and related apicomplexan (protozoan) parasites. *Trends in Parasitology* 17: 525–31.

West, S. A., Smith, T. G., and Read, A. F. 2000b. Sex allocation and population structure in apicomplexan (Protozoa) parasites. *Proceedings of the Royal Society of London B* 267: 257–63.

Wetzel, E. J., and Shreve, E. W. 2003. The influence of habitat on the distribution and abundance of metacercariae of *Macravestibulum obtusicaudum* (Pronocephalidae) in a small Indiana stream. *Journal of Parasitology* 89: 1088–90.

Wharton, D. A. 1986. *A Functional Biology of Nematodes*. Croom Helm, London.

Wharton, D. A. 1999. Parasites and low temperatures. *Parasitology* 119: S7–S17.

Whiteman, N. K., Santiago-Alarcon, D., Johnson, K. P., and Parker, P. G. 2004. Differences in straggling rates between two genera of dove lice (Insecta: Phthiraptera) reinforce population genetic and cophylogenetic patterns. *International Journal for Parasitology* 34: 1113–19.

Whitfield, J. B. 1998. Phylogeny and evolution of host-parasitoid interactions in Hymenoptera. *Annual Review of Entomology* 43: 129–51.

Whitfield, P. J., Pilcher, M. W., Grant, H. J., and Riley, J. 1988. Experimental studies on the development of *Lernaeocera branchialis* (Copepoda: Pennellidae): population processes from egg production to maturation on the flatfish host. *Hydrobiologia* 167/168: 579–86.

Whittington, I. D., Cribb, B. W., Hamwood, T. E., and Halliday, J. A. 2000. Host-specificity of monogenean (Platyhelminth) parasites: a role for anterior adhesive areas? *International Journal for Parasitology* 30: 305–20.

Whittington, I. D., and Ernst, I. 2002. Migration, site-specificity and development of *Benedenia lutjani* (Monogenea: Capsalidae) on the surface of its host, *Lutjanus carponotatus* (Pisces: Lutjanidae). *Parasitology* 124: 423–34.

Wickler, W. 1976. Evolution-oriented ethology, kin selection, and altruistic parasites. *Zeitschrift für Tierpsychologie* 42: 206–14.

Wickström, L. M., Haukisalmi, V., Varis, S., Hantula, J., Fedorov, V. B., and Henttonen, H. 2003. Phylogeography of the circumpolar *Paranoplocephala arctica* species complex (Cestoda: Anoplocephalidae) parasitizing collared lemmings (*Dicrostonyx* spp.). *Molecular Ecology* 12: 3359–71.

Wikel, S. K. 1999. Modulation of the host immune system by ectoparasitic arthropods. *BioScience* 49: 311–20.

Williams, C. M., Poulin, R., and Sinclair, B. J. 2004. Increased haemolymph osmolality suggests a new route for behavioural manipulation of *Talorchestia quoyana* (Amphipoda: Talitridae) by its mermithid parasite. *Functional Ecology* 18: 685–91.

Williams, E. H. Jr., and Bunkley-Williams, L. 1994. Four cases of unusual crustacean-fish associations and comments on parasitic processes. *Journal of Aquatic Animal Health* 6: 202–208.

Williams, G. C., and Nesse, R. M. 1991. The dawn of Darwinian medicine. *Quarterly Review of Biology* 66: 1–22.

Wilson, D. S. 1977. How nepotistic is the brain worm? *Behavioral Ecology and Sociobiology* 2: 421–25.

Wilson, K., Bjørnstad, O. N., Dobson, A. P., Merler, S., Poglayen, G., Randolph, S. E., Read, A. F., and Skorping, A. 2002. Heterogeneities in macroparasite infections: patterns and processes. In *The Ecology of Wildlife Diseases* (eds. P. J. Hudson, A. Rizzoli, B. T. Grenfell, H. Heesterbeek, and A. P. Dobson), Oxford University Press, Oxford, pp. 6–44.

Windsor, D. A. 1998. Most of the species on Earth are parasites. *International Journal for Parasitology* 28: 1939–41.

Wojcicki, M., and Brooks, D. R. 2004. Escaping the matrix: a new algorithm for phylogenetic comparative studies of co-evolution. *Cladistics* 20: 341–61.

Woolaston, R. R., and Baker, R. L. 1996. Prospects of breeding small ruminants for resistance to internal parasites. *International Journal for Parasitology* 26: 845–55.

Woolhouse, M.E.J., Dye, C., Etard, J. F., Smith, T., Charlwood, J. D., Garnett, G. P., Hagan, P., Hii, J.L.K., Ndhlovu, P. D., Quinnell, R. J., Watts, C. H., Chandiwana, S. K., and Anderson, R. M. 1997. Heterogeneities in the transmission of infectious agents: implications for the design of control programs. *Proceedings of the National Academy of Sciences of the U.S.A.* 94: 338–42.

Worthen, W. B. 1996. Community composition and nested-subset analyses: basic descriptors for community ecology. *Oikos* 76: 417–26.

Worthen, W. B., and Rohde, K. 1996. Nested subset analyses of colonization-dominated communities: metazoan ectoparasites of marine fishes. *Oikos* 75: 471–78.

Wright, D. H., Currie, D. J., and Maurer, B. A. 1993. Energy supply and patterns of species richness on local and regional scales. In *Species Diversity in Ecological Communities: Historical and Geographical Perspectives* (eds. R. E. Ricklefs and D. Schluter), University of Chicago Press, Chicago, pp. 66–74.

Wright, D. H., Patterson, B. D., Mikkelson, G. M., Cutler, A., and Atmar, W. 1998. A comparative analysis of nested subset patterns of species composition. *Oecologia* 113: 1–20.

Wright, D. H., and Reeves, J. H. 1992. On the meaning and measurement of nestedness of species assemblages. *Oecologia* 92: 416–28.

Yan, G., Stevens, L., and Schall, J. J. 1994. Behaviorial changes in *Tribolium* beetles infected with a tapeworm: variation in effects between beetle species and among genetic strains. *American Naturalist* 143: 830–47.

Yoshikawa, H., Yamada, M., Matsumoto, Y., and Yoshida, Y. 1989. Variations in egg size of *Trichuris trichiura*. *Parasitology Research* 75: 649–54.

Zander, C. D. 1998. Ecology of host-parasite relationships in the Baltic Sea. *Naturwissenschaften* 85: 426–36.

Zavras, E. T., and Roberts, L. S. 1985. Developmental physiology of cestodes: cyclic nucleotides and the identity of putative crowding factors in *Hymenolepis diminuta*. *Journal of Parasitology* 71: 96–105.

Zelmer, D. A. 1998. An evolutionary definition of parasitism. *International Journal for Parasitology* 28: 531–33.

Zelmer, D. A., and Arai, H. P. 2004. Development of nestedness: host biology as a community process in parasite infracommunities of yellow perch (*Perca flavescens* (Mitchill)) from Garner Lake, Alberta. *Journal of Parasitology* 90: 435–36.

Zelmer, D. A., and Esch, G. W. 1999. Robust estimation of parasite component community richness. *Journal of Parasitology* 85: 592–94.

Zervos, S. 1988a. Population dynamics of a thelastomatid nematode of cockroaches. *Parasitology* 96: 353–68.

Zervos, S. 1988b. Evidence for population self-regulation, reproductive competition and arrhenotoky in a thelastomatid nematode of cockroaches. *Parasitology* 96: 369–79.

Index